Databricks Data Intelligence Platform

Unlocking the GenAI Revolution

Nikhil Gupta
Jason Yip

Apress®

Databricks Data Intelligence Platform: Unlocking the GenAI Revolution

Nikhil Gupta
Livingston, NJ, USA

Jason Yip
Redmond, WA, USA

ISBN-13 (pbk): 979-8-8688-0443-4
https://doi.org/10.1007/979-8-8688-0444-1

ISBN-13 (electronic): 979-8-8688-0444-1

Managing Director, Apress Media LLC: Welmoed Spahr
Acquisitions Editor: Shaul Elson
Development Editor: Laura Berendson
Coordinating Editor: Gryffin Winkler

Cover Photo by JJ Ying on Unsplash (unsplash.com)

Distributed to the book trade worldwide by Springer Science+Business Media New York, 233 Spring Street, 6th Floor, New York, NY 10013. Phone 1-800-SPRINGER, fax (201) 348-4505, e-mail orders-ny@springer-sbm.com, or visit www.springeronline.com. Apress Media, LLC is a California LLC and the sole member (owner) is Springer Science + Business Media Finance Inc (SSBM Finance Inc). SSBM Finance Inc is a **Delaware** corporation.

For information on translations, please e-mail booktranslations@springernature.com; for reprint, paperback, or audio rights, please e-mail bookpermissions@springernature.com.

Apress titles may be purchased in bulk for academic, corporate, or promotional use. eBook versions and licenses are also available for most titles. For more information, reference our Print and eBook Bulk Sales web page at http://www.apress.com/bulk-sales.

Any source code or other supplementary material referenced by the author in this book can be found here: https://www.apress.com/gp/services/source-code.

If disposing of this product, please recycle the paper

Table of Contents

About the Authors...**xv**

About the Technical Reviewers ....**xvii**

Chapter 1: Databricks Platform: From Lakehouse to Data Intelligence Platform **1**

Data Platforms: Historical Perspective..2

Emergence of the Lakehouse 3

What Is a Lakehouse? 4

What Is the Databricks Lakehouse?..6

Key Features of the Databricks Lakehouse Platform 8

Introducing the Databricks Data Intelligence Platform 9

Conclusion 13

Chapter 2: Databricks Platform Overview **15**

Key Terminology..15

Databricks Compute or Clusters 18

Interactive or All-Purpose Clusters..19

Job Cluster 19

SQL Warehouse 20

Databricks All-Purpose Cluster Setup..20

Policy..22

Access Mode 24

Databricks Runtime Version 27

Autoscaling and Autotermination 28

Tags 28

Spot Instances 29

Cluster Pools 29

Cluster Sizing Considerations and Best Practices 30

Databricks Notebooks 31

Debugging 33

Serverless in Notebook 35

Databricks Widgets 36

Library Management 38

External Databricks Connectivity 39

Databricks CLI 39

Databricks REST API 40

Databricks Terraform 41

Conclusion 42

Chapter 3: Data Ingestion in Lakehouse 45

Introduction 45

Cloud Ingestion 46

Delta Ingestion 54

Auto Loader 55

COPY INTO 58

Conclusion 60

Chapter 4: Delta Lake - Deep Dive 61

The Challenges of Other Formats 61

What Is Delta Lake? 62

Delta Lake: Medallion Architecture 65

Delta Lake Key Features 68

　Update, Delete, and Upserts in Delta Table 68

　Schema Evolution 68

Time Travel 70

Clone Delta Tables 71

Generated Column 73

Change Data Feed 74

Universal Format 76

Delta Optimization 80

Liquid Clustering 82

Working with Liquid Clustering 83

Current Limitations 84

Predictive I/O 85

　ML/AI to the Rescue 87

Conclusion 88

Chapter 5: Data Governance with Unity Catalog 89

What Is Databricks Unity Catalog? 90

Unity Catalog: Before and After 91

Unity Catalog Hierarchy 92

Unity Catalog Admin Roles 94

　Getting Started with Unity Catalog 94

　Create a Metastore 95

Organizing Data in Unity Catalog 97

Key Features of Unity Catalog 99

　Centralized Metadata and User Management 99

　Centralized Access Controls 101

Data Lineage 102

Data Access Auditing 104

Data Search and Discovery 104

Row-Level Security and Column-Level Masking 105

    Row Filters 105

    Create a Row Filter 105

    Apply the Row Filter to a Table 105

    Column Masks 106

    Dynamic Views vs. Row Filters and Column Masks 106

Delta Sharing 107

    An Open Standard for Data Sharing 108

    How Delta Sharing Works 108

Conclusion 111

Chapter 6: Data Engineering Part 1: Orchestrating Data Pipelines Using Databricks Workflows 113

Databricks Workflow Jobs 114

Databricks Jobs and Tasks 115

    Configure Databricks Job Tasks: Task-Level Parameters 116

    Configure Databricks Job Tasks: Job-Level Parameters 119

Advanced Workflow Features 124

Monitoring Data Pipelines 130

Conclusion 132

Chapter 7: Data Engineering Part 2: Delta Live Tables 133

What Is Delta Live Tables? 134

    Data Ingestion Using DLT 135

    Change Data Capture with DLT 137

    Delta Live Tables Expectations 139

Creating a DLT Pipeline 142

Logging and Monitoring 145

Enhanced Autoscaling.. 147

Runtime Channels 148

Example: A Retail Sales Pipeline 148

Streaming Pipeline 149

Data Validation... 149

Data Lineage.. 150

Validation Dashboard.. 151

Conclusion 152

Chapter 8: Data Warehousing with DBSQL..................................153

What Is Databricks SQL?.. 154

SQL Warehouses 155

Photon 157

SQL Editor.. 158

Introduction to AI/BI Dashboards... 159

Alerts 161

Query History and Profile.. 161

Serverless Compute 163

Constraints in DBSQL 164

Constraints on Databricks 164

Enforced Constraints 165

Informational Constraints: Primary Key Foreign Key 166

Streaming Tables and Materialized Views 168

Streaming Tables... 169

Materialized Views 170

Create a Materialized View.. 170

Refresh a Materialized View 171

Lakehouse Federation 171

AI Functions in DBSQL 173

Consume LLM Models in DBSQL 173

Custom Functions Backed by a Serverless Serving Endpoint 176

Integrate BI Tools with Databricks 176

Publish to PowerBI Online from Databricks 177

Connect Power BI Desktop to Databricks 178

Conclusion 179

Chapter 9: Machine Learning Operations Using Databricks 181

Machine Learning with Databricks 182

Experiments 183

What Is the Glass Box Approach to Automated Machine Learning? ... 184

Machine Learning Lifecycle: MLOps 185

ML Example: Predicting Flight Delays with Databrick's AutoML 186

Data Exploration at Scale 188

Feature Store 191

Model Building 195

Deploy Model 200

MLOps Best Practices 215

Conclusion 218

Chapter 10: Generative AI with Databricks 219

What Is Generative AI? 219

Databricks Generative AI 221

The GenAI Journey 223

Prompt Engineering 224

Mosaic AI Playground 226

Use Cases 228

Retrieval Augmented Generation 230

    Similarity Search: The Magic Behind the Scenes 235

    A Practical Example for RAG: Using Structured Data.................... 236

    Step 1: Feature and Function Serving 237

    Step 2: Calculate Embedding and Sync to a Vector Database 240

    Step 3: Create a LangChainTool to Perform Various Tasks 242

    Step 4: MLflow LLM Evaluation 242

Mosaic AI Fine-Tuning API... 247

    Fine-Tuning Example 248

Pre-Training 248

    A Case Study of AI2's OLMo, a Truly Open-Source Large Language Model 249

Gen AI Pricing... 250

    What Are Tokens and Tokenizers? 251

Conclusion 253

Chapter 11: Large Language Model Operations...............................255

Machine Learning Operations 255

Large Language Model Operations 256

Components of LLMOps 258

Deep Dive into Each Process 263

    Prompt Engineering.. 263

    Retrieval Augmented Generation 266

    Model Fine-Tuning... 270

    Model Pretraining 271

A Case Study of AI2's OLMo, a Truly Open-Source Large Language Model.......271

    Model Governance.. 273

    LLM as a Judge 277

    Model Packaging and Deployment.. 279

Conclusion 283

Chapter 12: Mosaic AI Agent Framework: Creating Quality AI Agents 285

Part 0: The Installations..287

Part 1: LangChain Parametrization...287

Part 2: MLflow Evaluation 289

Part 3: Model Development...294

Part 4: Deployment...302

Evaluation Example..305

Conclusion 308

    Beyond LangChain...309

Chapter 13: DBRX: Creating an LLM from Scratch Using Databricks 311

What Is DBRX? 312

The DBRX Benchmarks 315

DBRX Architecture..318

    Shortcomings of the Transformer Architecture..320

    Mixture of Experts 322

    MegaBlocks: Efficient Sparse Training with Mixture-of-Experts 323

    Fine-Grained MoE...324

The MosaicML Stack..325

Distributed GPU Training 326

Model Serving...327

Using DBRX on Databricks 328

Conclusion 330

Chapter 14: The Databricks Data Intelligence Platform 331

Databricks IQ..333

Deep Dive into Databricks IQ 334

Databricks Assistant...334

AI-Powered Governance 339

Search and Discovery..347

AI/BI Genie (Previous Data Rooms)..348

How to Set Up Genie...349

Conclusion 352

Chapter 15: Databricks CI/CD 353

What Is CI/CD? 353

Stages of CI/CD 356

Introduction to Databricks Repos..358

Databricks UI vs. Git Terminologies...361

Databricks Asset Bundles 364

Case Study: Databricks MLOps Stack 365

Conclusion 369

**Chapter 16: Databricks Pricing and Observability Using
System Tables...371**

Costs Associated with the Databricks Platform 371

Cloud Infrastructure Costs 372

Databricks Pricing..373

What Are Databricks Units?..373

SQL Warehouse Pricing 379

Databricks Cost Management Best Practices 380

Databricks Observability: System Tables 383

Introduction to System Tables 384

Common Schemas/Tables Available with System Tables 386

System Table: Billing Usage Example 387

Conclusion 388

Chapter 17: Databricks Platform Security and Compliance 389

Databricks Architecture 390

Azure Databricks Deployment.. 391

    Capacity Planning... 391

    VNET Injection or Bring Your Own VNET.................................... 392

    Secure Cluster Connectivity (No Public IP/NPIP) 394

    Azure Private Link for Back-End and Front-End Connections 396

    Encryption and Auditing 397

    Customer Managed Keys... 397

Identity and Access... 399

    SSO and Multifactor Authentication.. 399

    IP Access Lists... 399

    Role-Based Access Control 401

    Token Management API 402

Security Analysis Tool 404

Databricks Security Best Practices... 407

Conclusion 408

Chapter 18: Spark Structured Streaming: A Comprehensive Guide..409

Spark Streaming 410

Structured Streaming.. 414

What Is Continuous Processing?... 415

Triggers 416

Output Modes... 417

Windowed Grouped Aggregation... 418

State Management... 418

Late-Arrival Handling: Watermark... 420

Auto Loader 422

Project Lightspeed 423

    Advanced State Management 424

    Use Case: E-commerce Operation 424

Structured Streaming Best Practices 426

Conclusion 428

**Chapter 19: From Ideation to Creation: A Walk-Through of
Building a GenAI Application 431**

    The Problem Statement 432

    Data Generation: Source 433

    Data Ingestion: Ingest 435

    Data Transformation: Transform 435

        Using Serverless SQL for Transformation 436

    Machine Learning Model for Diabetes Complication Classification: Query
and Process 444

    Generative AI: Serve 445

        Where Do We Start? 448

    Monitoring Dashboard: Analysis 452

    Conclusion 455

Index 457

About the Authors

 **Nikhil Gupta** is a seasoned data professional with more than 18 years of experience in big data technologies, driving innovation and strategic growth in the field. As a solution architect at Databricks, he leverages his expertise to help customers across various industries (including retail, consumer packaged goods, financial services, banking, and manufacturing) to modernize their data and AI implementations on the Databricks platform. His expertise spans a range of big data technologies, including data warehousing, data lakes, and real-time data processing, making him a trusted advisor for Fortune 500 companies.

 **Jason Yip** is a data and machine learning architect. He currently serves as the director of data and AI at Tredence, a leading data science and analytics company. He advises Fortune 500 companies on implementing data and generative AI strategies on the cloud. He serves on multiple advisory boards at Databricks, including the Partner Product Advisory Board and the Solution Architect Champion Advisory Board. He is a top voice on Databricks and a former Microsoft employee who successfully led the Microsoft Corporate Finance big data transformation using Databricks.

About the Technical Reviewers

 **Soumendra Mohanty** has led key growth portfolios (IIOT, data, analytics, AI, intelligent RPA, digital integration, digital experience, platforms), bringing world-class capabilities, innovative solutions, and transformation-led, outcomes-led value propositions to his clients. Under his leadership, Tredence has established a wide range of digital and data analytics capabilities and an enviable client-centric innovation culture to solve problems at the convergence of physical and digital.

With a career spanning 25 years, Soum has held various executive and leadership roles at Accenture, Mindtree, and L&T Infotech, leading multifaceted profit-and-loss functions, including merger-and-acquisition advisory for technology growth strategies and start-up ecosystems.

He is an accomplished thought leader and has published several books. He regularly speaks at various global forums, CDAO advisory gatherings, and educational institutions. He is an advisor to the Harvard Business Review (Analytics Stream).

Vishal Vibhandik is a veteran data architect with more than 20 years of experience in designing and implementing robust data solutions.

CHAPTER 1

Databricks Platform: From Lakehouse to Data Intelligence Platform

The intensifying pace of digital transformation has led companies to amass increasing volumes of diverse data from various sources. This data explosion carries enormous potential for organizations to uncover transformative insights to guide innovation and decision-making through advanced analytics.

In this chapter, we will examine the evolution of data platforms over the last decade or so. Then, we will discuss why today's ideal data platform is a lakehouse and how Databricks established the lakehouse category. We will then go in-depth to understand the various facets of the lakehouse platform as it is on Databricks.

Finally, we will discuss how generative AI (GenAI) and large language models (LLMs) have revolutionized the entire artificial intelligence (AI) landscape and how Databricks has embraced this technology to create the Databricks Data Intelligence Platform.

Data Platforms: Historical Perspective

The data landscape has undergone rapid evolution in recent years, necessitated by the exponential growth in information from an ever-expanding variety and volume of data. As organizations deal with this big data surge, the existing infrastructure has struggled to harness its potential effectively. This has led architects and technology leaders to start conceptualizing new integrated systems that can adeptly consolidate the strengths of current data platforms.

Let's start with data warehouses. They provided immense value over decades for descriptive analytics and business intelligence use cases relying on predefined structured data. However, as the focus and needs expanded to predictive analytics and leveraging the latest machine learning advancements, the nature of workloads moved beyond what traditional warehouses could proficiently support. Descriptive analytics for business intelligence based on predefined datasets are no longer enough. Further varied data types such as unstructured, semi-structured, and streaming use cases require more extensive and agile processing than data warehouse infrastructures are designed for.

The data lake concept therefore gained interest as an alternative to data warehouses, given its natural ability to ingest raw multistructured data quickly. One of the more popular technologies that was forefront of this was Hadoop and its ecosystem. However, lack of transactionality, data quality, and mixing modes inhibited unlocking the benefits promised by data lakes. The flexibility therefore came at the cost of governance, reliability, and vital enterprise capabilities. Consequently, the data lakes quickly turned into "data swamps."

Despite all these drawbacks, organizations with no better alternatives began using both these technologies in their data architecture: data warehouses for descriptive and business intelligence (BI) use cases and data lakes for AI/machine learning (ML) use cases with a variety of processing tools thrown in the mix (sometimes even a single tool for one use case). However, with two completely different systems, solving for two critical types of workloads started to be problematic. First, it created data silos, which necessitated moving data across the platforms and thus maintaining multiple copies of the same data. Second, the governance model of these disparate platforms was incompatible, thus requiring separate governance models for different systems. Finally, organizations started using different tools for BI and ML workloads, increasing operational efficiency and costs. Over time, the complexity of maintaining different systems increased. This is becoming not only costly but also slowing innovation.

More than ever enterprises needed a unified data infrastructure capable of managing diverse information seamlessly through its entire lifecycle to serve exponentially expanding analytical use cases.

Emergence of the Lakehouse

Let's understand how organizations look at their modern data platforms. First, the platform should be able to store all sorts of data in a single storage location, preferably cost-effectively. Then, that data should have a single governance and access model and, last, a technology that helps them solve all their use cases without moving any data or code.

However as discussed earlier, organizations using both a data warehouse and a data lake in their architecture are essentially looking at two different piecemeal systems leading to disconnected data silos, complex integrations, and fragmented governance, severely hampering building enterprise-grade analytic solutions that could positively impact the business.

This reality has catalyzed the emergence of an evolutionary new paradigm pioneered by Databricks: the *lakehouse*. The lakehouse architecture aims to bring together the most impactful capabilities of data warehouses and data lakes into an integrated whole on the cloud. Reliable

3

support for varied workloads using consistent data, managed securely under standard governance policies, holds the promise to finally harness big data comprehensively.

With its seismic potential to reshape the analytics landscape, the lakehouse undoubtedly constitutes one of most pivotal recent data platform innovations.

What Is a Lakehouse?

Let's dig a bit deeper and understand what a lakehouse is. A *lakehouse* is a data architecture paradigm aiming to bridge the gaps between data lakes and data warehouses. The goal is to provide the flexibility and scalability of a data lake as well as to provide the performance, reliability, and governance typically associated with a data warehouse. A lakehouse seeks to implement some of the managed data capabilities seen in warehouses directly on top of object stores or cloud-based storage. Figure 1-1 compares the three.

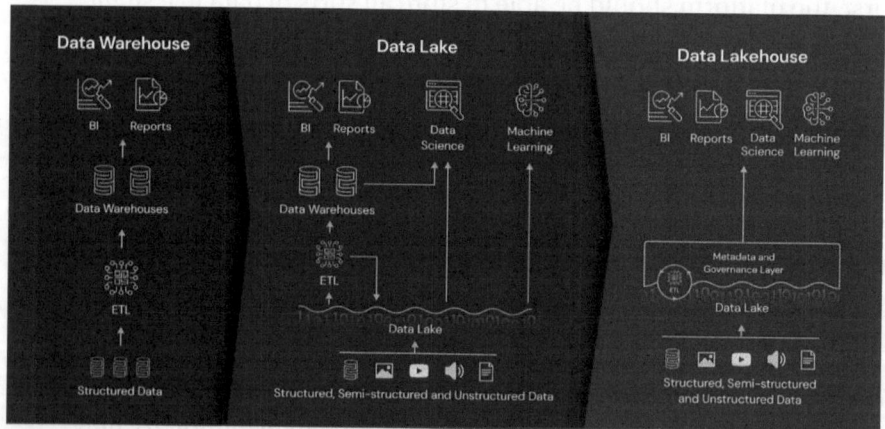

Figure 1-1. *Data warehouse versus data lake versus data lakehouse*

The data lakehouse construct addresses these gaps by consolidating the capabilities of data warehouses and lakes:

- Natively manages both structured and varied unstructured data

- Leverages cloud-scale object storage as the foundational data repository

- Provides reliability, security, and governance across storage and processing

- Provides high performance through technologies such as caching, indexing, and partitioning

- Supports real-time and batch workloads via unified streaming architecture

- Provides open extensibility to accommodate rapidly evolving analytics needs

The lakehouse breaks down data silos and enables simplified management by converging workloads on the same platform under standard governance policies. This makes it possible to get a single view of information at scale to power advanced analytics. With cloud infrastructure adding unlimited elasticity, lakehouses finally make it feasible to ask bigger questions of data than ever before possible.

If you were to design a new-generation analytical data management system using cheap distributed storage as a foundation, you would end up with something resembling a lakehouse: flexible schemas but faster queries. The goal is real-time insights without compromising governance.

What Is the Databricks Lakehouse?

Now that you understand the lakehouse paradigm, let's move on to see how a lakehouse is implemented on Databricks. Published in *Conference on Innovative Data in 2021*, Databricks researchers Michael Armbrust et al. wrote "Lakehouse: A New Generation of Open Platforms that Unify Data Warehousing and Advanced Analytics" (https://www.cidrdb.org/cidr2021/papers/cidr2021_paper17.pdf).

Figure 1-2 shows Databricks lakehouse platform.

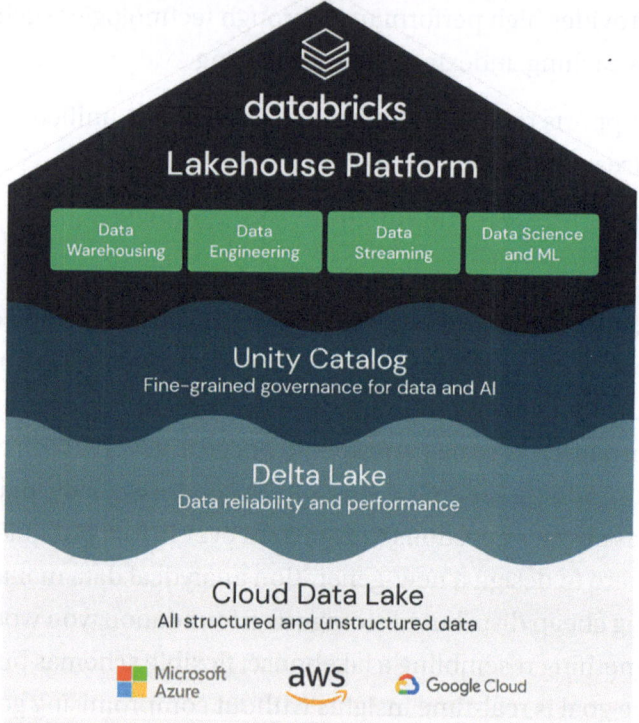

Figure 1-2. *Databricks lakehouse platform*

Databricks with the lakehouse architecture presented a potential solution to consolidate disparate data sources into a single location while avoiding some of the limitations of existing architectures. Databricks provides one of the most mature enterprise-scale implementations of a lakehouse architecture through its integrated data and AI platform. Built on open source and open standards, the Databricks lakehouse architecture simplifies your data estate by eliminating the silos that historically complicate data and AI.

Let's decode this a bit and do a deep dive into a Databricks lakehouse. Databricks leverages cloud object storage (S3-AWS, ADLS-Azure, and Google Cloud Storage [GCP]) as a central data store at its foundation. This enables enormous volumes of structured, unstructured, and semi-structured data to be housed in native formats in one of the cheapest storage available on the cloud. This is what constitutes the "lake" in the lakehouse. Once the data lands in the cloud in raw format, it is moved to Delta Lake format. Please note that data is still in your cloud storage but in Delta Lake format. Delta Lake is an open-source storage layer that brings performance, reliability, and governance to the data lakes. Delta Lake applies atomic transactions, caching, indexing, and time travel to make large-scale storage reliable and performant for mission-critical workloads. Basically, Delta Lake gives as the "warehouse" type capabilities to the data stored in your cloud storage. This constitutes the "house" in the lakehouse architecture.

As shown in Figure 1-2, Unity Catalog provides unified data governance for all data within the lakehouse. It manages all data assets, including tables, schemas, views, and even AI models, centrally.

Finally, the Databricks platform provides features that enable all data personas within your organization to build a variety of use cases be it data engineering, data science, streaming, or data warehousing.

To conclude, Databricks provides a unified lakehouse platform built on open-source technologies that is cloud-agnostic and able to handle diverse use cases at any scale. This platform makes data available for multiple analytics use cases, from business intelligence to machine learning.

Key Features of the Databricks Lakehouse Platform

Databricks' enterprise data cloud provides a leading implementation of the lakehouse paradigm. The following are some core concepts and capabilities:

Delta Lake: This open format optimizes the storage of massive volumes of structured and semi-structured data for reliability, performance, and governance.

Unified batch and streaming: Databricks processes batch and real-time data via the same platform using Spark structured streaming. This enables new ways to combine historical with streaming data.

Unity Catalog: Unity Catalog captures metadata and usage information across diverse data types and storage systems for unified discovery and governance.

Multilanguage support: The platform natively integrates languages like SQL, Python, R, Java, and Scala to support various analytics use cases on the same data.

Cloud-native architecture: By leveraging managed cloud infrastructure, Databricks automates resource management and scaling to meet the needs of the most demanding workloads.

Secure and governed access: Comprehensive access controls, encryption, and data masking enforce strict oversight and granular auditing.

Autoscaling and collaboration: Data scientists can quickly scale their work to production while closely collaborating with business users via sharing dashboards, reports, and applications.

Introducing the Databricks Data Intelligence Platform

If you look back in the technology world, 2023 was a groundbreaking year. It is when the world saw the power of GenAI LLMs and the potential they hold. Almost instantaneously organizations could imagine the future use cases that could be built by leveraging them. GenAI became the talk of every boardroom, and everybody was looking at using the technology to take a lead on their competitors.

Databricks with its Databricks lakehouse platform was uniquely placed to utilize this technology to not only enhance its platform but also help enterprises build their GenAI use cases. Let's talk about these two in detail.

First, Databricks enhanced its lakehouse platform by merging it with GenAI capabilities; this is called the Databricks data intelligence platform. Databricks used LLMs in almost every part of its platform, from assisting developers in troubleshooting coding errors to automatically generating insights from your data. We will discuss each of these features in detail in Chapter 12. The overall platform became more and more intelligent and thus enhances the user experience.

Second, Databricks built capabilities and features inside the platform that allow organizations to build their own GenAI use cases. Features like Vector Search, the Fine Tuning API, and RAG Studio enable organizations to productionalize their GenAI use cases from RAG applications to even building their own model from scratch. We will discuss these features in detail in Chapter 9.

Thus, Databricks enhanced its platform using LLMs and allowed users to create their GenAI applications on the platform.

To understand the data intelligence platform on Databricks, let's look at this analogy. Figure 1-3 shows the most powerful spaceship ever built—SpaceX's Starship. It is important to note that at the core, sitting underneath, is the Super Heavy booster, which is capable of withstanding 2.8 million pounds of weight while standing and, when in flight, propelling the second stage to space with its raptor engine.

Figure 1-3. *SpaceX spaceship*

With this concept in mind, it is not hard to understand the Databricks data intelligence platform (see Figure 1-4). It is also comprised of two major components. At the core, it is powered by the lakehouse platform and GenAI, which makes the platform much more intelligent to user needs and requirements.

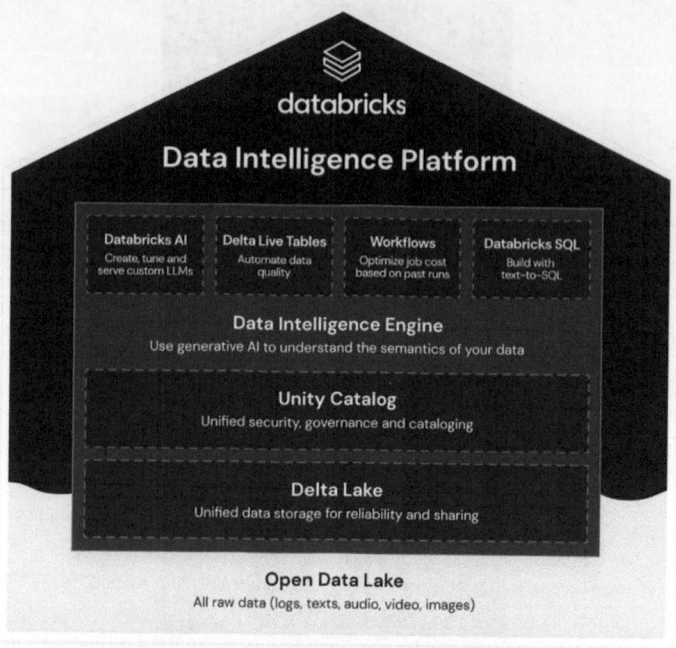

Figure 1-4. *Databricks data intelligence platform*

To conclude, fueled by the latest development of generative AI, Databricks has integrated the Data Intelligence Engine into the core of its offering. This is equivalent to SpaceX's Starship. In short, Databricks has leveraged the latest GenAI models and technology to create the Data Intelligence Engine (Databricks IQ), which fuels all parts of the platform.

With Mosaic ML and Databricks IQ, developers can seamlessly create their workload like they are working with a data subject-matter expert (SME) like never before. Databricks AI can also allow data scientists to leverage large language models as they are, refreshing their domain-specific knowledge with RAG, fine-tuning with more specialized knowledge, or even training a brand new LLM from scratch. This powerful second stage can propel the Databricks platform to a new era, enabling organizations to create the next generation of data and AI applications with quality, speed, and agility.

Conclusion

In this chapter, looked at the evolution of data platforms. Data warehouses are excellent for BI use cases, and data lakes with their open storage are used for ML use cases. However, by using both, these incompatible systems in their architecture created data silos, and hence businesses could not utilize their full data for business decisions. The Databricks Lakehouse Platform enables organizations to store all their data in one place. Whether it is structured, semi-structured, or unstructured data, it is stored in an open data lake. Then the raw data is moved into Delta Lake format, which provides reliability and improves performance. Unity Catalog provides a single governance layer, and the Databricks platform offers features to do use cases from data engineering, data warehousing, streaming, and data science. Finally, we discussed how Databricks built intelligence into their platform by utilizing GenAI and LLMs to create the Databricks data intelligence platform. In the next chapters, we will deep dive into various parts of the Databricks platform.

CHAPTER 2

Databricks Platform Overview

In this chapter, you will learn various aspects of the Databricks data intelligence platform. This chapter will provide a brief overview of the Databricks platform and set the stage for deep dives into various product features in later chapters. Initially, you will learn about the most common terms unique to the Databricks platform. After that you will learn about Databricks compute (clusters) and Databricks notebooks. Again, this chapter acts as a foundation for the rest of the chapters and features we will cover in them.

Key Terminology

The Databricks platform delivers three services catering to the specific needs of various personas: Data Engineering, Machine Learning, and SQL. Let's first look at the key Databricks terminology used throughout this book. Most of these terms will also be explained in detail in subsequent chapters.

- **Account:** A Databricks account allows admins to centrally manage and control access to their Databricks resources such as workspaces, users, and metastore. Billing and support are handled at the account level. A Databricks account can have multiple workspaces.

© The Editor(s) (if applicable) and The Author(s),
under exclusive license to APress Media, LLC, part of Springer Nature 2024
N. Gupta and J. Yip, *Databricks Data Intelligence Platform*,
https://doi.org/10.1007/979-8-8688-0444-1_2

- **Workspace:** Databricks workspaces provide a collaborative environment for data teams to access all Databricks assets. Workspaces are accessed via a web app and help users organize their work on Databricks. Users can create, manage, and share notebooks, clusters, and libraries within workspaces.

- **Databricks file system (DBFS):** DBFS is a storage location provisioned when creating a Databricks workspace. It is important to note that DBFS should not be used to store production data, libraries, or scripts.

- **Cluster (compute):** A Databricks cluster is a group of virtual machines (VMs) that process your data workloads. They allow you to execute code from notebooks, libraries, or custom code. Clusters can be created, scaled, and managed using the Databricks UI or application programming interface (API) or command-line interface (CLI), and they provide features like autoscaling and spot instances. Clusters do not store data. Data is always stored in your cloud storage account and other data sources.

- **Notebooks:** Notebooks are a collaborative IDE that allows you to write and execute code in Scala, Python, R, SQL, or Markdown and visualize results in real time. They come with features such as version history, co-editing, providing comments, and even scheduling as a job. Notebooks need to be connected to a cluster to execute commands. Users can share notebooks via the Web or download them to a local machine.

- **Databricks Git folders (formerly known as repos):**
  A folder is a feature of Databricks that allows users to source-control their data and AI projects by integrating with Git providers like GitHub, GitLab, Azure DevOps, etc. A folder also enables users to work directly with Git repo-backed folders from the Workspace UI.

- **Catalog:** A catalog is a centralized metadata browser that provides a single source of truth for all data assets in an organization. It allows users to discover, manage, and govern data across multiple workspaces, clusters, and teams. We will discuss this more in Chapter 5.

- **Workflows:** Databricks workflows enable you to orchestrate and schedule your code and data pipelines. Workflow jobs allow the code execution to occur either on an already existing cluster or on a cluster of its own. Jobs can be run from code in notebooks, JAR files, or Python scripts. They can be created manually through the UI or the REST API or the CLI.

- **Libraries:** Libraries are packages or modules that provide additional functionality to solve your business problems. These may be custom-written Scala or Java JARs, Python egg or wheel files, or custom-written packages. You can write and upload libraries manually through the UI, use the Libraries API, or install them directly via package management utilities like PyPi, Maven, or CRAN.

- **Databricks runtime (DBR):** Databricks Runtime is a set of core components that run on clusters. Databricks constantly updates the runtime with newer versions, and each version includes updates that improve the usability, performance, and security of big data analytics.

- **Databricks Unit (DBU):** DBU is the unit of processing capability and is billed per second. This is how Databricks charges users for the compute they use.

- **Delta Lake:** Delta Lake is an open-source storage layer that provides ACID transactions, scalable metadata management, and unified data management across data pipelines. It allows users to manage large datasets and provides a reliable and secure way to store and organize data.

After reviewing the key terminologies, we will dive into two topics: clusters and Databricks notebooks. As Databricks users, these are the two elements you will start working with when you first use the platform.

Databricks Compute or Clusters

Databricks is a fully managed PaaS offering that requires no infrastructure administration, management, or maintenance. Users and processes run code on clusters of VMs for data engineering, data science, and data analytics workloads. This includes batch and real-time production ETL pipelines, streaming analytics, ad hoc analytics, machine learning, deep learning, and graph analytics.

Databricks clusters consist of one or more virtual machine instances over which computation workloads are distributed. In the typical case, a cluster has a driver node alongside one or more worker nodes. During processing, the driver distributes workloads across available worker nodes. The driver program takes care of the job execution within the cluster. A job is split into multiple tasks distributed over the worker nodes. Clusters can be fixed-size clusters or autoscaling; by default, they auto-terminate after 120 minutes of inactivity (this is configurable). Databricks can also provide a single-node cluster option, typically limited to development or testing with small workloads.

Databricks has three main cluster types, and depending on the use case you are running, you can select one to improve efficiency and manage costs.

Interactive or All-Purpose Clusters

All-purpose compute is best suited for interactive analytics using notebooks, dashboards, or IDEs that require fast responses for an interactive user experience. They are best for ad hoc analysis, data exploration, or development. They can be either single user or shared by multiple users and can be terminated and restarted (manually, API or cluster setting).

Job Cluster

Job clusters should be utilized when running Databricks jobs. As a best practice, all production jobs or ETL pipelines should be run on job clusters, as they provide a fully isolated environment. Job clusters are pure ephemeral compute, as they terminate themselves when the job ends, thus reducing resource usage and costs. In later chapters, we will learn more about job clusters while discussing Databricks workflows. Now we have the option to run job clusters in serverless mode.

SQL Warehouse

SQL warehouses are meant to run SQL workloads and queries, primarily in the DBSQL part of the platform. If you are writing SQL queries, creating visualizations/dashboards, or connecting your favorite tool to Databricks, SQL warehouses is the way to go.

After defining a cluster and the types of clusters present in the Databricks environment, let's examine how to set one up using the Databricks UI.

Databricks All-Purpose Cluster Setup

This section will discuss the various attributes that need to be selected and how to configure them. Figure 2-1 shows the cluster creation interface.

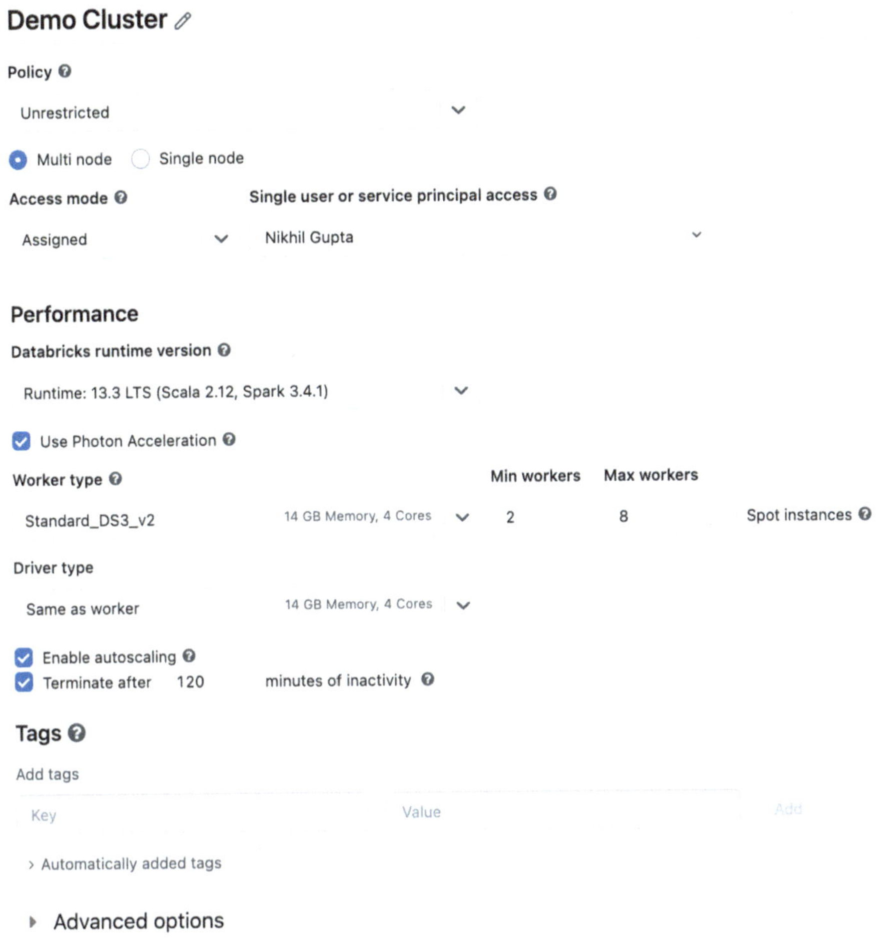

Demo Cluster ✏

Policy ❓

Unrestricted   ⌄

🔵 Multi node   ⚪ Single node

Access mode ❓   **Single user or service principal access** ❓

Assigned   ⌄   Nikhil Gupta   ⌄

Performance

Databricks runtime version ❓

Runtime: 13.3 LTS (Scala 2.12, Spark 3.4.1)   ⌄

☑ Use Photon Acceleration ❓

Worker type ❓		Min workers	Max workers	
Standard_DS3_v2	14 GB Memory, 4 Cores ⌄	2	8	Spot instances ❓

Driver type

Same as worker   14 GB Memory, 4 Cores ⌄

☑ Enable autoscaling ❓
☑ Terminate after   120   minutes of inactivity ❓

Tags ❓

Add tags

Key	Value	Add

› Automatically added tags

▸ Advanced options

Figure 2-1. *Creating an interactive cluster*

Next, we will look at some of the important parameters on this page.

Policy

Cluster policies enable admins to limit the attributes available for cluster creation. Users can select a cluster policy from the policy drop-down on the cluster configuration page. You can configure ACLs that limit cluster policies to specific users and groups.

For example, in Figure 2-2, the cluster policy allows users to create a cluster with the defined configurations as given in the JSON file. Only the configurable fields are visible when the user uses this policy, and the rest are hidden. This allows admins to control the clusters that the users can create. Further, only admin users can create, edit, and delete cluster policies. Admin users also have access to all policies.

```
1   {
2     "node_type_id": {
3       "type": "allowlist",
4       "values": [
5         "Standard_DS3_v2",
6         "Standard_DS4_v2",
7         "Standard_DS5_v2",
8         "Standard_NC4as_T4_v3"
9       ],
0       "defaultValue": "Standard_DS3_v2"
1     },
2     "spark_version": {
3       "type": "unlimited",
4       "defaultValue": "auto:latest-ml"
5     },
6     "runtime_engine": {
7       "type": "fixed",
8       "value": "STANDARD",
9       "hidden": true
0     },
1     "num_workers": {
2       "type": "fixed",
3       "value": 0,
4       "hidden": true
5     },
6     "data_security_mode": {
7       "type": "allowlist",
8       "values": [
9         "SINGLE_USER",
0         "LEGACY_SINGLE_USER",
1         "LEGACY_SINGLE_USER_STANDARD"
2       ],
3       "defaultValue": "SINGLE_USER",
4       "hidden": true
5     },
6     "driver_instance_pool_id": {
7       "type": "forbidden",
8       "hidden": true
9     },
0     "cluster_type": {
1       "type": "fixed",
2       "value": "all-purpose"
3     },
```

Figure 2-2. *Sample cluster policy*

Cluster policies present three main benefits. First, it helps control costs as these policies prevent individuals from spinning up unnecessarily large and enforce specific configurations such as auto-termination. Second, cluster policies help improve governance as admins can enforce cluster tags to track usage by team or project or control cluster access to users/groups. Finally, as more and more users are onboarded on the Databricks workspace, disruption is minimized by standardizing the cluster creation process.

Before we move further, let's look into a particular Databricks-managed cluster policy that is available: Personal Compute. This policy allows users to create single-machine easy compute resources for an individual user to start running workloads immediately, minimizing compute management overhead for admins. Some of the properties of Personal Compute are that the clusters created are single-node, single-user (Unity Catalog enabled), and all-purpose clusters with the latest Databricks runtime.

The next configuration we want to look into is Access Mode.

Access Mode

Cluster access modes are divided into three distinct types (see Figure 2-3).

- Standard single-user clusters

- Shared clusters (for multiple users) with User Isolation data access mode

- No isolation shared clusters

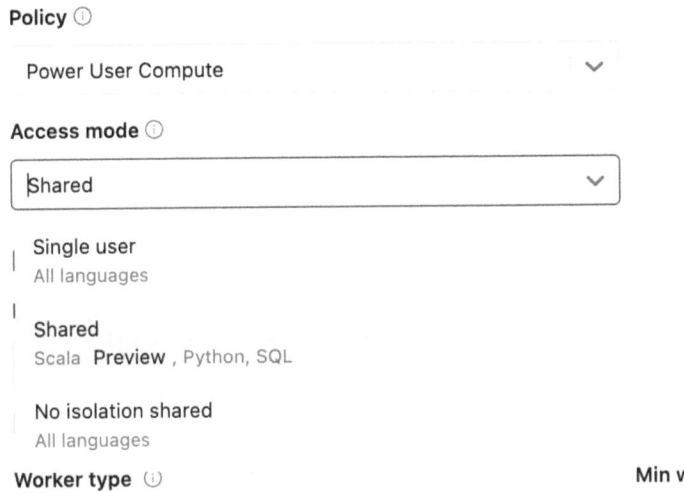

Figure 2-3. *Cluster access mode*

Standard single-user clusters: Standard single-user clusters are Unity Catalog (UC)–enabled clusters for a single user designated while creating or editing the cluster. Other users cannot attach to the cluster, regardless of the cluster permission settings. Standard clusters can run workloads developed in any language such as Java, Python, R, Scala, or SQL, and they can be fixed size or autoscaling.

Shared (for multiple users) clusters: Shared clusters are UC-enabled clusters ideal for multiple users accessing a single cluster to run interactive or automated jobs. These clusters only support SQL, Python, and R. The key benefits of shared clusters are that they provide Apache Spark–native fine-grained sharing for maximum resource utilization and minimum query latencies so that all users on the cluster can run jobs by sharing total compute resources (CPU and RAM) among all the users on the cluster. Shared clusters can help reduce costs for a shared user work environment, as well as experimentation, testing, and execution of some production workloads.

The "No isolation shared" option for legacy support and does not support Unity Catalog. This access mode is generally recommended for new clusters only if there is a specific need.

Table 2-1 summarizes the access modes along with Unity Catalog support. Databricks recommends using Unity Catalog for fine-grained access controls.

Table 2-1. *Databricks Access Modes*

Access Mode	Visible to User	UC Support	Supported Languages
Single user	Always	Yes	Python, SQL, Scala, R
Shared	Always (Premium plan or above required).	Yes	Python (on Databricks Runtime 11.3 LTS and above), SQL, Scala (on Unity Catalog–enabled compute using Databricks Runtime 13.3 LTS and above)
No Isolation Shared	Admins can hide this access mode by enforcing user isolation in the admin settings page.	No	Python, SQL, Scala, R
Custom	Hidden (for all new compute).	No	Python, SQL, Scala, R

Databricks Runtime Version

The Databricks runtime is a collection of core software components running on the clusters of machines managed by Databricks. You can select this setting in an all-purpose compute, but in SQL warehouses, it is auto-selected. The Databricks runtime version includes Spark but also adds several components and updates that substantially improve big data analytics' usability, performance, and security. As a best practice, select the most recent runtime version. Long-Term Support (LTS) versions are released every six months and supported for two years (see Figure 2-4).

Performance

Databricks runtime version ⓘ

Runtime: 14.3 LTS (Scala 2.12, Spark 3.5.0)	⌄

Standard	›	15.3	Scala 2.12, Spark 3.5.0
ML	›	15.2	Scala 2.12, Spark 3.5.0
		15.1	Scala 2.12, Spark 3.5.0
		14.3 LTS	Scala 2.12, Spark 3.5.0
		14.2	Scala 2.12, Spark 3.5.0
		14.1	Scala 2.12, Spark 3.5.0
		13.3 LTS	Scala 2.12, Spark 3.4.1
		12.2 LTS	Scala 2.12, Spark 3.3.2
		11.3 LTS	Scala 2.12, Spark 3.3.0
		10.4 LTS	Scala 2.12, Spark 3.2.1
		9.1 LTS	Scala 2.12, Spark 3.1.2

Figure 2-4. *Databricks runtime*

Apart from the Standard runtime version, there is also an augmented machine learning (ML) runtime version. This runtime version caters to ML workloads and is optimized for them. Further, many ML libraries come pre-installed and optimized with this runtime.

Finally, there is a checkbox for Photon, Databricks' vectorized execution engine for optimizing performance and costs. We will discuss Photon more in Chapter 8.

Autoscaling and Autotermination

A lot of times, the compute capacity is unknown, say, for example, during the development phase when the data engineer is writing and developing a pipeline. If the "Enable autoscaling" is checked, you can define the minimum and maximum number of workers to be added to the cluster. Thus, Databricks will allocate the necessary number of workers according to its needs during job execution. For example, you can create a cluster with a minimum of two workers and a maximum of eight workers. The cluster at start time will have two workers. As the user starts to process data and if need be (say for a big join between two tables) more workers will be added until a maximum of eight workers is reached. When there is no more need for eight workers, the cluster will scale down to two workers. This also is a huge cost-saving mechanism as you do not always need big clusters running.

You may also enable autotermination (Terminate After) for a cluster. During cluster formation, you can choose an inactivity time in minutes after which the cluster should terminate. If the difference between the current time and the last command issued/executed on the cluster exceeds the chosen inactivity interval, Databricks terminates the cluster automatically.

Tags

Cluster tags allow you to monitor costs and attribute Databricks usage/costs to your organization's entities, such as business units and teams. So, it is important to set tags on your clusters. These tags propagate down to VMs, which helps you charge back costs to your departments or business units.

A few default tags are created, such as Vendor (Databricks), ClusterID, ClusterName, and Creator. You can also create up to 20 custom tags.

Spot Instances

Spot instances are unused computes in the respective cloud environment. They are massively discounted compute compared to traditional virtual machines. You can use spot VMs to run your clusters, thus saving on costs.

One key point is that cloud providers can terminate or recall the spot instances when there is demand from Azure. However, Databricks automatically terminates spot VMs by starting the pay-as-you-go VMs to guarantee job completion. Databricks clusters are resilient to interruptions and well-suited for enterprise data and AI use cases.

Cluster Pools

Pools are pre-reserved VM instances, so when users request new clusters, Databricks can pull from an existing pool instead of acquiring from the cloud provider. You can set up spot instances for the pool and allocate them as clusters start. When creating a pool, select the desired instance size and Databricks Runtime version; then choose All Spot from the On-demand/Spot option; see Figure 2-5.

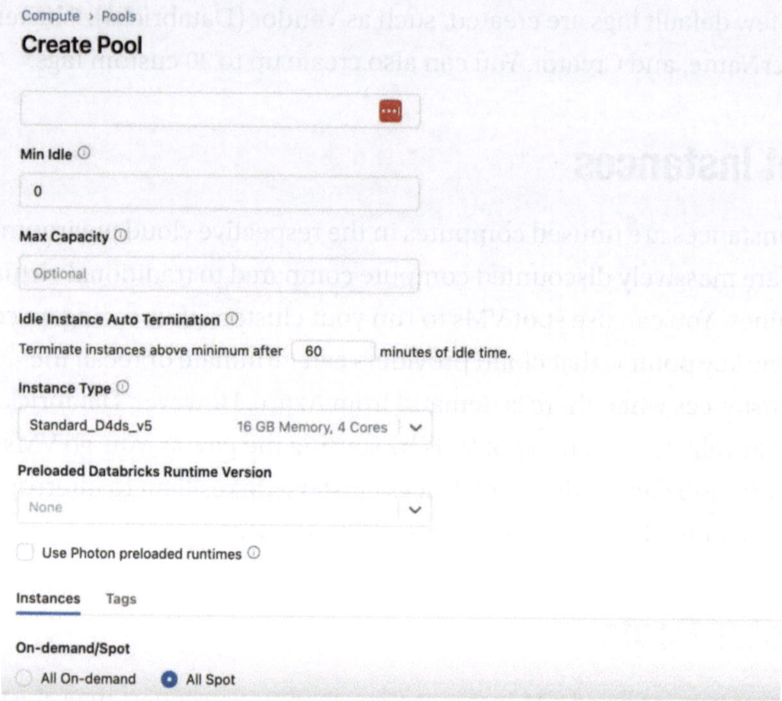

Figure 2-5. *Databricks compute pools*

Cluster Sizing Considerations and Best Practices

Here are some cluster best practices:

- Use autoscaling clusters when the compute capacity required is unknown.

- Set automatic termination when applicable.

- Use the latest Databricks Runtime version for recent features and performance optimizations.

- Use cluster tags for project- or team-based chargeback.

- Use the cluster event log and Spark UI to analyze cluster activities and submitted job performance.

- Configure cluster log delivery to deliver Spark driver and worker logs to cloud storage.

- Use cluster access control to configure permissions for users and groups.

- User cluster policies limit cluster types that users can launch.

After learning about clusters, we will learn about another important feature: Databricks notebooks.

Databricks Notebooks

If you are familiar with Jupiter notebooks, Databricks notebooks share the same concept. However, Databricks notebooks don't use the same back end as Jupyter notebooks, so if you clone the notebook from source control to your local environment, you must first convert it to `.ipynb` format.

A Databricks notebook is a code-first development tool that enables conversational data interaction by developing code and visually presenting results. With it, you can iteratively explore and visualize your data, create ETL pipelines, write reports or prototypes, and train ML models. Databricks notebooks provide capabilities like real-time co-authoring, support for multiple languages, automatic versioning, and built-in data visualizations. Figure 2-6 provides a view of a sample Databricks notebook.

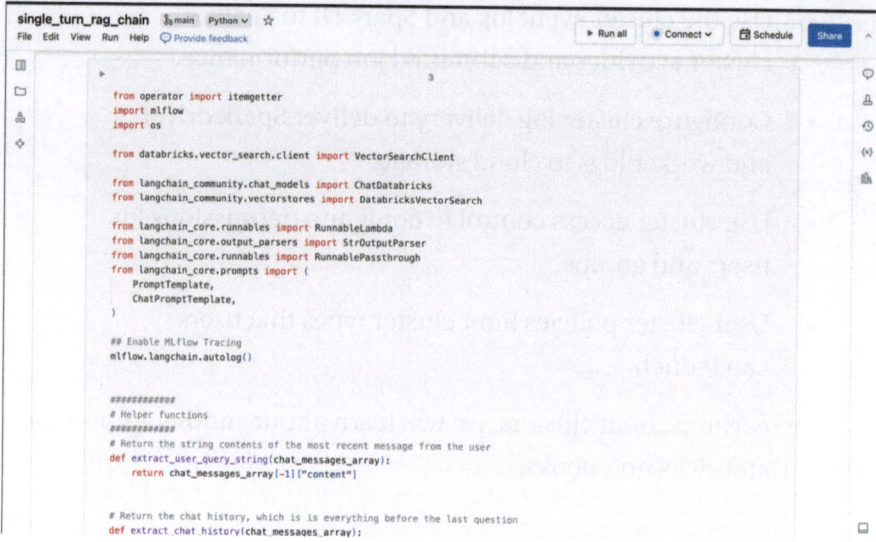

Figure 2-6. *Databricks notebook*

Now let's see some of the capabilities of Databricks notebooks:

- **Multiple language support:** Databricks notebooks allow you to develop code in multiple languages, such as Python, SQL, Scala, and R. This gives developers the flexibility to develop in the language of their choice or even use multiple languages within the same notebook using the magic command (e.g., % SQL or % Python). Notebooks also provide Markdown capabilities so you can maintain documentation along with the code itself.

- **Collaborative:** Notebooks allow developers to co-author or work on the same notebook in real time similar to working in your Google Docs environment. Further, users can collaborate by writing and leaving comments for their team members, which can then be worked upon later.

- **Reproducible:** Notebooks automatically track changes you made and store the version in version history, allowing you to look back at a previous version easily and compare what changed in the current notebook. Further, you can also integrate notebooks into your Git repositories for your CI/CD.

- **Visualizations:** Databricks notebooks have built-in visualizations, including bar, line, pie, scatter, map, and more. Users can create one or more visualizations for each command's result. Notebooks also allow you to bring external libraries like ggplot (R), matplotlib (Python), and Plotly for more advanced figures. Visualizations are automatically refreshed and updated whenever commands are rerun.

- **Scheduled:** In addition to interactive features, you can quickly create automated jobs from the same notebook and schedule them at specific intervals as per the use case. Thus, you can make your notebook run a job. In Chapter 6, we will see how you can orchestrate a pipeline using multiple notebooks.

Debugging

Debugging your Python code has never been easier with a Databricks notebook. You can set breakpoints and step into your Python code with a debugger. Use the Debug cell button to start debugging, as shown in Figure 2-7 and Figure 2-8.

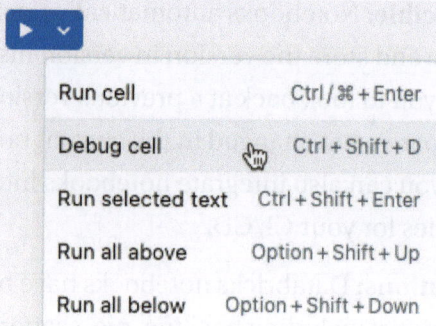

Figure 2-7. *Notebook cell debugging*

```
|| Stop   00:16                                    Debug ▶ ⤴ ↓ ↑                        Python  🗑 ◇ ⟨⟩ ⋮
    # Chain configuration
    # We suggest using these default settings
●   rag_chain_config = {
        "databricks_resources": {
            # Only required if using Databricks vector search
            "vector_search_endpoint_name": VECTOR_SEARCH_ENDPOINT,
            # Databricks Model Serving endpoint name
            # This is the generator LLM where your LLM queries are sent.
            "llm_endpoint_name": "databricks-dbrx-instruct",
        },
        "retriever_config": {
            # Vector Search index that is created by the data pipeline
            "vector_search_index": destination_tables_config["vectorsearch_index_name"],
            "schema": {
```

Figure 2-8. *Python debugger*

Variable Explorer also allows you to see all the variables in your notebook, greatly helping with the debugging experience. It supports Python, Scala, and R in this view, making the notebook a real-time cross-language compiler. Figure 2-9 shows a view of the explorer. The Variable Explorer provides a convenient one-click action to inspect all variables, including DataFrames. You can click any DataFrame to explore it in a new notebook cell, allowing you to visualize or profile it easily.

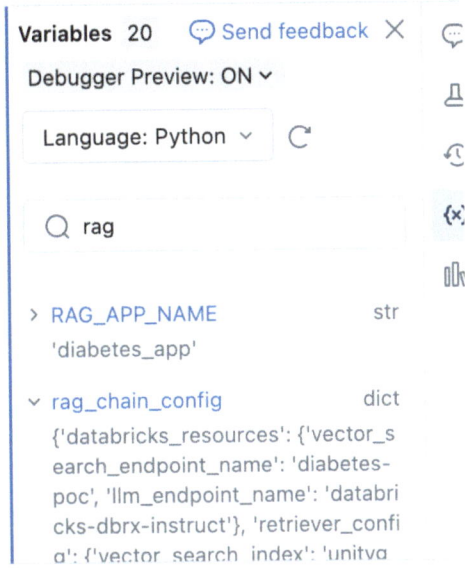

Figure 2-9. *Variable Explorer*

Serverless in Notebook

To enable rapid development experience from end to end, Databricks now enables users to use serverless SQL warehouse in notebooks. This allows SQL developers to continue to collaborate with other team members in the same environment. See Figure 2-10.

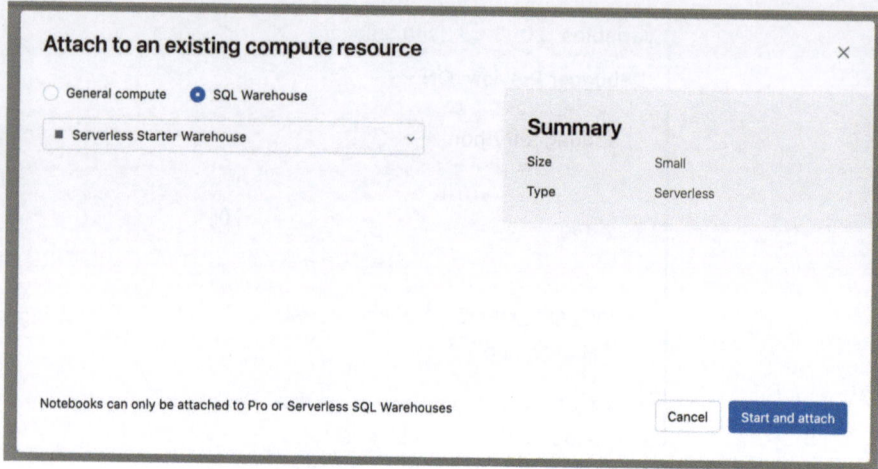

Figure 2-10. *Attaching a SQL warehouse to a notebook*

Databricks Widgets

Databricks widgets are input elements that allow you to parameterize your notebooks. Consider a scenario where you want to use the same notebook code but with multiple different inputs. One way could be to create multiple static notebooks by hard-coding values. Still, a more elegant and preferred way would be to add input elements to your notebook, making the same notebook more reusable. In short, Databricks widgets allow you to parameterize your notebooks by creating input widgets that can be adjusted to pass different values into the same notebook code.

There are four types of widgets for use with Databricks notebooks (see Figure 2-11).

- **Text Input:** Allows users to enter a text value in an
  input box:

  dbutils.widgets.text("widget_name", "Value", "Label")

- **Dropdown:** This provides a drop-down menu to select from a list of options. It is useful for predefined categories and options.

  dbutils.widgets.dropdown("widget_name", "Value", ["option1", "option2", "option3"], "Label")

- **Combo box:** This is a combination of a text box with a drop-down. Users can either type a custom value or select an option.

- **Multiselect:** This allows users to select multiple values via checkboxes.

  dbutils.widgets.checkbox("widget_name", True, "Label")

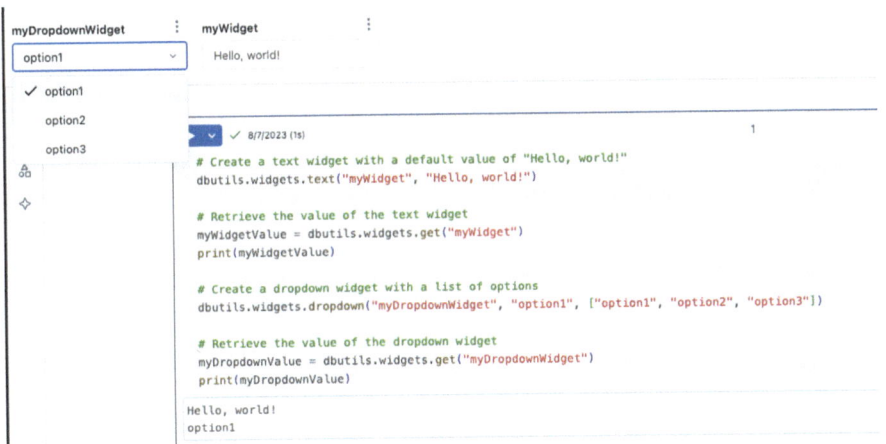

Figure 2-11. *Sample widgets in Databricks*

Once the widget is created, its value can be accessed using `dbutils` `widgets.get()` or via `:filter_value` or `${filter_value}` in SQL for DBR 15.1 or below. The value can then be used as input parameters in your code to customize data processing, visualization, or analysis.

Library Management

Libraries could be either third-party or prewritten custom code that must be available to Databricks notebooks or clusters to execute your code/jobs successfully. Libraries can be written in multiple languages and reused as needed by developers. Further, they could be stored locally in DBFS or cloud storage or called from external repositories such as PyPI, Maven, or CRAN.

Databricks Runtime includes many commonly used libraries installed on the cluster. The release notes give a list of libraries for your runtime version. However, you may need to install more custom or specific libraries at the time of code execution. Databricks provides two main options for library installation: cluster-scoped and notebook-scoped libraries.

Cluster-scoped libraries provide the ability to install libraries on specific clusters so that they can be used by all notebooks/jobs running on that cluster (Figure 2-12).

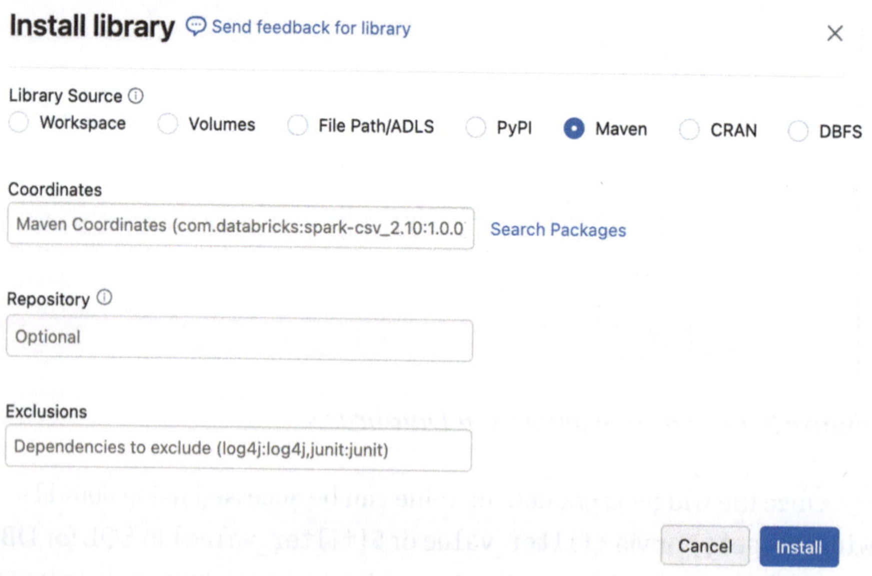

Install library 💬 Send feedback for library   ✕

Library Source ⓘ

○ Workspace   ○ Volumes   ○ File Path/ADLS   ○ PyPI   ● Maven   ○ CRAN   ○ DBFS

Coordinates

| Maven Coordinates (com.databricks:spark-csv_2.10:1.0.0 | Search Packages

Repository ⓘ

| Optional |

Exclusions

| Dependencies to exclude (log4j:log4j,junit:junit) |

Cancel   **Install**

Figure 2-12. *Library installation page*

There are several sources, including workspace files, cloud object storage, UC volumes, paths on local machines, or external repositories like PyPI, Maven, or CRAN.

Only Python and R allow you to install notebook-scoped libraries and create an environment scoped to a notebook session. Notebook-scoped libraries are used only when needed for your notebook and can be installed using the %pip magic command. These libraries do not persist and must be re-installed after each session.

External Databricks Connectivity

In this section, we will discuss how you can connect to Databricks beyond the browser, like Databricks CLI and API. While these are for administrative purposes in the beginning, the ecosystem has evolved a lot so we do day-to-day development in our favorite IDE offline.

Databricks CLI

The Databricks command-line interface (aka Databricks CLI) provides an easy-to-use tool for automating the Databricks platform from your terminal command prompt. From the CLI, you can start/stop a cluster, run Databricks jobs, and more.

To connect Databricks CLI to the Databricks workspace, you need to generate a Databricks personal access (PAT) token. To do so, browse to User Settings ➤ Developer ➤ Access Tokens (see Figure 2-13). The token will be visible to you only once and by default is valid for 90 days; you will need to regenerate it afterward.

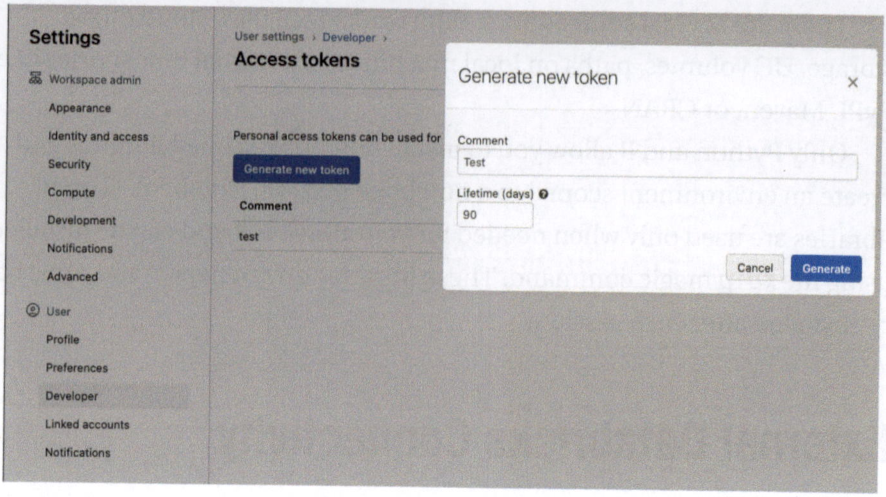

Figure 2-13. *PAT token generation*

Once you have the PAT token, you can quickly connect to the workspace by using the following and providing the PAT token when prompted:

```
databricks configure --host <workspace-url> --profile
<configuration-profile-name>.
```

Finally, you can run the following to create clusters:

```
databricks clusters create --cluster-name my-cluster --node-
type-id Standard_D2_v2 --num-workers 4
```

Databricks REST API

The Databricks REST API allows users to interact programmatically with their Databricks workspace. More or less anything that can be done via the UI can be done via the REST APIs. Users can interact with the Databricks REST APIs via curl requests, Python requests, Postman applications, or the databricks-api Python package. Here again, you would require a PAT token to authenticate to the Databricks workspace.

The Databricks documentation includes a REST API reference Guide that details both the workspace and account-level APIs for all three cloud platforms (`https://docs.databricks.com/api/azure/workspace/introduction`).

Databricks Terraform

The Databricks Terraform provider allows you to interact with almost all of Databricks' resources. Behind the scenes, it is powered by the Databricks SDK. Both Databricks SDK and Terraform providers are official Databricks open-source projects and are actively supported by Databricks.

Administrators often use the Databricks Terraform provider for automated deployment and disaster recovery. Figure 2-14 illustrates the vast scope that it supports for Databricks management. The ultimate meaning of DevOps is leveraging infrastructure as code (IaC) to manage operations and not depend on the user interface or a series of predocumented commands.

The Databricks Terraform provider can be found at the Terraform website:

`https://registry.terraform.io/providers/databricks/databricks/latest`

You can find the full source code of Databricks Terraform at Databricks GitHub, which is one of the top trending repos:

`https://github.com/databricks/terraform-provider-databricks`

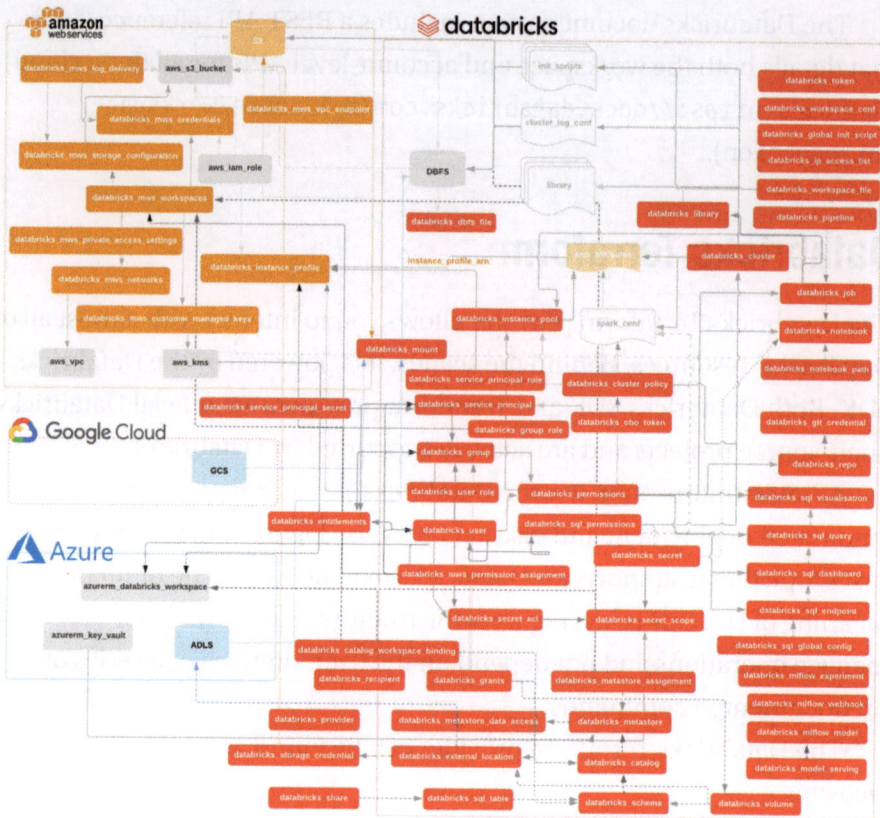

Figure 2-14. *Databricks Terraform provider*

Conclusion

In this chapter, we discussed the basic terminology associated with the
Databricks data intelligence platform. This formed the foundation for the
concepts and features we will learn throughout the book. Databricks has
evolved as not only a management tool on top of Spark but also provides
lots of features and toolings to manage your data and AI assets, be it tables,
jobs, policies, and development environments. Everything comes out of
the box. The open-source repo also contains countless useful tools that
Databricks is maintaining on behalf of the community.

In addition to key terms, we looked at two commonly used services: clusters and notebooks. Clusters form the compute on the Databricks and are now available as serverless. A notebook is the IDE where you write your code and execute it on your data using clusters. We concluded the chapter by looking at external connectivity to Databricks via Databricks CLI, the REST API, and Terraform.

CHAPTER 3

Data Ingestion in Lakehouse

Organizations have a wealth of information siloed in various data sources. It could be relational databases, on-prem data warehouses, big data storage like Hadoop systems, ERP/CRM systems, or real-time streaming sources. A significant number of analytics use cases need to not only process this data efficiently but also do it in a unified manner to produce meaningful reports and predictions. So to start this journey, organizations need to ingest data from different sources to a single location. In this chapter, we will look into how you can ingest data from various sources incrementally and efficiently into your Delta Lake.

Introduction

In a Databricks lakehouse, organizations can ingest data from a variety of sources to create a "single source of truth" for their data, enabling comprehensive analytics and data science capabilities across all their data. To break down the ingestion process, especially for batch data, it is mostly a two-step process, as shown in Figure 3-1.

The first step is to upload raw data from a variety of sources be it on-prem or other systems into your cloud storage (S3, ADLS, or Google Cloud Storage). This is normally referred to as *cloud ingestion*. Once it lands in your cloud storage, the second step is to move it into your Delta Lake layer. This is referred to as *delta ingestion*. Now for Delta ingestion there are two popular and efficient techniques: the Auto Loader and the COPY INTO command. Later in the chapter we will discuss both in detail.

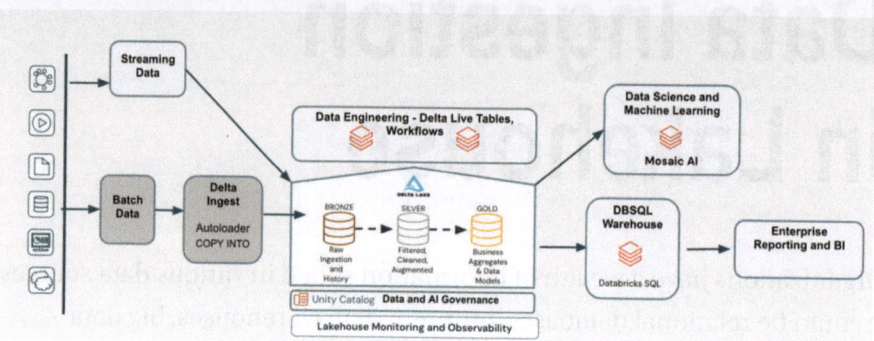

Figure 3-1. *Databricks reference architecture: ingestion*

We will discuss Delta Lake in length in Chapter 4, but we'll touch on it here. Databricks' integration with Delta Lake ensures reliability and performance at scale, providing ACID transactions and a unified process for batch and streaming data. This unification of data not only simplifies data management but also empowers organizations to derive more valuable insights, make data-driven decisions, and, ultimately, drive business growth.

Now, let's move in and learn the various methods used for both cloud and Delta ingestion.

Cloud Ingestion

As a first step, we need to get data into the cloud and, more specifically, into your cloud data storage. Usually, we call this layer the *landing zone* where the data lands from various sources and can be stored in any format, be it CSV, Parquet, JSON, etc. This layer is a source for Delta ingestion into the Delta bronze layer.

There are a number of alternatives that can be used to bring data to the cloud. The first method is via the built-in Databricks connectors that ingest data from sources such as Workday, MySQL, Salesforce, etc. Moreover, the Databricks UI provides an intuitive way to move the data directly to Delta Lake. Next are native cloud tools like Azure Data Factory for Azure Cloud. Finally, ingestion can happen via third-party tools such as Fivetran via Partner Connect.

Next, we will look into these three options in much more detail:

- **Databricks Native Connectors, Add Data and File Upload:** The Databricks' File Upload UI and Add Data UI (see Figure 3-2) allow you to easily move data for ingestion into Delta tables with Unity Catalog. It enables you to ingest data from a wide range of data sources in a secure manner via notebook templates or drag-and-drop functionality.

  Add Data UI: The Add Data UI acts as a central location for all your ingestion needs from various data sources into the Databricks lakehouse.

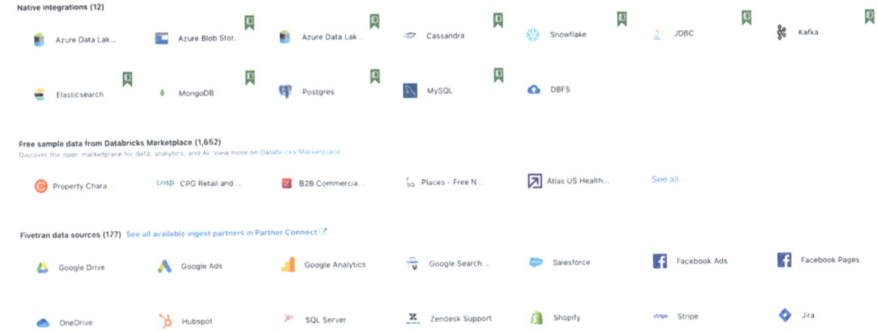

Figure 3-2. The Add UI Interface

Now developers can click any data source they want to ingest data from and then follow the UI flow or generated Databricks notebook with instructions to finish data ingestion step-by-step directly into Delta Lake.

Databricks supports several integrations, such Azure Data Lake Storage or Amazon S3 as the destination. Further, there are built-in connectors to support data transfers from data sources such as Snowflake, Kafka, MySQL, etc. Once you click the source, a notebook gets generated wherein you can give the source and target parameters (Figure 3-3).

Load data from MySQL to Delta Lake

This notebook shows you how to import data from JDBC MySQL databases into a Delta Lake table using Python.

Step 1: Connection information

First define some variables to programmatically create these connections.

Replace all the variables in angle brackets <> below with the corresponding information.

```
driver = "org.mariadb.jdbc.Driver"

database_host = "<database-host-url>"
database_port = "3306" # update if you use a non-default port
database_name = "<database-name>"
table = "<table-name>"
user = "<username>"
password = "<password>"

url = f"jdbc:mysql://{database_host}:{database_port}/{database_name}"

print(url)
```

Step 2: Reading the data

Now that you've specified the file metadata, you can create a DataFrame. Use an option to infer the data schema fro

First, create a DataFrame in Python, referencing the variables defined above.

```
remote_table = (spark.read
  .format("jdbc")
  .option("driver", driver)
  .option("url", url)
  .option("dbtable", table)
  .option("user", user)
  .option("password", password)
  .load()
)
```

Step 3: Create a Delta table

The DataFrame defined and displayed above is a temporary connection to the remote databa

To ensure that this data can be accessed by relevant users througout your workspace, save it

```
target_table_name = "<target-schema>.<target-table-name>"
remote_table.write.mode("overwrite").saveAsTable(target_table_name)
```

Figure 3-3. *Sample notebook, MySQL to Delta table*

Once these notebooks are run using Databricks clusters, the data is transferred from the source directly to the Delta tables.

Further, you can leverage more than 150 other connectors in the UI that are supported by Fivetran.

File upload UI: The file upload UI allows you to drag and drop local files seamlessly and enables the secure uploading of these files to create a Delta table. The UI is accessible across all personas through the navigation bar (Figure 3-4) or from the Catalog UI by clicking the + Add icon. The file upload UI offers the option to create a new table or overwrite an existing table.

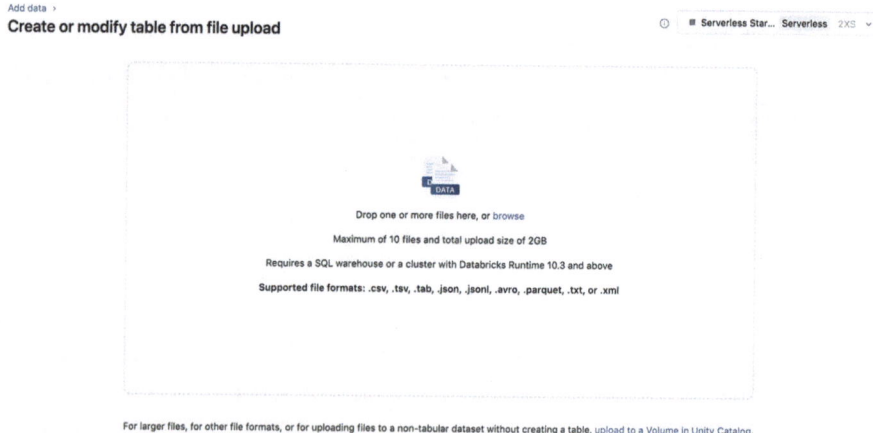

Figure 3-4. *Data ingestion, file upload UI*

You can use the File Upload UI to ingest via the following features:

- Select or drag and drop one or multiple files (CSV or JSON, etc.)

- Preview and configure the resulting table and then create the Delta table

- Autoselect default settings such as column types

- Modify various format options and table options

Therefore, both the Add UI and File Upload UI provide user-friendly interfaces to ingest data, which could be local or in other data storage platforms, into the Databricks lakehouse platform. Next we move into the cloud data ingestion via cloud-native tools.

- **Ingestion via cloud-native tools:** Another popular way to ingest data into the cloud is via cloud-native technologies. For example, for batch ingestion, we can use ADF (Azure), Glue (AWS), or Data Fusion (GCP). For stream ingestion, EventHub (Azure), Kinesis (AWS), Google Pub/Sub, or Kafka are popular choices.

  Let's look into an example of using Azure Data Factory (ADF) in Azure Cloud. ADF has more than 90 built-in data source connectors that can ingest data from various sources in the Azure cloud. Further, ADF seamlessly orchestrates Azure Databricks notebooks to connect and ingest all of your data sources into a single data lake. It also has a Delta format connector that can read and write Delta format into the lakehouse, providing seamless integration with Databricks.

- **Ingestion via third-party tools:** The next ingestion method is to leverage the extensive Databricks partner ecosystem and especially ingestion partners such as Fivetran, Hevo, Rivery, etc. To make this a seamless process, Databricks has closely worked with them and not only validated their technology but also aligned with them to build integrations that enable you to load data into cloud storage. These integrations enable low-code, scalable data ingestion from various sources into a Databricks lakehouse. These partners are featured in Databricks Partner Connect (Figure 3-5), which provides a UI interface that simplifies connecting third-party tools to your lakehouse for data ingestion.

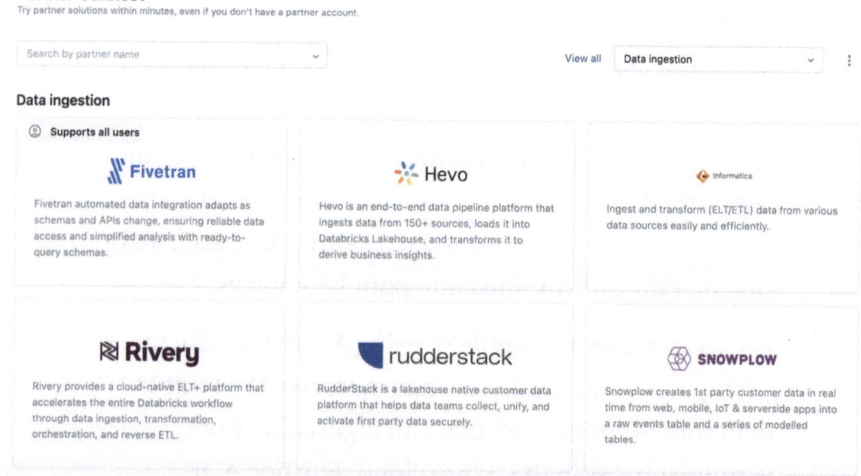

Figure 3-5. *Databricks Partner Connect*

Partner Connect lets you create trial accounts with select Databricks technology partners and lets you connect your Databricks workspace to partner solutions from the Databricks UI. With just a few clicks, Partner Connect will automatically configure resources such as clusters, tokens, and connection files for customers to connect with data ingestion, prep and transformation, and BI and ML tools.

Fivetran is a popular third-party data ingestion Databricks partner that offers simple no-code connectors that can ingest more than 150 data sources (e.g., MySQL, DynamoDB, SFTP) into destination data stores such as Databricks Delta Lake. Fivetran's ingestion solution helps customers avoid setting up manual or open-source connectors that might be less performant when managing the ingestion process. The connector for Fivetran works as follows:

- Set up a Databricks connection with an interactive cluster (jobs clusters are not available for Fivetran ingestion).

- Specify the data source in the connection as well as the schedule (takes just five minutes).

- Once complete, Fivetran will run a Databricks job and use the COPY INTO or MERGE command to append or update Delta Lake tables, which will contain the data from the source as scheduled.

Therefore, Databricks with its vast Partner ecosystem allows you to use the third-party technology to move data from a variety of sources into the lakehouse.

Delta Ingestion

The data has now landed in your cloud storage, or the landing zone. Here, the data could be in any format, such as CSV, JSON, Parquet, etc. The next step is to move that data into Delta Lake (the bronze layer) to complete your second-step data ingestion process (Figure 3-6).

Now this might sound simple, but there are a couple of ways where things could go wrong. For example, you could accidentally miss some files to process, which leads to missing data or could ingest previous ingested files, leading to duplicates and reverting or deleting those files would be even more complicated. Further, if there is a schema change in the source system, it could lead to failed jobs or even lost or corrupted fields in your data files.

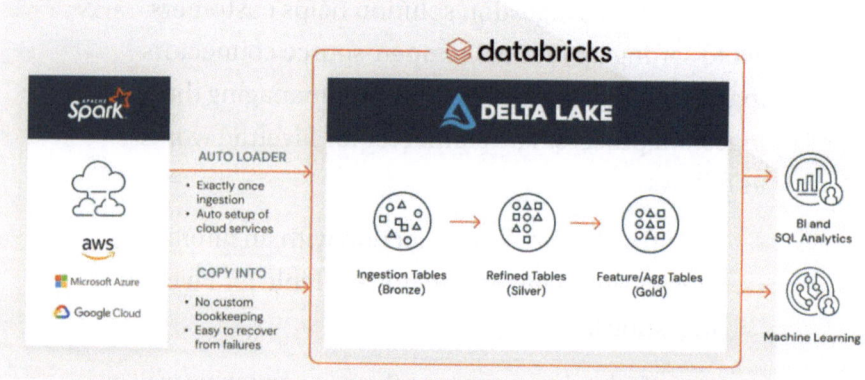

Figure 3-6. *Delta ingestion via the Auto Loader and COPY INTO*

To overcome these challenges, two common methods developed by Databricks are recommended: the Auto Loader and COPY INTO. Now, let's look into both in detail.

Auto Loader

The Auto Loader provides a highly efficient way to incrementally and efficiently process large amounts of data as it arrives in cloud storage. It also guarantees that each data file is processed exactly once. This is important because processing only new files incrementally solves the missing or duplicate data problem, which in turn helps save processing times and lowers cost for data ingestion.

The Auto Loader is designed for structured, semi-structured, and unstructured data. The Auto Loader can ingest JSON, CSV, XML, Parquet, Avro, ORC, text, and BINARYFILE file formats into Delta Lake.

Under the hood, the Auto Loader provides a structured streaming source called cloudFiles. Given an input directory path on the cloud file storage, the cloudFiles source automatically processes new files as they arrive, with the option of also processing all existing unprocessed files in that directory. The Auto Loader can be set up easily using the following syntax:

```
Df = Spark.
    readStream.
    format("cloudFiles") \
    .option("cloudFiles.format", "json") \
    .load("<path-to-source-data>") \
    .writeStream \
    .option("maxFilesPerTrigger", "2000") \
    .trigger("availableNow", "True") \
    .option("mergeSchema", "true") \
    .option("cloudFiles.inferColumnTypes", "true") \
    .option("checkpointLocation", "<path-to-checkpoint>") \
    .start("<path_to_target")
```

Let's look into the previous code and discuss a few important parameters. In the first part we are creating a readStream to read in input JSON files that have landed in the raw folder. In the second part, we do a writeStream and ingest the data into Delta Lake. The following are some noteworthy options in the previous syntax:

- **Checkpoint:** In the case of failures, Checkpoint helps the Auto Loader to resume the processing from where it left off by using the information stored in the checkpoint location and continuing to provide exactly-once guarantees when writing data into Delta Lake. You don't need to maintain or manage any state yourself to achieve fault tolerance or exactly-once semantics.

- **Trigger.AvailableNow:** The Auto Loader can be scheduled to run in Databricks Jobs as a batch job by using Trigger.AvailableNow. The AvailableNow trigger will instruct the Auto Loader to process all the files that arrived before the query start time. New files that are uploaded after the stream has started are ignored until the next trigger. Let's assume that the incoming data is spiky and instead of processing continuously, you want to process the data nightly in as a batch job. Trigger.AvailableNow allows you to do that without changing your code/architecture.

- **mergeSchema:** The mergeSchema option tells the Auto Loader to detect dynamically the evolution of the schema, for example, new fields added to the data. This prevents users from tracking and handling these changes

- **manually.inferColumnTypes:** The schema inference
  has always been expensive and slow at scale, especially
  with dynamic JSON. The Auto Loader efficiently
  samples data to infer the schema and stores it under
  `cloudFiles.schemaLocation` in your bucket.

- **Rescue_Data:** The source system often sends data that
  might be malformed and not fit in the table structure.
  The Auto Loader automatically adds the `_rescued_`
  `data` column, which stores the new columns that can
  be processed later.

Let's look under the hood as to how the Auto Loader discovers files.
When you begin to scan hundreds of files and millions of rows, it becomes
an expensive operation leading to ingestion challenges and higher
storage costs.

Scanning folders with many files to detect new data is expensive,
leading to ingestion challenges and higher cloud storage costs. To solve
this issue and support an efficient listing, Databricks Auto Loader offers
two modes: Direct Listing and File Notification (Figure 3-7).

- **Directory Listing:** This is the default mode in which
  the Auto Loader identifies new files by periodically
  listing the contents of the input directory on the
  cloud storage. This mode allows you to quickly start
  without any additional permission configurations as
  long as you have access to the data on cloud storage.
  To ensure eventual completeness of data, the Auto
  Loader automatically triggers a full directory listing
  after completing a configured number of consecutive
  incremental listings. Directory Listing mode is suitable
  for small to medium-sized directories or when the
  volume of incoming files is moderate.

- **File Notification:** In this mode, the Auto Loader sets up a managed cloud notification and queue service that subscribes to file events from the input directory. This requires additional cloud permissions to set up. File notification is more performant and scalable for very large input directories or a high volume of files, say millions/hr.

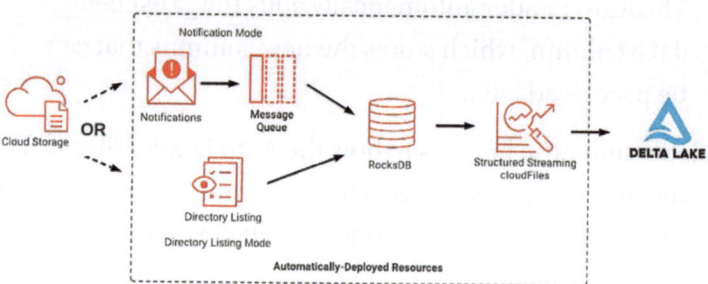

Figure 3-7. *Auto Loader modes: direct listing and file notification*

To conclude, the Auto Loader is a scalable solution that handles the incremental ingestion of billions of files and guarantees only once processing. Further, it comes with features like schema inference and schema evolution and rescues data that would have been otherwise ignored or lost. Next, let's look into the second option, COPY INTO command.

COPY INTO

COPY INTO is a SQL command that lets you load data from cloud storage into a Delta table. It supports many common file formats, including JSON, CSV, Parquet, Avro, and text files. COPY INTO is idempotent by default, so files are processed only once. This saves time and cost as your ETL pipeline processes every file only once instead of a full load each time. Now, the COPY INTO command is perfect for scheduled or ad-hoc ingestion use cases in which the data source location has a small number of files, which we would consider in the thousands of files. It is recommended that for a larger number of files the Auto Loader is suitable. COPY INTO supports target schema evolution, merging, mapping, and inference.

Let's look into a quick example.

COPY INTO requires a table to exist as it ingests the data into a target Delta table. If the ingestion is for the first time, you create an empty Delta table.

```
DROP TABLE IF EXISTS test_table;
CREATE TABLE test_table;
```

Once the table is created, you can ingest the data from a cloud storage location to the Delta table.

```
COPY INTO test_table
FROM 's3://my-bucket/exampleData'
FILEFORMAT = CSV
VALIDATE
FORMAT_OPTIONS ('header' = 'true', 'inferSchema' = 'true',
'mergeSchema' = 'true')
COPY_OPTIONS ('mergeSchema' = 'true')
```

Let's look into a few specifics from the previous code.

VALIDATE: The COPY INTO validate mode (runtime 10.3 and above) lets you preview and validate your source data before you write or ingest the files. Some of the validations are to see if the schema matches that of the target table or it needs to change, if all nulls and constraints are met, and if the data is parseable. The result of validate mode is a sample table that you can view.

If you find inconsistencies, such as nonmatching column names, format issues, etc., you can go back and fix them in the code.

Now once you are satisfied with the preview table, you can remove the VALIDATE keyword and rerun the COPY INTO command.

```
COPY INTO test_table
FROM 's3://my-bucket/exampleData'
FILEFORMAT = CSV
FORMAT_OPTIONS ('header' = 'true', 'inferSchema' = 'true',
'mergeSchema' = 'true')
COPY_OPTIONS ('mergeSchema' = 'true')
```

To conclude, the COPY INTO SQL command lets you load data from a file location into a Delta table. This is a retriable and idempotent operation; files in the source location that have already been loaded are skipped.

Conclusion

In this chapter, we covered how to ingest data in a Databricks lakehouse. Data ingestion is usually a two-step process for batch data. The first is to bring the data in any format to the cloud storage. The source system could be varied from on-prem data warehouses/data lakes to cloud databases. This is usually termed *cloud ingestion* and can be done in several ways such as Databricks native connectors, cloud ingestion tools, or third-party tools.

Thereafter, the data is then moved into a Delta lake, and the two most recommended approaches here are the Auto Loader and the COPY INTO command, both of which incrementally process data but also ensure that the data files are processed only once, helping data engineers to efficiently manage data ingestion into Delta Lake.

CHAPTER 4

Delta Lake - Deep Dive

In this chapter, we will examine a crucial aspect of the lakehouse paradigm: the storage format for your data. As discussed in Chapter 1, the ideal storage format for a lakehouse is one that provides similar data management and performance features of a data warehouse but is an open format and built on top of cloud data lakes. Delta Lake is a storage protocol that exactly fits the requirements. Delta Lake is an open, performant storage format that enables organizations to build data lakehouses, allowing data warehousing and machine learning directly on the data lake.

We will focus on understanding why we need Delta Lake as the storage protocol in the lakehouse architecture. Thereafter, we will discuss the medallion architecture and some key features of Delta Lake, including merge capabilities, liquid clustering, and optimizations. We will end with some best practices when working with Delta Lake.

The Challenges of Other Formats

Before we start looking into Delta Lake, let's first understand some of the challenges of data lakes and other storage formats. To be honest, data lakes and standard storage formats, such as CSV, Parquet, JSON, etc., have been around for quite some time. However, there have been inherent challenges in terms of reliability and performance while storing data in these traditional storage formats. Let's discuss some of these in a bit of detail.

N. Gupta and J. Yip, *Databricks Data Intelligence Platform*,
https://doi.org/10.1007/979-8-8688-0444-1_4

First, when you use formats like Parquet and CSV, etc., it is extremely difficult to roll back the data to its original state if your ETL job fails, leaving data corrupt. Not only that, it is hard to apply inserts, updates, and deletes to data stored in traditional storage formats. Next, there is a lack of schema enforcement, which leads to lower data quality, which was one of the main reasons for the low adoption of data lakes. Another important reason was that the performance of data in traditional data lakes was way behind that of warehouses due to issues such as small file problems (a large number of very small files slowing the processing) and no ability to cache queries or input data.

All these issues prevented the large-scale adoption of data lakes with common file formats, such as CSV, Parquet, etc., from becoming the de facto storage layer. However, the introduction of Delta Lake was a game-changer. Let's look into why.

What Is Delta Lake?

Delta Lake is an open-source storage layer that sits on top of data lakes and provides reliability, data governance, and performance. At its core, Delta Lake, like transactional databases, provides ACID compliance to data lakes with schema enforcement.

Figure 4-1 shows the core components of a Delta table.

Contents of a Delta table

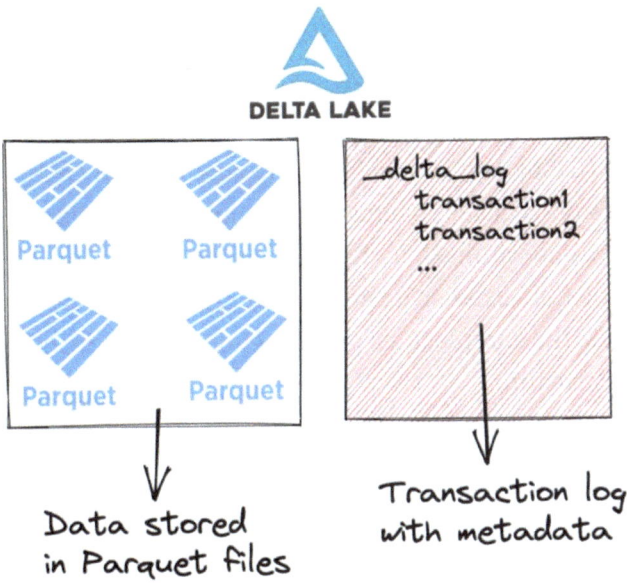

Figure 4-1. *Delta Lake components*

The components are as follows:

- **Parquet files:** Parquet, which organizes data in a highly efficient columnar format, has been the de facto format for storing big data for quite some time. Delta is built on top of Parquet, as the actual data is stored in Parquet format, which ensures data compression and encoding optimizations.

- **Delta log:** The Delta log is the transactional log that acts a ledger and stores all the edits made to the Delta table. It acts as the single source of truth for the Delta table. The Delta Log is found in the _delta_log subdirectory within the Delta folder, which contains the

Parquet data files for the table. The Delta log enables the most common features such as ACID transactions, time travel, scalable metadata handling, etc.

- **Cloud Object Storage Layer:** It is important to note that the data is always in your cloud object storage layer (S3 for AWS, ADLS for Azure, and GSC for GCP). This storage layer ensures the durability and scalability of the data within Delta Lake, enabling users to store and process extensive datasets without the need to handle the complexities of managing the underlying infrastructure.

After looking at what composes the Delta Lake, let's look into some key features of Delta Lake.

- **Schema enforcement:** Delta enforces the schema by default and blocks bad writes to the data. However, it provides the flexibility to evolve the schema as needed.

- **ACID transactions:** ACID transactions ensure reliability and consistency, even during failures.

- **Version control:** As discussed earlier, the Delta log acts as a ledger and tracks all the changes made to tables. If required, say when your job fails midway, the older version can be easily restored.

- **Unified batch and streaming:** Delta provides the unique capability of a unified source and sink for streaming and batch processing. For example, you can stream and add batch data to the same Delta table.

- **Time travel:** The transaction log gives Delta the ability to time travel, enabling users to revert or access any version of the table as they were at a specific point in time.

- **Compliance:** Delta logs help improve data governance, security, and regulatory compliance needs.

Delta Lake: Medallion Architecture

The Medallion architecture, sometimes referred to as the *multihop* architecture, is the concept of logically separating the data in a lakehouse into multiple layers with each layer having specific properties. A standard medallion architecture consists of three main layers: Bronze, Silver, and Gold. It is best practice to curate your data by using a layered architecture approach, as it allows data teams to structure the data according to quality levels and define roles and responsibilities per layer. See Figure 4-2.

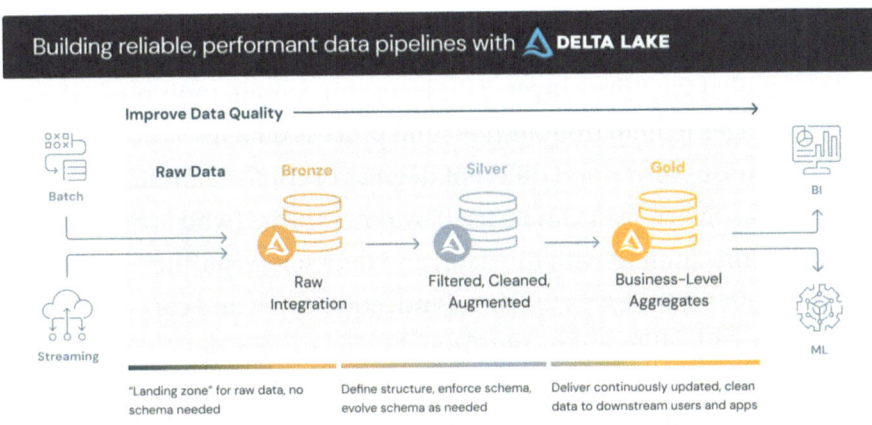

Figure 4-2. *Delta Lake medallion architecture*

Let's look into the three layers and see what they mean.

- The first layer is the raw layer, often called the **Bronze layer**. This layer will preserve the data as close as possible to the original data. It is always a best practice to maintain a copy of your source system data for the following reasons:

  - The source system copy helps to back out a production workload in case of any error.

  - The bronze layer helps to reprocess a data pipeline.

  - The Bronze layer loads preserve historical data for analytical processing and enable insights and trend analysis.

  - A source system copy helps hydrate a data lake to enable new use cases and is often required by data scientists so they have access to nontransformed/unbiased data.

- The second layer is the staging layer, which can be called the **Silver layer**. This layer can contain multiple stages to help troubleshoot and process data in various forms and different degrees of conformation. The Silver layer can be used by power users (who are more familiar with the data) and data scientists, but with some risk as the data is not conformed and can provide different results from the one usually open to all business users. A silver layer load typically consists of the following:

- Filtered and augmented data typically formatted per business requirements.

- For data scientists, the data in this layer often is free from class imbalance problems and enables faster model development in neural networks and other approaches.

- The third layer is called the refined layer, or the **Gold layer**. This layer is open to all business users. It will contain the confirmed (agreed upon) data and will be treated as the one true version for the business. This layer can contain smaller subsets of the data for a specific purpose (sometimes called data marts). The Gold layer often does the following:

  - Answers very specific business questions

  - Most likely is fully aggregated data

  - Is the data that is ready for the presentation layer for BI tools to slice and dice this information (an OLAP cube)

  - Summary data and quality checked (dimensions serve as single source of truth)

Now after understanding the inner workings of Delta and the medallion architecture, let's look at some of the key features of Delta Lake.

Delta Lake Key Features

The following sections cover the key features of Delta Lake.

Update, Delete, and Upserts in Delta Table

Delta supports both Update and Delete commands, both of which are not supported by traditional Parquet format. Further, it provides the ability to upsert using the MERGE SQL Command.

Let's examine how the MERGE SQL operation can be used to upsert data into a Delta table from a source table, view, or DataFrame.

```
MERGE INTO target
USING source
ON source.key = target.key
WHEN MATCHED THEN
  UPDATE SET *
WHEN NOT MATCHED THEN
  INSERT *
WHEN NOT MATCHED BY SOURCE THEN
  DELETE
```

These are important operations and can be easily done in traditional databases, but now you can also do so within your Delta Lake layer.

Schema Evolution

It is important to note that Delta enforces the schema by default. This prevents users from adding data that does not conform to the existing schema, avoiding unwanted data additions to your table and maintaining data quality. Any new write to a table is checked for compatibility with the target table's schema before it is committed. If the data is not compatible, Delta Lake cancels the transaction altogether (no data is written) and raises an exception to let the user know about the mismatch.

However, data sources evolve over time due to changing requirements, which might involve adding or dropping new fields to existing tables. So, to fulfill this use case, although Delta, by default, enforces schema, it also supports schema evolution.

Therefore, schema evolution allows users to easily change a table's current schema to accommodate changing data such as including one or more new columns while performing an append or overwrite operation. Therefore, schema evolution can be used when you intend to change the schema of your table by either setting the option "mergeSchema" to "true or setting the property spark.databricks.delta.schema.autoMerge. enabled *to* true.

By including the mergeSchema option in your query, any columns present in the DataFrame but not in the target table are automatically added to the end of the schema as part of a write transaction. Nested fields can also be added, and these fields will be added to the end of their respective struct columns as well.

From Spark 3.0 onward, explicit DDL (using ALTER TABLE) is fully supported. The following code snippets provide some examples of how this can be utilized:

- Adding new columns (at arbitrary positions)

```
ALTER TABLE table_name ADD COLUMNS (col_name
data_type [COMMENT col_comment] [FIRST|AFTER
colA_name], ...)
```

- Reordering existing columns

```
ALTER TABLE table_name ALTER [COLUMN] col_name
(COMMENT col_comment | FIRST | AFTER colA_name)
```

- Renaming existing columns

```
ALTER TABLE table_name RENAME COLUMN old_col_
name TO new_col_name
```

To conclude, Delta supports both schema enforcement, which prevents adding data that does not conform to the existing schema, and schema evolution, which gives users the flexibility to make intended changes to the table.

Time Travel

Delta Lake's time travel feature allows users to access and query historical versions of data stored in Delta tables. This is important because it eliminates the need to maintain point-in-time copies of data, which is cumbersome and costly. Delta Log acts as a transaction log that maintains a granular view of changes made to data over time.

Some of the most common use cases where you might need to access previous versions of data are auditing data as it changes over time, reproducing ML experiments or reports, or rolling back to the earlier version in case of job failures.

Let's move and see this in action. As explained earlier, every operation that executes on Delta table is automatically versioned in the Delta log (Figure 4-3).

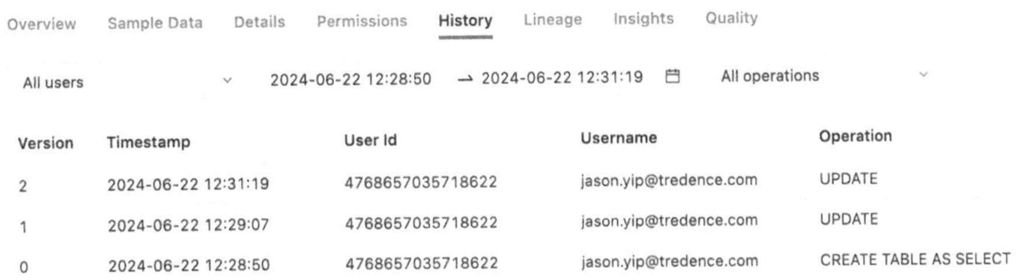

	Overview	Sample Data	Details	Permissions	**History**	Lineage	Insights	Quality

All users	˅	2024-06-22 12:28:50	→ 2024-06-22 12:31:19	📅	All operations	˅

Version	Timestamp	User Id	Username	Operation
2	2024-06-22 12:31:19	4768657035718622	jason.yip@tredence.com	UPDATE
1	2024-06-22 12:29:07	4768657035718622	jason.yip@tredence.com	UPDATE
0	2024-06-22 12:28:50	4768657035718622	jason.yip@tredence.com	CREATE TABLE AS SELECT

Figure 4-3. *Delta Log snapshot*

You can query the previous versions of the Delta table by doing the following:

1. Using a timestamp

   ```
   SELECT count(*) FROM my_table VERSION AS OF 5238
   ```

2. Using a version number

   ```
   SELECT * FROM employee_delta VERSION AS OF 2
   ```

A key question is how far back one can go to query previous versions of the Delta table. By default, you can query historical versions of the table for 30 days. Now, depending on the use case, one can increase or decrease the time by using the command `delta.logRetentionDuration`. This gives users the flexibility to manage storage costs versus the need to go back and access historical data.

Clone Delta Tables

When you clone a table, you are basically creating a replica of a table at a given point in time. As the name suggests, clones have metadata as the source table but behave as a separate table with a separate lineage or history. Therefore, any changes made to clones affect only the clone and not the source. Further, if the source data changes after the clone is created, those changes are not reflected in the cloned table automatically.

You can create a copy of an existing Delta Lake table on Databricks at a specific version using the `clone` command. Also, clones have a separate independent log history from the source table. Time travel queries on your source table and clone may not return the same result.

There are two types of clones that can be created: deep clones and shallow clones. Let's look into both of these:

> **Deep clone:** A deep clone makes a full copy of the source table's metadata and data files. This is similar to copying a table with a CTAS command (`CREATE TABLE... AS... SELECT...`). Since the metadata is being copied from the source table, you do not need to re-specify partitioning, constraints, and other information as you have to do with CTAS.
>
> Deep clones are helpful when creating a completely independent copy of a Delta table for use cases like archiving specific tables or do transformations on a new copy to test some transformations
>
> Deep clones can be quickly created using the following syntax:
>
> ```
> CREATE OR REPLACE TABLE db.target_table CLONE
> db.source_table --
> ```
>
> **Shallow clone:** A shallow (*also known as Zero-Copy*) clone duplicates only the metadata of the source table. The data files of the table itself are not copied, so another physical copy of the data is not created,

which helps save storage costs. These clones are not self-contained and depend on the source from which they were cloned as the source of data.

Shallow clones are useful when you want to perform experiments on a new table, such as testing new code on production data, without affecting the production tables.

Shallow clone can be created using the following syntax:

```
CREATE OR REPLACE TABLE my_test SHALLOW CLONE
my_prod_table;
```

One point to remember is that shallow clones are not self-contained tables like deep clones. If the data is deleted from the source table for any reason, your shallow clone may not be usable.

Generated Column

Generated columns are a special type of columns whose values are automatically generated based on user-specified functions over the columns in the Delta table.

When you write to a table with generated columns and you do not explicitly provide values for them, Delta Lake automatically computes the values. If you explicitly provide values for them, the value must satisfy the constraint (`<value>` `<=>` `<generation expression>`) IS True or write will fail.

Change Data Feed

One of the important functionalities of working in a medallion architecture is what we call *change data capture*. Change data capture basically refers to the process of capturing only incremental changes to a source table and merging only those changes with the target table.

Within the medallion architecture, as the data moves from bronze to silver to gold, you can implement the CDC functionality by using the change data feed (CDF) in Delta Lake. See Figure 4-4.

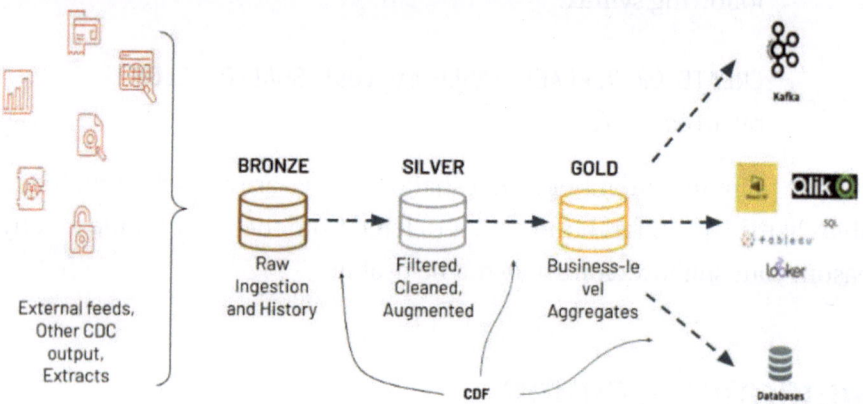

Figure 4-4.   *CDF*

The Delta CDF captures the row-level changes between versions of a Delta table. When CDF is enabled on a Delta table, the Databricks runtime records "change events" for all the data written into the table into a separate folder alongside the Delta log. The captured includes both the row data and corresponding metadata indicating whether the specified row was inserted, deleted, or updated.

It is important to note that CDF only provides the CDC capability within the medallion architecture and not for data ingested from source systems, e.g., databases to Delta Lake. To take advantage of the CDF functionality, bring your external data sources to the Bronze layer and then enable CDF from that point forward. This will allow you to use the Change Data Feed in moving to the Silver or Gold layers or feeding out to an external platform.

Change Data Feed can be easily enabled on all *new tables* by setting the property `spark.databricks.delta.properties.defaults.enableChangeDataFeed = true;` either in cluster settings or in the notebook.

You can set this property on the `CREATE TABLE` command as well.

```
CREATE TABLE student (id INT, name STRING, age INT)
TBLPROPERTIES (delta.enableChangeDataFeed = true)
```

If the table already exists, use `ALTER TABLE` to set the property.

```
ALTER TABLE myDeltaTable SET TBLPROPERTIES (delta.
enableChangeDataFeed = true)
```

Once the CDF feature is enabled on the table, a `_change_data` folder gets created under the table directory and records the change data for `UPDATE`, `DELETE`, and `MERGE` operations (Figure 4-5).

```
1  %sql
2  SELECT * FROM table_changes('silverTable', 2, 4) order by _commit_timestamp

▸ (1) Spark Jobs
```

Country	NumVaccinated	AvailableDoses	_change_type	_commit_version	_commit_timestamp
1	Australia	100	3000	insert	2	2021-04-12T20:48:05.000+0000
2	USA	10000	20000	update_preimage	3	2021-04-12T20:48:08.000+0000
3	USA	11000	20000	update_postimage	3	2021-04-12T20:48:08.000+0000
4	UK	7000	10000	delete	4	2021-04-12T20:48:11.000+0000

Showing all 4 rows.

Figure 4-5. *CDF change log*

This table can now be used to update only the changes, say, from the Silver table to the Gold table.

CDF can be useful in a number of use cases. For example, you can now update only the changes from your Silver table to Gold tables with substantially less processing cost. Another use case might be when you want to transmit data incrementally from Gold tables to external systems that can ingest change data output to reduce the processing overhead. Finally, for audit and compliance purposes, it might be necessary to keep a record of when, where, and how data has been changed. CDF with its change log helps to maintain the logs.

Universal Format

As enterprises move toward building their lakehouse architectures, one of the decisions they need to make is to choose the data format. Ideally they want to store data in an open-source format but one that gives then data warehouse capabilities. There are three open-source formats that meet this criteria: Delta, Iceberg, and Hudi.

Now if we go one level deeper in these formats, we see that all three are built on top of Parquet with the difference in the metadata layer. But these differences make these formats incompatible to be read by the same reader. The problem is further complicated when different departments within the organizations try to use these different formats within the lakehouse architecture. See Figure 4-6.

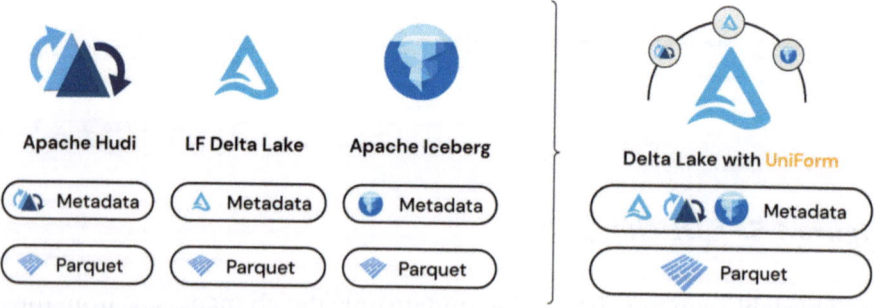

Figure 4-6. UniForm decoded

To solve this problem, Databricks announced Databricks UniForm (Universal Format) with the Delta 3.0 release. As discussed earlier, all three formats are built on top of Parquet. UniForm takes advantage of this fact and is able to make Delta tables accessible as Iceberg or Hudi tables to respective readers without any data duplication or additional costs. When a table is created with UniForm activated, the metadata for the additional formats (e.g., Iceberg) is automatically instantiated and subsequently updated in response to any data mutation.

Note that prior to the release of Delta UniForm, the ways to switch between open table formats were copy- or conversion-based and only provided a point-in-time view of the data.

Let's see an example there is a Delta reader and an Iceberg reader that is trying to read the Delta tables that are written by a Delta writer. Uniform in this scenario will generate Iceberg metadata asynchronously along with Delta metadata, thereby allowing both readers to read from the same Delta table. It is important to note that this is possible only with Unity Catalog (discussed in the next chapter), which essentially acts as an Iceberg catalog as well and is compatible with Iceberg APIs. See Figure 4-7.

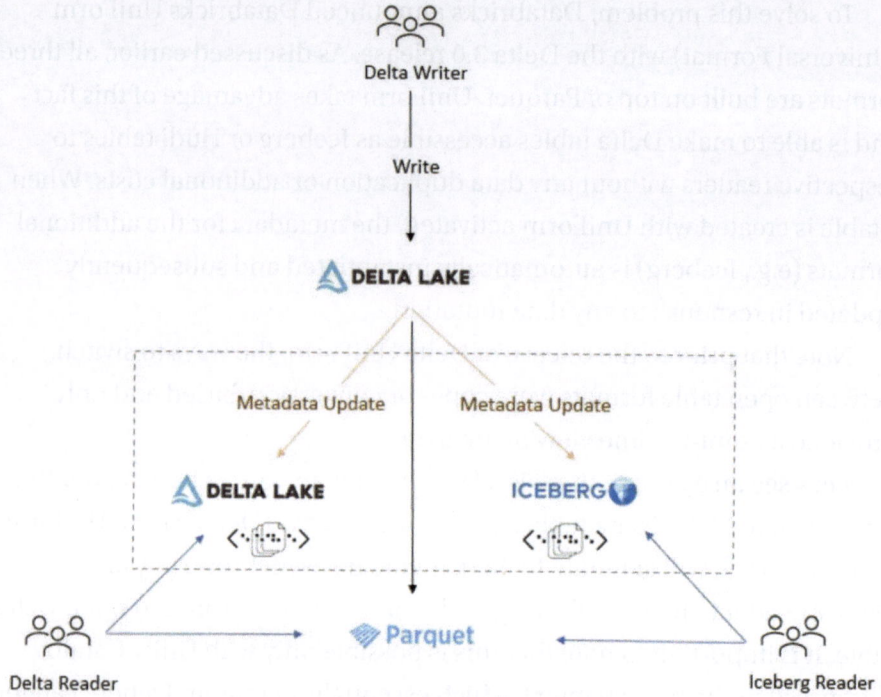

Figure 4-7. *Inner workings of Uniform*

You can enable Uniform on a new table by running the following command. Please note that Uniform is available only for UC-enabled tables.

```
CREATE TABLE uniform.test.T(name string , age int) TBLPROPERTIES(
 'delta.enableIcebergCompatV2' = 'true',
 'delta.universalFormat.enabledFormats' = 'iceberg');
```

Let's add some data.

```
INSERT INTO uniform.test.T VALUES ('Mark', 35), ('Tom', 42)
```

If we look into the table properties for the data, we can see something like Figure 4-8.

	A^B_C key	A^B_C value
1	delta.columnMapping.maxColumnId	2
2	delta.columnMapping.mode	name
3	delta.enableIcebergCompatV2	true
4	delta.feature.columnMapping	supported
5	delta.feature.icebergCompatV2	supported
6	delta.minReaderVersion	2
7	delta.minWriterVersion	7
8	delta.universalFormat.enabledForma...	iceberg

Table ⌄ +

↓ 8 rows | 1.35 seconds runtime

Figure 4-8. *Properties of Delta table with UniForm enabled*

Thus, with open table formats, organizations experience seamless data management, ensuring data integrity and enabling smooth transactions across multiple users and processing engines.

In the next part of the chapter, we will discuss some of the most common performance optimization techniques, such as vacuum, optimize, partitioning, and z-order. These techniques are not only optimization tools but also help slash storage costs, enhance parallelism, and reduce operating load on the infrastructure.

Delta Optimization

It is important to have clean and optimized Delta tables to enhance query performance and build efficient pipelines. As discussed earlier, tables can grow very large over time and then run into issues like small file problems or file layouts that do not support the query patterns. These techniques aim to alleviate some of the issues discussed.

- **Partitioning:** As the name suggests, partitioning refers to grouping of data files under the same column based on the partition key. Partitioning data can significantly enhance query performance as it will help Spark to skip a lot of unnecessary data partition (i.e., subfolders) during scan time. Partitioning works best with low-cardinality columns, and one can choose columns that are commonly used in queries for partitioning.

```
CREATE TABLE table_name
USING delta
PARTITIONED BY (column_name)
-- OR --
ALTER TABLE table_name ADD PARTITION
(column_name = 'value')
```

  As a best practice, do not partition tables under 1TB in size and partition data by a column if you expect each partition to be at least 1GB. Further, always choose a low-cardinality column—for example, year or date—as a partition key.

- **Optimize:** As discussed earlier, Delta folders might accumulate a very large number of small files (small file problem), which has an impact on query performance. Optimize compacts and pack these small files to a configurable size, which is optimum to maximize the performance of big data processing engines. Optimize keeps all the data as is, but table statistics are recalculated, and metadata is cleaned up by removing unnecessary entries. The target file (1GB default) size of the new command can be changed by tweaking the following:

```
spark.databricks.delta.optimize.maxFileSize
```

You can run Optimize on a Delta table by simply running this command:

OPTIMIZE table_name

As a practice, OPTIMIZE (with or without ZORDER) should be done on a regular basis, say once a day or weekly, to maintain a good file layout for better downstream query performance. Also, run Optimize on a separate job cluster because with compute-intensive VMs, it is a compute-intensive operation.

- **Z-order:** Z-ordering reorganizes data within Delta tables to improve query performance. It rearranges the data based on specified columns, allowing Delta Lake to skip irrelevant data during query execution. In short, the entire table is rewritten according to the columns mentioned in the z-order command.

  As a best practice, always choose high-cardinality columns (for example, customer_id in an orders table) for z-ordering. This is the opposite of partitioning, where low-cardinality columns are chosen. Further, choose the columns that are most frequently used in filter clauses or as join keys in the downstream queries. Finally, it is best to limit the columns to four or fewer because more than that and the effectiveness of z-order degrades.

- **Vacuum:** Vacuum deletes files that are redundant in in the Delta folders. By default Delta retains older files up to 7 days and can be configured using the property `delta.deletedFileRetentionDuration`.

Vacuum is not reversible, so it should be used with caution. Further, once it is done on the table, your ability to use time travel is limited, but the vacuum saves on storage costs as unnecessary files are deleted. So, depending on the use case, you can consider whether you want to vacuum a particular table.

After learning the fundamental optimization techniques, let's move on to two of the newer optimizations: liquid clustering and predictive I/O.

Liquid Clustering

Liquid clustering is a new feature introduced for the Delta table in Runtime 13.1 and above. Let's examine how you can utilize this feature to enhance the performance of your Delta tables without much manual intervention.

As discussed, two of the most common techniques used to optimize your Delta tables for efficient storage and data retrieval are table partitioning and z-order.

When done right, these techniques help users increase the performance of their queries. But both require careful consideration. For example, you need to use the right column to partition your data, and z-order needs to be done each time new data is added to your table. Therefore, data engineers need to constantly work to keep the tables optimized.

Liquid clustering aims to replace both these features with much less manual intervention, thus reducing data management and tuning overhead. It's flexible and adaptive to data pattern changes, scaling, and data skew.

With liquid clustering, keys (columns) can be chosen purely based on the query access pattern. You do not need to consider things like cardinality, key order, file size, potential data skew, and future access pattern change. Further, the keys can be changed without rewriting the files in the table; thus, over time, as the query pattern changes, the data layout adapts accordingly.

As a best practice, you should enable liquid clustering for all your new Delta tables. Some of the scenarios where liquid clustering are highly useful is when tables have significant data skew, when they are growing rapidly in size with new data, and when queries involve frequent filtering by high cardinality columns.

Let's see how liquid clustering works internally.

Working with Liquid Clustering

Liquid clustering is enabled during the creation of a Delta table by using the command CREATE BY and defining the clustering keys. Once enabled, run OPTIMIZE jobs to cluster data incrementally.

```
-- Create an empty table
CREATE TABLE table1(col0 int, col1 string) USING DELTA CLUSTER
BY (col0);

-- Using a CTAS statement
CREATE EXTERNAL TABLE table2 CLUSTER BY (col0)  -- specify
clustering after table name, not in subquery
LOCATION 'table_location'
AS SELECT * FROM table1;
```

```
--Trigger the Liquid clustering job
OPTIMIZE table2;
```

Some of the other useful use cases and commands are as follows:

```
-- Using a LIKE statement to copy configurations
CREATE TABLE table3 LIKE table1;
```

```
--Change the Cluster Key
ALTER TABLE table_name CLUSTER BY (new_column1, new_column2);
```

```
--disable the cluster Key
ALTER TABLE table_name CLUSTER BY NONE;
```

Another important aspect of liquid clustering is determining how to choose the right clustering keys. To start, choose columns that are frequently used in queries regardless of their cardinality. You can begin with one column and add up to four columns when needed. Finally, as the queries and workload evolve, use ALTER TABLE tbl CLUSTER BY to change the clustering keys as often as you want. The best part is that there is no need to rebuild the table.

Current Limitations

According to the Databricks documentation, the following limitations exist:

- You can only specify columns with statistics collected for clustering keys. By default, the first 32 columns in a Delta table have statistics collected.

- You can specify up to four columns as clustering keys.

- Structured streaming workloads do not support clustering-on-write.

Predictive I/O

Predictive I/O is a collection of Databricks ML-powered optimizations that improve the performance for your data interactions. Its accelerated reads reduce the time to scan and read data, while accelerated updates reduce the amount of data that needs to be rewritten. Predictive I/O is enabled by default on serverless SQL and Pro SQL warehouses and clusters with runtime 14.0 and above.

Let's move into and see how predictive I/O works with a simple analogy. Imagine all the data transactions are no more than read and write. Think of the Windows defragmentation function, which has existed all the way back to our Windows 95. File systems are often represented by data blocks, just like containers or buckets, but over time, there will be some room left in each block, regardless of the size of the block. Everyone who has done some packing for a trip would understand this concept. How many bags do we bring? Below is an illustration of a simple file system.

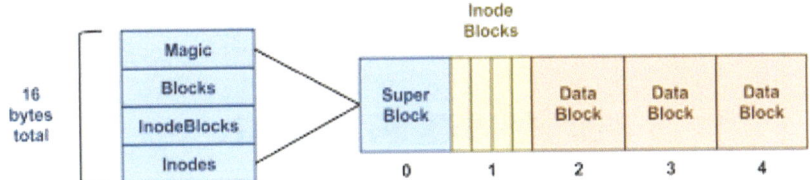

Figure 4-9. *A Simple File System*

Source: `https://www3.nd.edu/~pbui/teaching/cse.30341.fa18/project06.html`

In terms of file systems, there is a concept of defragmentation, which is simply reorganizing all the files into proper blocks to optimize storage and read and write efficiency. You would agree that anything organized would be more efficient to retrieve. The idea is simple.

In terms of the Delta format, there are three concepts that we need to consider.

- **File size:** The Delta format is organized by Parquet files, which is similar to the block size discussed in a file system.

- **Copy on write:** The Delta format supports ACID transactions. To perform updates or deletes, the analogy is similar to taking out something (delete) and placing it back into the file system. In the form of Parquet files, it must be written back onto the disk to be able to read again. So anything that's changed will need to be rewritten. Even the slightest change would affect the whole file, making the write operation very expensive if frequent updates are required.

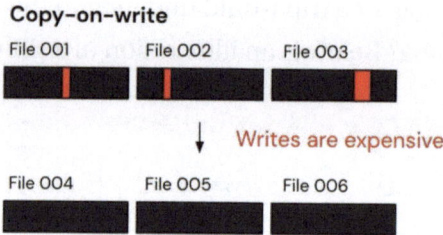

Figure 4-10. *Copy-on-write operation*

- **Merge on read**. To avoid expensive writes, the Delta format created a _delta_log folder, which keeps track of transactions, like update, delete, and insert. Similar to files in a file system, these log files can be either large or small and will become fragmented over time. While the write-to-the-log file is a cheap operation, the read will become expensive because it requires handling a large amount of operations in real time during reads.

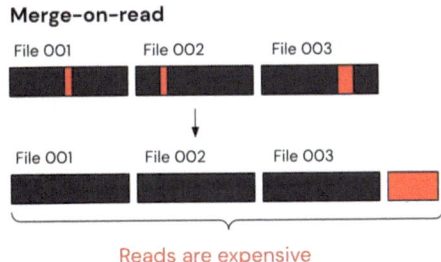

Figure 4-11. *Merge-on-read operation*

ML/AI to the Rescue

By now, you may wonder: how do we tune these settings? What is an optimal file size? Do we need to choose between copy on write or merge on read?

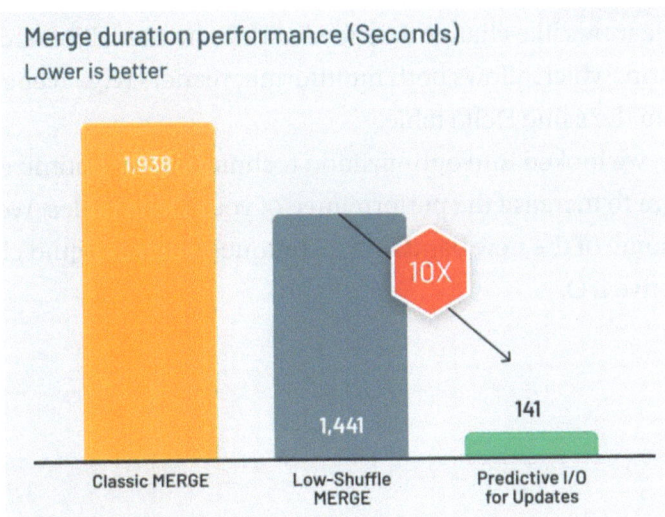

Figure 4-12. *Predictive I/O for Updates makes MERGE up to 10x faster than Low-Shuffle Merge (LSM)*

Source: https://www.databricks.com/blog/announcing-public-preview-predictive-io-updates

Databricks, with its vast experience, has developed machine learning models to optimize these settings. Developers no longer need to worry about the what, when, and how. The result is a 10x gain in update, merge, and delete.

Conclusion

In this chapter, we looked into one of the building blocks of the Databricks lakehouse architecture: Delta Lake. This format provides both reliability and performance to your data. Delta Lake is the most critical part of your lakehouse as it gives all the warehouse-type capabilities to your data, like ACID transactions, updates/deletes and merge functionality, schema enforcement and evolution, time travel, etc. We also looked into some advanced features like change data feed within the medallion architecture, and UniForm, which allows both multiformat readers (e.g., Iceberg reader) to read from the same Delta table.

Finally, we looked into optimization techniques like optimize, z-order, and vacuum to increase the performance of your Delta tables. We also reviewed some of the new hands-off techniques, such as liquid clusters and predictive I/O.

CHAPTER 5

Data Governance with Unity Catalog

Data is one of an organization's most significant assets. An important determinant of a company's performance and growth is how well its data is handled regarding quality, management, and ownership. Organizations today, especially with ever-expanding use cases for GenAI, face expanding data privacy regulations. Nonetheless, the reliance on data is increasing as organizations look to help optimize operations and drive business decision-making. Therefore, they are looking for data governance on their data platforms to ensure that not only their data assets but, more importantly, their AI products are consistently developed and maintained and their precise guidelines and standards are adhered to.

In this chapter, we will look at Unity Catalog—Databricks' data governance solution. We will introduce the concept of Unity Catalog and how it differs from traditional Databricks' hive metastore. Further, we will look at how you can enable Unity Catalog in your workspace and architect your data estate. Finally, we will deep dive into some of the key features of Unity Catalog, like centralized management, data lineage, and Delta Sharing.

© The Editor(s) (if applicable) and The Author(s),
under exclusive license to APress Media, LLC, part of Springer Nature 2024
N. Gupta and J. Yip, *Databricks Data Intelligence Platform*,
https://doi.org/10.1007/979-8-8688-0444-1_5

What Is Databricks Unity Catalog?

Unity Catalog is Databricks' governance solution and is a unified system for managing its data assets (see Figure 5-1). It is a central storage repository for all metadata assets, accompanied by tools for governing data, access control, auditing, and lineage.

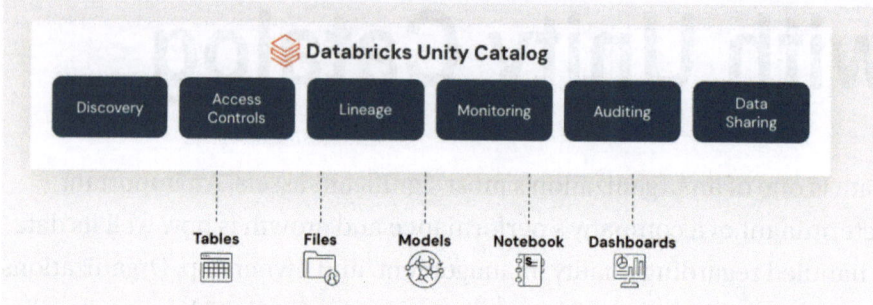

Figure 5-1. *Databricks Unity Catalog*

It maintains an extensive audit log of actions performed on data across all Databricks workspaces in your account. It provides capabilities such as effective data discovery, centralized metadata and user management, data lineage, and much more. It offers views and controls across all structured, semi-structured, and unstructured streaming data, AI models, notebooks, workplaces, files, tables, and dashboards.

In short, it brings all your Databricks workspaces together, offering fine-grained management of data assets and access. This streamlines operations by reducing maintenance overheads, accelerates processes, and increases efficiency and productivity.

Unity Catalog is the foundation of the Databricks Data Intelligence Platform, which understands the uniqueness of your data. If you are looking to build your next GenAI application, it is essential to enable Unity Catalog in your Databricks environment.

Unity Catalog: Before and After

Before Unity Catalog, Databricks workspaces were separate and independent units (see Figure 5-2). Each workspace had its own metastore, user management (adding/removing users), and Table ACL store. A simple example is that if a user created a table in one Databricks workspace, it would not automatically be available in another workspace. This led to data and governance isolation boundaries between workspaces, and if you wanted to bring consistency between your workspaces, it would mean duplication of effort. Some users handled this by developing pipelines or code to synchronize their metastores and ACLs, while others set up their self-managed external metastores to use across workspaces. However, these solutions added more complexity and maintenance.

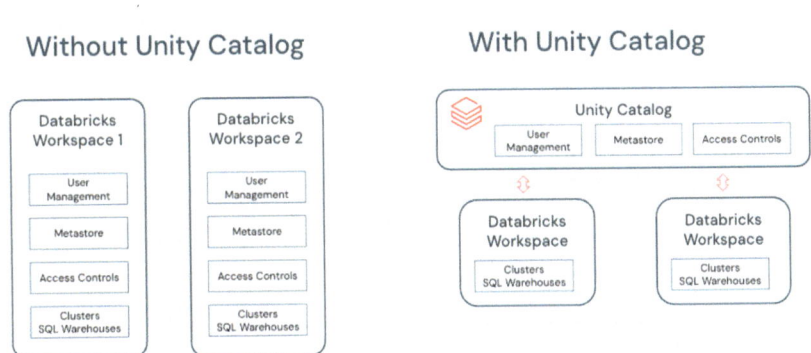

Figure 5-2. *Before and after Unity Catalog*

With Unity Catalog, Databricks has moved all three (User Management, Metastore, and Access Controls) out of workspaces to an *account* that works across all workspaces. The account, including the Account Console, which is a user interface to control the account, lives purely in the control plane. As a best practice, there should be one account per organization (i.e., your entire company) per cloud provider. A Databricks account lets you set up data, controls, and user management in one place and use them across multiple Databricks workspaces.

Unity Catalog Hierarchy

Now let's move on and understand some key concepts with Unity Catalog (UC) such as the metastore, catalog, etc. See Figure 5-3.

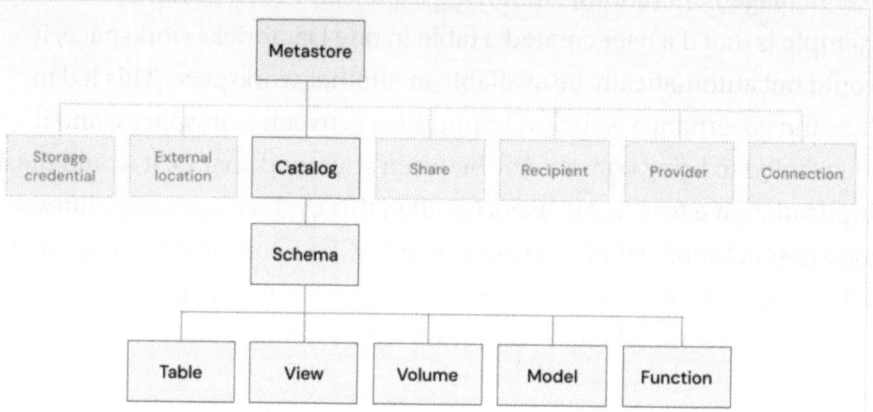

Figure 5-3. *Unity Catalog hierarchy*

- **Metastore:** A metastore stores metadata about data and AI assets and the permissions that govern access to those assets. UC metastore is a container in your cloud storage managed by Databricks. You can enable UC for a workspace by attaching it to a UC metastore. There should be one metastore per region, and all workspaces should be assigned to that metastore in that region. The metastore has a three-level hierarchy: catalog, schema, and tables.

- **Catalog:** A catalog serves as the top-level container in the three-level namespace hierarchy. It organizes the data assets and contains schemas (databases), tables, views, volumes, models, and functions.

- **Schema (database):** This is the second level in the three-level namespace and contains tables and views.

- **Tables:** Tables are defined within a schema and provide governance for tabular data. There are two types of tables: External and Managed.

  - **External tables:** In external tables, data is stored outside the managed storage location(s) for the associated schema/catalog/metastore. It is used when direct access to the data outside Databricks is required. Unity Catalog governs access to "external" tables but does not manage the underlying data. This means when you drop a table it deletes only the metadata and not the underlying data. You can use Delta and other file formats (CSV, JSON, etc.) while creating an external table.

  - **Managed tables:** These are the default way to create tables in UC. They are stored in a managed storage location (at the schema, catalog, or metastore-level storage location). Unity manages the data life cycle and file and folder layouts for these tables. The underlying data format is Delta. When a table is dropped, the underlying data is deleted from cloud storage within 30 days.

- **Volumes:** Volumes are defined within a schema and provide governance for nontabular data (e.g., image files, etc.). They can store and access files in any format (unstructured, semi-structured, structured) but cannot store tables. Volumes can be "managed" by defaulting to the schema's managed storage location or "external" by specifying an external storage location.

Unity Catalog Admin Roles

It is important to understand various admin roles associated with Unity Catalog.

- **Account admin:** Account admins administer and control anything at the account level, including SCIM, SSO, Metastore creation/deletion, and assignment of metastores to workspaces and create credentials for external location access. Account admins can query all data or perform grants on all data objects.

- **Metastore admin:** Metastore admins can create catalogs and assign their ownership (via grants) to groups or individuals. They can also create external locations. Metastore admins have visibility to all securable objects within the metastore they are admin of.

- **Data owners:** Data owners can perform grants on data objects they own and create new nested objects. For example, a catalog owner can create a schema and then a table within that schema.

- **Workspace admin:** Workspace admins are similar to cloud administrators and, as the name suggests, manage the workspaces. They can define cluster policies on workspaces, add/remove user assignments, elevate user permissions within a workspace of various objects like notebooks, etc., and change job ownership.

Getting Started with Unity Catalog

In this section, we will quickly review how to get started with Unity Catalog by creating a metastore and assigning users and groups to workspaces via the account console.

Create a Metastore

The steps to create a metastore are detailed in Databricks documentation (`https://docs.databricks.com/en/data-governance/unity-catalog/create-metastore.html`). As a quick overview, account admins can log into Databricks Account Console and create a metastore. See Figure 5-4.

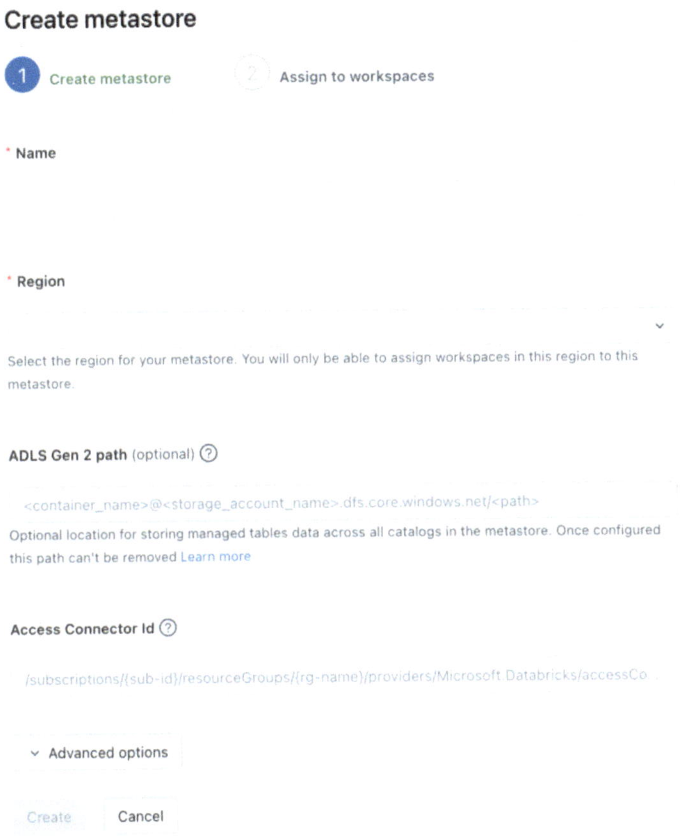

Figure 5-4. *Unity Catalog metastore setup interface*

The key inputs required are as follows:

1. Name of the metastore.

2. The region in which the metastore is created. One needs to remember that we can have one metastore per region.

3. ADLS Gen 2 Path or the S3 bucket, which will be the root bucket for the metastore.

4. Access Connector ID (Azure): An Access connector in Azure allows you to use Managed Identity to access storage containers on behalf of Unity Catalog users.

    IAM role ARN (AWS): Amazon Resource Name for the bucket that was setup in #2 (`https://docs.aws.amazon.com/IAM/latest/UserGuide/reference_identifiers.html#identifiers-arns`).

Once the metastore is created, you can assign the metastore to a workspace and thereby enable Unity Catalog (Figure 5-5) for the particular workspace.

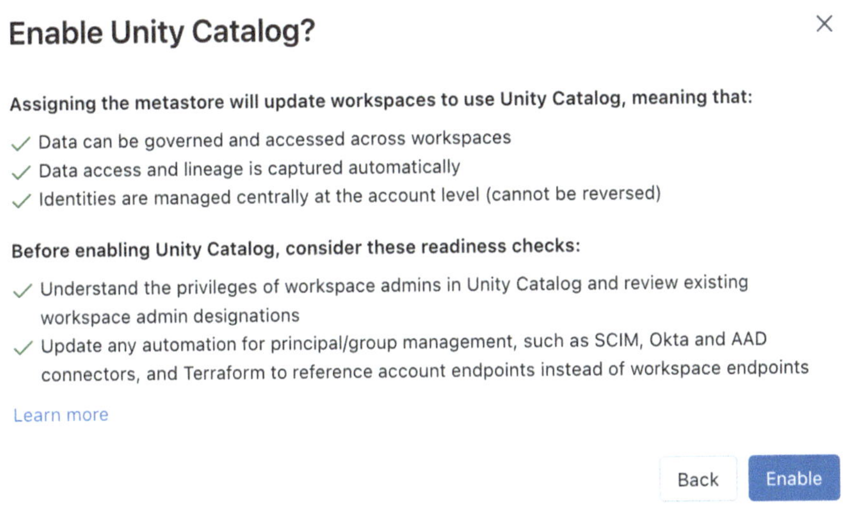

Enable Unity Catalog? ✕

Assigning the metastore will update workspaces to use Unity Catalog, meaning that:

✓ Data can be governed and accessed across workspaces
✓ Data access and lineage is captured automatically
✓ Identities are managed centrally at the account level (cannot be reversed)

Before enabling Unity Catalog, consider these readiness checks:

✓ Understand the privileges of workspace admins in Unity Catalog and review existing workspace admin designations
✓ Update any automation for principal/group management, such as SCIM, Okta and AAD connectors, and Terraform to reference account endpoints instead of workspace endpoints

Learn more

Back   Enable

Figure 5-5. *Final screen before enabling Unity Catalog*

Organizing Data in Unity Catalog

As discussed earlier, the catalog is the top-level container in the three-level namespace. As a best practice, you should use catalogs to segregate your organization's information architecture. This simply means catalogs can correspond to a department, team, business unit, or development environment scope (Dev, UAT, Prod), as shown in Figure 5-6.

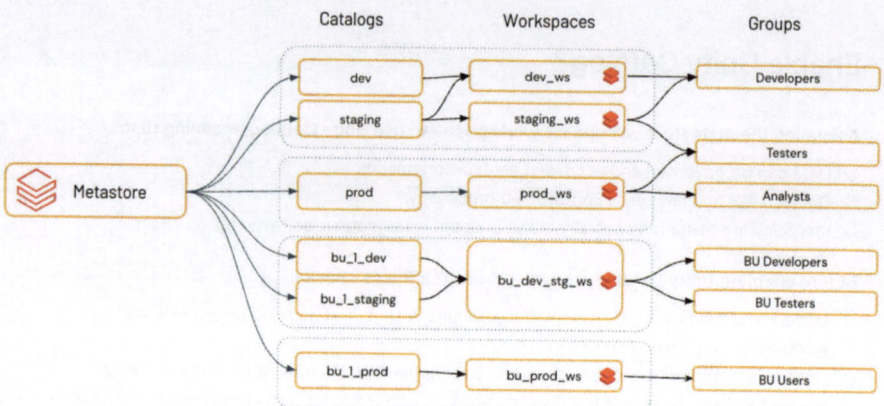

Figure 5-6. *Sample Unity Catalog structure*

Another typical pattern is that developers use workspaces as a data isolation tool—for example, using different workspaces for prod and dev environments or a specific workspace for processing sensitive data.

Therefore, while working in that specific workspace, they want to see only that specific catalog. For example, while working in a dev environment, you want only the dev catalog visible, not prod. Unity Catalog has a feature that allows you to **bind a catalog to specific workspaces**. This ensures that all specified data processing is handled in the appropriate workspace. These environment-aware ACLs allow you to ensure that only specific catalogs are available within a workspace, regardless of a user's individual ACLs. This means the metastore admin or the catalog owner can define the workspaces that a data catalog can be accessed from.

To learn more, please go to the following website:

```
https://docs.databricks.com/en/data-governance/unity-catalog/
create-catalogs.html#optional-assign-a-catalog-to-
specific-workspaces
```

Figure 5-7 illustrates the workspace setup architecture.

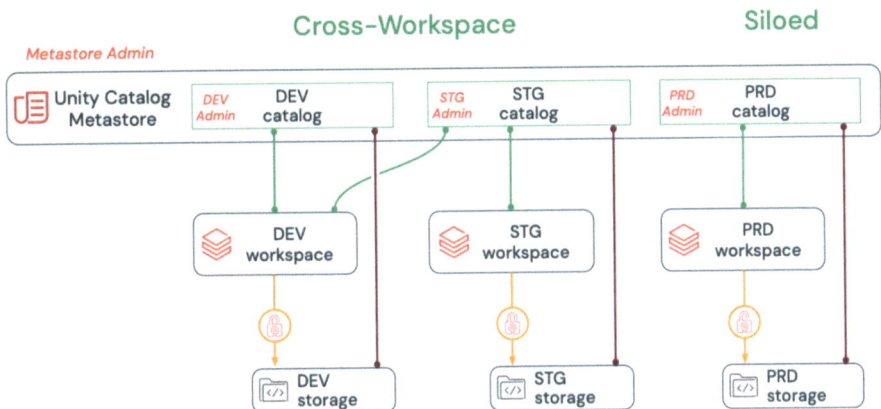

Figure 5-7. *With Unity Catalog, we can attach a catalog to SDLC workspaces*

Key Features of Unity Catalog

Let's talk about the key features in more detail.

Centralized Metadata and User Management

As explained earlier, Unity Catalog provides a single metastore across all workspaces in an account. This enables users to create and access tables, views, etc., across workspaces. Now, you can create multiple catalogs; set up schemas, tables, and views in one place; and access them across workspaces.

It is important to note that when multiple metastores are set up in an organization, the catalogs cannot be attached to the workspaces in other metastores. The solution is to use Delta Sharing, which will be discussed later in this chapter. See Figure 5-8.

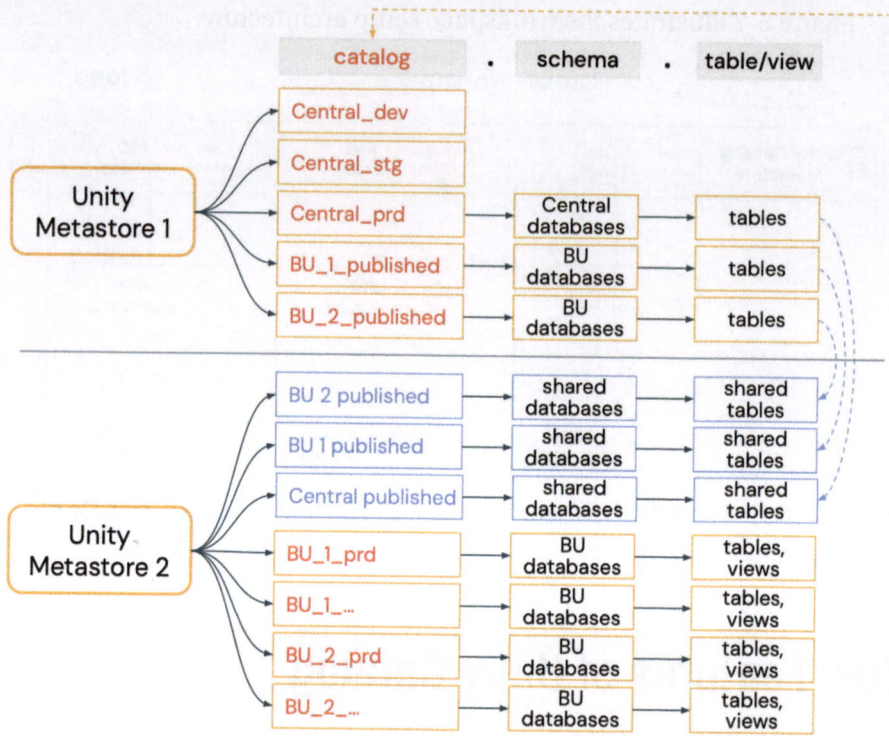

Figure 5-8. *Delta Sharing strategy with multiple metastores*

Another important feature of Unity Catalog is centralized user management. Before, UC admins had to add users to each new workspace either manually or through some SCIM synchronization and maintain those workspaces by workspace. With Unity Catalog, once you have synced your identity provider, say Azure AAD, via SCIM to Databricks Account Console, you can assign users/groups to all different workspaces via the account console, hence centrally managing users across workspaces. As a best practice, you should enable SCIM integration at the account level and sync users to workspaces with Identity Federation. Do not use SCIM at the workspace level at all.

Centralized Access Controls

One of the main requirements in any data platform is strict control over access to data to safeguard it and adhere to various data protection policies within your organization. Unity Catalog provides a centralized management method as data access policies are applied across all relevant workspaces and data assets.

The access control mechanisms use identity federation, allowing Databricks users to be service principals, individual users, or groups. In addition, SQL-based syntax, the Databricks UI, or even Terraform and APIs can be used to provide and control fine-grained access across a wide range of resources, including schemas, tables, views, clusters, notebooks, and dashboards.

Let's look into how you can use ANSI SQL to grant permission scopes on *securable objects* like tables or locations to *principals* like groups, users, or service principals. As a best practice, use groups for securing access to tables and owning securable objects. If a group owns an object, then any users in that group are owners.

```
GRANT <privilege> ON <securable_type> <securable_name> TO
'<principal>'
GRANT SELECT ON iot.events TO engineers
```

The same functionality is also available via Databricks UI in an easy-to-use point-and-click manner, which helps for easy access and auditing on the spot, as shown in Figure 5-9.

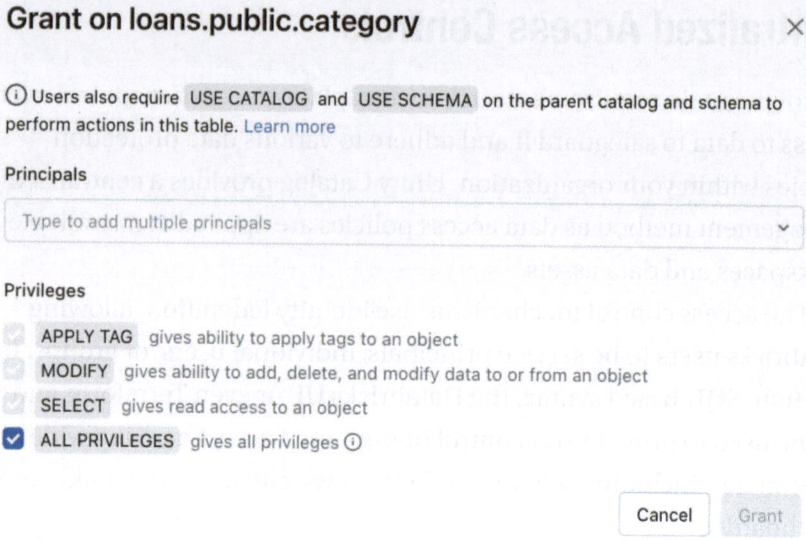

Grant on loans.public.category   ✕

ⓘ Users also require USE CATALOG and USE SCHEMA on the parent catalog and schema to
perform actions in this table. Learn more

Principals

Type to add multiple principals   ⌄

Privileges

☑ APPLY TAG   gives ability to apply tags to an object

☑ MODIFY   gives ability to add, delete, and modify data to or from an object

☑ SELECT   gives read access to an object

☑ ALL PRIVILEGES   gives all privileges ⓘ

[Cancel]   [Grant]

Figure 5-9. Granting permissions on a UC table

In addition, Databricks offers the ability to set these ACLs on objects via REST API or CLI, which means that Unity Catalog can support and power anything, from legacy entitlement request processes to modern dev/sec/ops initiatives.

Data Lineage

Data lineage is the process of tracking data flows from their source to their destination. It has gained significance due to the large volume of data processed through complex transformations and serves various purposes, including auditing and debugging. Thus, data lineage has become vital in understanding data movement, tracking, monitoring jobs, debugging failures, and tracing transformation rules.

Unity Catalog has end-to-end data lineages for all workloads, giving visibility into how data flows are consumed. Data lineage is automatically aggregated across all workspaces connected to a Unity Catalog metastore, which means that the lineage captured in one workspace can be seen in any other workspace that shares the same metastore.

Unity Catalog provides users with both table- and column-level lineage in a single lineage graph, giving users a better understanding of what a particular table or column is made up of and where the data is coming from. Users can easily follow the data flow through different stages, gaining insight into the tables and fields' relationships.

Further, the Unity Catalog tracks lineage for notebooks, workflows, ML models, and dashboards. This improves end-to-end visibility into how data is used in your organization and allows you to understand the impact of any data changes on downstream consumers. See Figure 5-10.

Figure 5-10.   *Unity Catalog lineage*

Data lineage holds critical information about the data flow and uses Unity Catalog's common permission model. This means that users with appropriate permissions can view the lineage data flow diagram, thus adding an extra layer of security.

Finally, Unity Catalog also offers rich integration with various data governance partners, such as Collibra and Purview, via Unity Catalog REST APIs, enabling easy export of lineage information to these partner catalogs.

Data Access Auditing

Unity Catalog automatically captures user-level audit logs and records the data access activities. These logs encompass various events associated with the catalog, such as creating, deleting, and altering multiple components within the metastore, including the metastore itself. Additionally, they cover actions related to storing and retrieving credentials, managing access control lists, handling data-sharing requests, and more.

The built-in system tables let you easily access and query the account's operational data, including audit logs, billable usage details, and lineage information.

We will do a deep dive into system tables later in the book when discussing observability in Chapter 9.

Data Search and Discovery

Unity Catalog offers a unified UI across the platform with enhanced search capabilities. Further, it leverages a common permissioning model to ensure security, enabling users to access assets they have access to. It allows tagging and documenting data assets, offers a comprehensive search interface, and utilizes lineage metadata to represent relationships within the data.

As we will discuss in Chapter 14, Databricks has greatly enhanced the platform's search and discovery capabilities by using LLMs and GenAI capabilities.

Row-Level Security and Column-Level Masking

Organizations are continuously striving to protect and secure their data, and one important way they are looking to do so is through row and column-level security. This feature is now available in the Unit Catalog–enabled Databricks workspaces.

Row Filters

Row filters allow you to apply a filter to a table so that subsequent queries only return rows for which the filter predicate evaluates to true. A row filter is implemented as a SQL user-defined function (UDF). A row filter accepts zero or more input parameters where each input parameter binds to one column of the corresponding table.

Create a Row Filter

```
CREATE FUNCTION <function_name> (<parameter_name>
<parameter_type>, ...)
RETURN {filter clause whose output must be a boolean};
```

Apply the Row Filter to a Table

```
ALTER TABLE <table_name>
SET ROW FILTER <function_name> ON (<column_name>, ...);
```

Let's look at an example. We want to create a function to filter data for the U.S. region. If the function is called by a user in the admin group, the RETURN_IF condition will be passed and all the data; otherwise, RETURN_IF will return the rows with region='US'.

```
CREATE FUNCTION us_filter(region STRING)
RETURN IF(IS_MEMBER('admin'), true, region="US");

ALTER TABLE sales SET ROW FILTER us_filter ON region;
```

Column Masks

Column masks let you apply a masking function to a table column. The masking function gets evaluated at query runtime, substituting each reference of the target column with the results of the masking function. For most use cases, column masks determine whether to return the original column value or redact it based on the identity of the invoking user. Column masks like row filters are expressions written as SQL UDFs.

```
CREATE FUNCTION <function_name> (<parameter_name>
<parameter_type>, ...)
RETURN {expression with the same type as the first parameter};
ALTER TABLE <table_name> ALTER COLUMN <col_name> SET MASK
<mask_func_name> [USING COLUMNS <additional_columns>];
```

In this example, if the user, the query results will mask the SSN numbers for nonadmin users.

```
CREATE FUNCTION ssn_mask(ssn STRING)
RETURN IF(IS_MEMBER('admin'), ssn, "****");

ALTER TABLE users ALTER COLUMN table_ssn SET MASK ssn_mask;
```

Dynamic Views vs. Row Filters and Column Masks

Now, an important question is why row-level filters are needed when Databricks already has dynamic views, which help users create abstracted, read-only views of one or more source tables. Further, dynamic views, row filters, and column masks let you apply complex logic to tables and process their filtering decisions at query runtime.

Let's discuss an important distinction between the two. Creating a dynamic view defines a new table name that must not match the name of any source tables. This abstraction layer ensures data integrity and prevents unintentional alterations to the core data. As a best practice, use dynamic views if you need to apply transformation logic such as filters and masks to read-only tables and if it is acceptable for users to refer to the dynamic views using different names than the source tables.

On the other hand, row-level filters and column masks apply logic directly to the table itself, and users don't have to deal with new or different table names or aliases. Again, use row filters and column masks if you want to filter or compute expressions over specific data but still provide users access to the tables using their original names.

Delta Sharing

Organizations seek to securely exchange data with their customers, suppliers, or partners to unlock further business value. However, a key requirement is that data sharing should happen securely to establish trust in data quality, security, and privacy. Some of the most common use cases for data sharing are data monetization with customers, B2B sharing with partners, suppliers, or intra-company data sharing among various departments.

Data sharing is not a new concept, and traditionally, organizations have deployed two main methods to do so. The first is via self-built solutions or tools via APIs, JDBC/ODBC, or file transfers via SFTP. The second is via commercial software vendors. The problem with the first is around scalability and infrastructure maintenance, while the problem with the second is costs and a lack of flexibility in terms of data access.

An Open Standard for Data Sharing

Databricks Unity Catalog comes with Delta Sharing, an open protocol for securely sharing data internally and across organizations in real time. Delta Sharing is fully integrated with Unity Catalog and allows you to centrally manage and audit the shared data across organizations.

Some of the differentiators or benefits of delta sharing include accessing data where it resides without creating any copies or moving to other platforms. Further, you can integrate Delta Sharing with either open-source clients (e.g., Pandas, Spark) or commercial clients (Power BI, Databricks and others) that support the protocol.

Delta Sharing is a transformative solution to access and share data. Let's see how Delta Sharing works.

How Delta Sharing Works

There are three main ways to share data with Delta Sharing:

- **The Databricks-to-Databricks sharing** in which both the provider and the recipient are on Unity Catalog–enabled workspaces. It has some advanced features like notebook sharing, AI Model Sharing, data governance, auditing, and usage tracking for both providers and recipients.

- **The Databricks open sharing protocol** allows a provider with a Unity Catalog–enabled workspace to share data with a recipient on any computing platform.

- **A user-managed implementation of the open-source Delta Sharing server,** which lets you share from any platform to any platform, whether Databricks or not. This is open sourced with instructions at https:// github.com/delta-io/delta. See Figure 5-11.

Figure 5-11. *Delta Sharing in action*

Let's examine how to set up Delta Sharing. The first step is for the data provider to register a Delta Lake table with the Delta Sharing server. This is done by creating a share (a read-only collection of data objects like tables, views, etc.) in a UC-enabled workspace.

```
CREATE SHARE IF NOT EXISTS test_share
```

```
ALTER SHARE test_share
 ADD TABLE test_table
```

The next step is to create a recipient, basically an individual or organization gaining access to a share.

```
CREATE RECIPIENT IF NOT EXISTS, recipient;
```

Once the recipient is created, each recipient gets an activation link that the recipients can use to download their credential. The Delta server identifies and authorizes the recipient/consumer based on these credentials. See Figure 5-12.

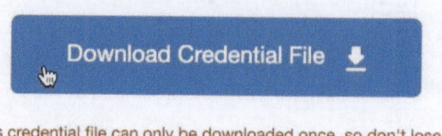

Figure 5-12. *Confirmation of Delta Sharing*

The recipient can download the credential file and then use it to access data. One key point to note is that the credential file is a single download only. The recipient can now use the file to authenticate and access data using various methods, such as Pandas, Java, or even Power BI.

Many open-source and commercial partners trust Delta Sharing, and Databricks also works with data providers to share data across the ecosystem. See Figure 5-13.

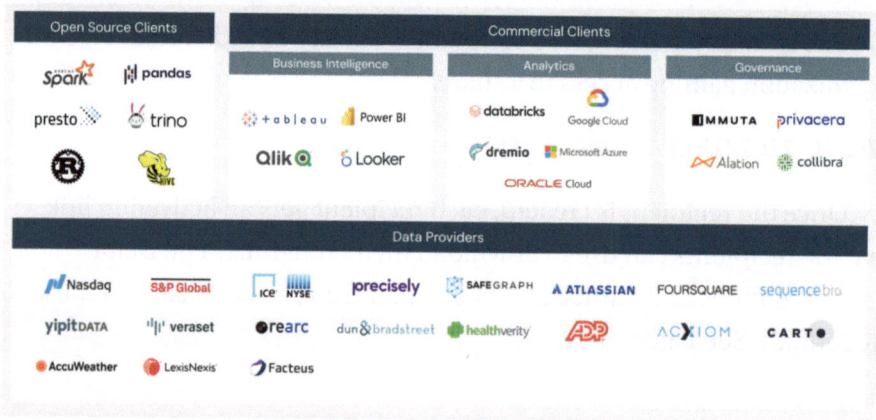

Figure 5-13. *Customers who use Delta Sharing*

To summarize, Unity Catalog unlocks Delta Sharing, which allows you to do secure in-place data sharing for data in Delta Lake to any tool that supports Delta Lake.

Conclusion

Databricks recognized two critical areas that needed attention: discovery and governance. Before Unity Catalog, data cataloging and governance were disjointed and cumbersome. With Unity Catalog, Databricks created an in-house solution that would seamlessly integrate with its ecosystem. Unity Catalog is the foundation for the Data Intelligence Platform and all the GenAI use cases that organizations are looking to deploy.

Unity Catalog serves as a central repository for all data assets, including files, tables, views, dashboards, and more. It provides a robust data governance framework, ensuring proper control and oversight. An extensive audit log records all actions performed on data stored in a Databricks account. Finally, Unity Catalog seamlessly ties in with other components of the Databricks ecosystem.

Data Engineering Part 1: Orchestrating Data Pipelines Using Databricks Workflows

The goal of orchestration is to configure multiple tasks into one complete end-to-end process or job. The orchestration service also needs to react to events or activities throughout the process and make decisions based on outputs from one automated task to determine and coordinate the next tasks. Finally, orchestration tools must provide full monitoring and observability capabilities to enable data engineers to have full visibility of their pipelines. See Figure 6-1.

© The Editor(s) (if applicable) and The Author(s),
under exclusive license to APress Media, LLC, part of Springer Nature 2024
N. Gupta and J. Yip, *Databricks Data Intelligence Platform*,
https://doi.org/10.1007/979-8-8688-0444-1_6

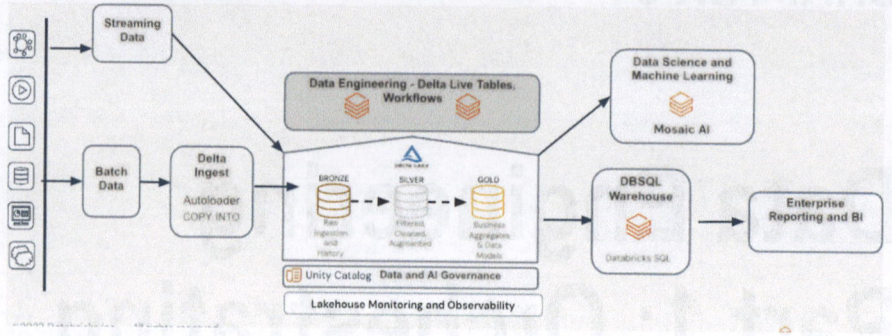

Figure 6-1. *End-to-end architecture of Databricks workflows*

Databricks workflows provide simple, reliable, and easy-to-use functionality that enables organizations to tackle the challenges of data orchestration efficiently. Data teams can easily create and manage multistep pipelines that transform and refine data and train machine learning algorithms within Databricks workspaces, thus saving time and effort of managing a separate tool. In this chapter, we will explore some key concepts of Databricks workflow jobs and examine the features that make it the orchestration platform of choice for Databricks lakehouses.

Databricks Workflow Jobs

Databricks workflows offer a unified and streamlined approach to orchestrating your data, BI, and AI workloads. You can define data workflows through the workflow user interface or programmatically using APIs, making them accessible to both technical and nontechnical teams.

Databricks workflows are similar to Azure Data Factory or Airflow, some popular orchestration services. Although these tools provide features for complete orchestration services, they do have a learning curve. They are an additional tool in your data stack, adding more maintenance and cost of ownership to your data platform.

Databricks workflows have evolved significantly over the last few years, not only with basic functionality such as scheduling, managing dependencies, Git integration but also by adding advanced-level features like retires, duration threshold, and repair and conditional tasks. These features give data engineers the capabilities to orchestrate their entire workload on the Databricks platform. Further, it is essential to note that there are no extra costs using workflows. The cost is for the underlying clusters/compute that the jobs use while executing.

In the next section, we will look at some of the building blocks and features of Databricks workflows.

Databricks Jobs and Tasks

Let's first understand the concept of a Databricks *job*. A Databricks job is a unit of orchestration within Databricks workflows. Basically, it is a method for running data processing and analysis applications in the Databricks workspace.

A job can consist of one or many tasks, each representing a specific unit of work such as an individual step or action. A job can consist of a single task or a large workflow with multiple tasks chained together by complex dependencies. For example, a data project might consist of ingesting data from various sources, transforming that data through the medallion architecture, and serving it via both as a SQL dashboard and an ML model. The entire flow can be a single job, and each activity is a task joined via dependencies.

Next, we will look into how you can create your first job.

Configure Databricks Job Tasks: Task-Level Parameters

As discussed, a task is the building block of a Databricks job. Figure 6-2 provides a snapshot of how to create a task. Let's examine some of the required parameters that you need to provide to do so:

- **Type:** Different "types" of tasks can be executed within a job. You can execute Databricks notebooks, JARs, Python Scripts and Wheels, SQL Queries, Delta Live Tables (DLT) Pipelines, or even DBT jobs. You can select the Type parameter depending on where your code resides.

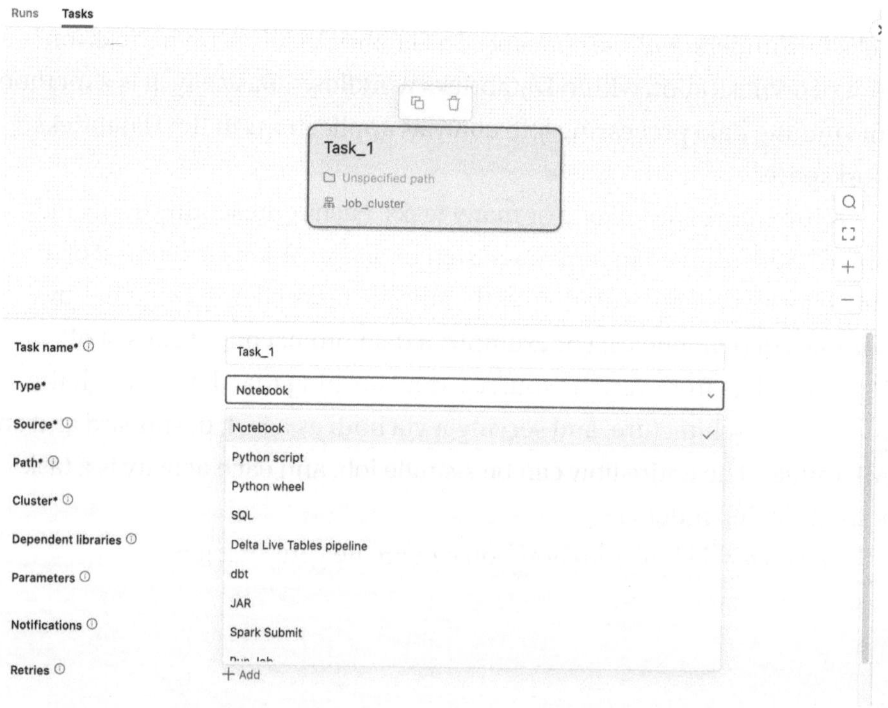

Figure 6-2. *Creating a task*

- **Source:** This parameter is primarily used when Notebook is selected as the type. A Databricks notebook is one of the most common utilities that data engineers utilize as a source of their pipeline code. These notebooks can either be in a Databricks workspace or stored in a Git repository like GitHub. Databricks workflows provide the functionality of calling these notebooks both directly from the workspace and from a Git repository via Git integration. Therefore, in the source, users select the path of the reference notebook in this parameter.

- **Cluster:** This field allows us to define the type of compute that would be used to run the job. Three options are available.

  - **Job cluster:** Job clusters have been the most common/preferred way to run your jobs in a production environment. They are pure ephemeral clusters, which means they get spun up once the job starts, execute the job, and then terminate when the job ends. Further, job clusters are around 50% cheaper than interactive/all-purpose clusters. It is highly recommended that one uses job clusters for production workloads. As a best practice, use the latest LTS version of cluster runtime for your workloads.

  - **Interactive clusters:** Interactive or all-purpose clusters are best used for developing ETL pipelines, testing jobs, and ad hoc queries. Interactive clusters should ideally not be used in production as they are not cost-efficient.

- **Serverless:** Serverless workflows are fully managed services that are operationally simpler and more reliable. Serverless compute provides capabilities like auto-optimization, selecting appropriate compute resources, automatic retires to job failures, etc. Serverless jobs were released very recently, and we believe they provide excellent compute for short and frequently running jobs.

- **Dependent libraries:** The configuration allows users to specify any libraries required for a task's successful execution. All libraries specified in the configuration are installed when the clusters start and are available when the job runs. These libraries could be installed from a public repo like Maven/Cran or from an ADLS/S3 folder.

- **Parameters:** Parameters provide values to a parameterized notebook. Developers often design parameterized notebooks for abstraction, so one notebook can be reused in multiple tasks with different parameter values rather than creating copies of the same task. This configuration lets you dynamically set and retrieve parameter values across tasks to build more mature and sophisticated data pipelines.

- **Notifications:** Notifications allow users to receive automatic updates when the task starts, succeeds, fails, or runs beyond the defined duration thresholds. Users can configure alerts to be notified via email or other communication channels like Slack, Teams, PagerDuty (and more), providing real-time observability of your task's execution.

- **Task retries:** Task retries determine when and how many times failed runs are retried. This feature enhances your workflows' reliability and fault tolerance by automatically attempting to recover from transient issues.

- **Duration threshold:** This configuration helps you define the execution time limits for a task. It either warns you if the task runs longer than expected or alerts you and terminates the task if it runs beyond the maximum set completion time.

After defining the parameter values for some or all configurations, you can hit the Create Task button to create your first task. Similarly, going back to our example, you can create tasks for all the other steps that need to be completed. Once your tasks have been created, let's move on to some of the job-level parameters you can define.

Configure Databricks Job Tasks: Job-Level Parameters

One of the key capabilities of any orchestration service is to run the jobs on a schedule, and the job-level Schedules & Triggers parameter does exactly that. Users can configure Databricks jobs to run either at a predefined time (schedule) or on a trigger (event-based or continuous).

Let's explore workflow triggers further and see different scenarios where they could be best used.:

- **Scheduled:** The Scheduled trigger enables you to automate the execution of your job by defining a specific time for it to run. It is important to note that this batch-based scheduling is not intended for low-latency use cases, as Databricks enforces a minimum of 10 seconds between subsequent runs. See Figure 6-3.

119

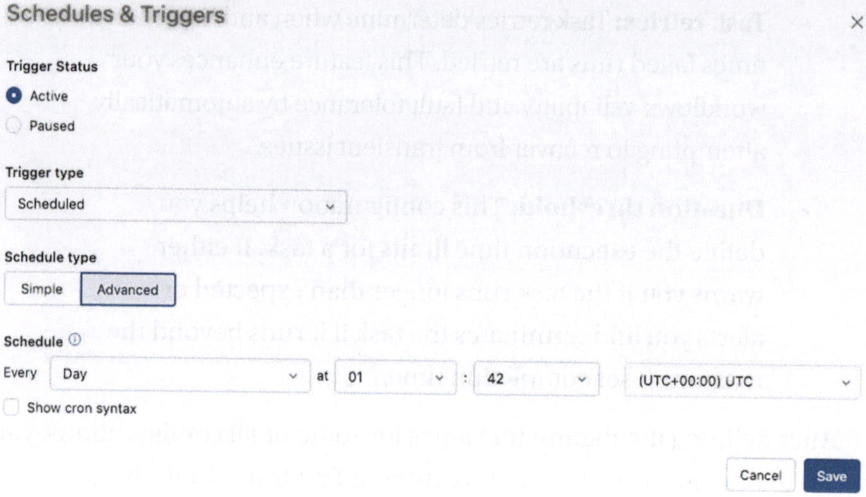

Schedules & Triggers   ✕

Trigger Status
◉ Active
◯ Paused

Trigger type
| Scheduled | ∨ |

Schedule type
| Simple | Advanced |

Schedule ⓘ
Every | Day | ∨ | at | 01 | ∨ | : | 42 | ∨ | (UTC+00:00) UTC | ∨ |
☐ Show cron syntax

Cancel   **Save**

Figure 6-3. *Scheduled trigger*

- **File Arrival:** File arrival triggers a Databricks workflow
  when a new file arrives in a particular configured
  cloud storage folder. This is useful when the file arrival
  schedule is irregular and you do not want a cluster
  to be always up and running to monitor the folder.
  One important thing to note is that you can only use
  this trigger in a Unity Catalog–enabled workspace.
  Further, one must use an external location added in
  the UC metastore and have READ permissions to the
  folder and Can Manage permissions on the job. See
  Figure 6-4.

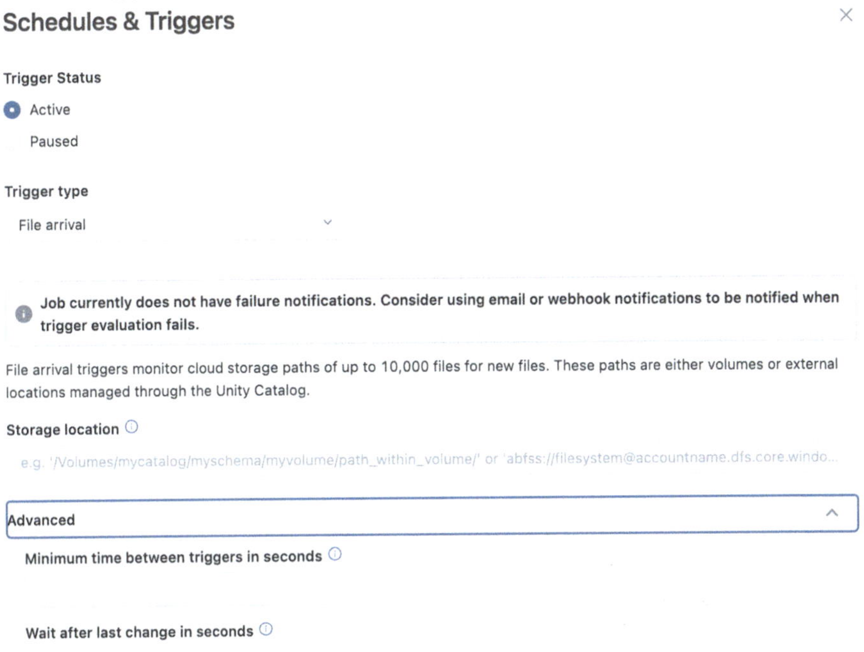

Schedules & Triggers   ✕

Trigger Status

🔵 Active

    Paused

Trigger type

    File arrival   ⌄

ⓘ **Job currently does not have failure notifications. Consider using email or webhook notifications to be notified when
    trigger evaluation fails.**

File arrival triggers monitor cloud storage paths of up to 10,000 files for new files. These paths are either volumes or external
locations managed through the Unity Catalog.

Storage location ○

    e.g. '/Volumes/mycatalog/myschema/myvolume/path_within_volume/' or 'abfss://filesystem@accountname.dfs.core.windo...

    Advanced   ⌃

    Minimum time between triggers in seconds ○

    Wait after last change in seconds ○

Figure 6-4. *File arrival trigger*

There are two other parameters one can set in the
file arrival trigger:

- **Minimum time between triggers in seconds:** This
  is the minimum time to wait before another run is
  triggered after a run is completed.

- **Wait after the last change in seconds:** This is the
  time to wait after a new file arrives before a run
  triggers. If another file arrives within this time
  frame, the timer will be reset.

- **Continuous:** As the name suggests, this trigger is for jobs running continuously until stopped. By setting the trigger type, Databricks will always ensure one active run of that job. A new job run will automatically kick off if the previous run completes or fails. You cannot use task dependencies with a continuous job, nor can you set retry policies.

- **Table update:** This trigger monitors for changes such as update, delete etc., in a Unity Catalog table (managed or external). See Figure 6-5.

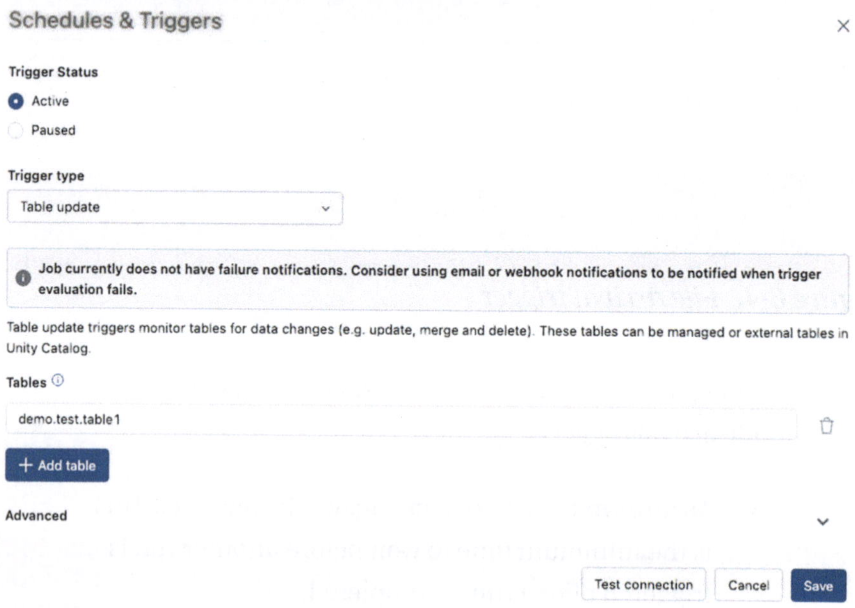

Schedules & Triggers   ✕

Trigger Status
- ● Active
- ○ Paused

Trigger type

Table update   ⌄

ⓘ Job currently does not have failure notifications. Consider using email or webhook notifications to be notified when trigger evaluation fails.

Table update triggers monitor tables for data changes (e.g. update, merge and delete). These tables can be managed or external tables in Unity Catalog.

Tables ⓘ

demo.test.table1   🗑

＋ Add table

Advanced   ⌄

Test connection   Cancel   Save

Figure 6-5. *Table update trigger*

Job Tags: Job tags allow users to easily identify and locate jobs by ownership, topic, and department. Job tags propagate to the job cluster and underlying VMs, which helps users assign charge-backs to a particular business unit. Furthermore, applying tags simplifies the process of filtering and identifying clusters based on specific criteria. This makes tracking, monitoring, and optimizing resources within your Databricks environment easier.

Job parameters: Job parameters give users more flexibility and control over their tasks in the workflows. They provide an easy way to add granular configurations to a pipeline, which is useful for reusing jobs for different use cases, a different set of inputs, or running the same job in different environments (e.g., dev staging and prod environments).

Job parameters allow users to provide both static values and dynamic values (that are provided by the system at runtime). An example of the dynamic value would be, say, the job ID is defined as {{job.id}} on the Parameter tab, which is the unique identifier assigned to the job. See Figure 6-6.

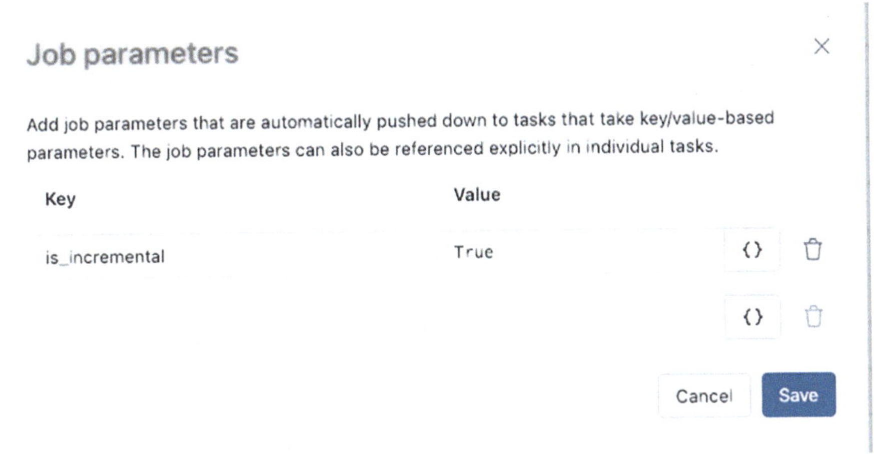

Figure 6-6. *Job parameters*

Table 6-1 identifies different parameter types for different task types.

Table 6-1. *Different Task Types in a Databricks Workflow*

Task Type	Parameter
Notebook	Key value pairs that set a value of a notebook widget
JAR	Array of strings passed to the Java main method
Spark Submit	Array of strings passed in for additional spark-submit arguments
Python Script	Array of strings retrievable using argument parse in Python
Python Wheel	Can specify positional arguments as array of strings; or keyword arguments as key-value pairs

To summarize, with **job parameters** you can parameterize your tasks that will give you more reusability of your jobs.

With this, we looked into some key features and configurations while setting up your Databricks jobs and tasks. In the next section, we will discuss some of the more advanced and newer features of Databricks workflows.

Advanced Workflow Features

In this section, we will look into some of the advanced features of Databricks workflows such as cluster reuse, conditional execution, etc.

Cluster Reuse: This feature allows users to utilize a job cluster across multiple tasks. Let's understand why it is such a useful feature.

Consider that there is a job that consists of five tasks. Without this feature (as it used to happen earlier), a new cluster would spin up when each task started and terminated when it ended. This led to five clusters being spun up and terminated, thus leading to more time to execute the entire job. But with the cluster reuse feature, you can configure only one cluster to spin up to run all the tasks and then terminate. This reduces the cluster initialization time for each task, leads to efficient cluster utilization, and decreases overall job latency.

Another important aspect of this feature is that the user still has the flexibility, if needed, to configure a particular cluster for a specific task. Continuing on our previous example, if a particular task requires a cluster of different configurations, for example, a compute-intensive task, one can configure a bigger cluster specifically for the task. Therefore, the different/ bigger cluster would spin up for this particular task.

Repair and Re-run: This feature enables users to repair/rerun failed or canceled jobs by running only the subset of failed tasks and any dependent tasks with the job. Because successful tasks are not run again with this feature, it reduces the time and resources required to recover from unsuccessful job runs.

Now, continuing from our previous example, suppose Task_3 (the third task) was unsuccessful. After fixing the cause of the failure, you can rerun the workflow starting from Task_3 instead of running all the tasks. This feature is particularly useful if the tasks prior to the failure were long or expensive to run. This eliminates the need to rerun those tasks, again reducing redundancy.

Conditional Execution of Tasks: Conditional execution helps build a dependency chain between two tasks within a job based on a condition. This is an important feature that helps orchestrate multistage data pipelines as it allows users to better control over complex workflows and implement advanced orchestration scenarios. In conditional execution, a task is executed only if the status of upstream tasks meets the specified condition.

Conditional execution consists of two main capabilities, the "If/else condition task type" and "Run if dependencies," which together enable users to create not only a branching logic in their workflows but also more sophisticated dependencies between tasks in a pipeline thereby giving them more flexibility into their workflows.

- **If/else condition task type:** The if/else Task type, as
  name suggests, enables users to add branching logic
  to their jobs. The *If/else condition* task is used to run
  a part of a job DAG based on the results of a Boolean
  expression given as a condition.

- **"Run if" dependencies:** The "Run if" dependencies
  are task-level configurations that provide users with
  more flexibility in defining task dependency. Take for
  example a task that has several dependencies over
  multiple prior tasks; users can now define what are
  the conditions that will determine the execution of
  the dependent task. These conditions are referred
  to as "Run if dependencies." One can now define
  whether the dependent task will run if all dependencies
  succeed, at least one succeeded, all finished regardless
  of status, etc.

The following are the task-level "Run if" dependencies available:

- All succeeded (all dependencies are executed and
  succeeded)

- At least one succeeded (at least one of the
  dependencies has succeeded)

- None failed (none of the dependencies have failed and
  at least one has executed)

- All done (all dependencies completed and at least one
  has executed)

- At least one failed (at least one dependency has failed)

- All failed (all dependencies have failed)

Late Jobs: Data teams usually have hundreds of jobs running in production. When you run a large number of jobs on the platform, users find it challenging to monitor all these jobs in real time and usually know about the status of the individual jobs only once they have been completed. This could be problematic, especially for long-running jobs, as it could lead to missed SLAs and prove costly.

For example, let's assume a particular job takes around 40 minutes to complete, and, for some reason, on a particular day, it took more than 3 hours to complete. First, there is no way to know that the job is running way over its usual runtime, and this would lead to higher costs as well.

The late job feature in Databricks workflows enables users to manage this use case efficiently. It allows users to define a "soft timeout" after which they receive a warning that a job or task run is taking longer than expected. Additionally, users can set the "timeout duration" after which the job will be stopped. See Figure 6-7.

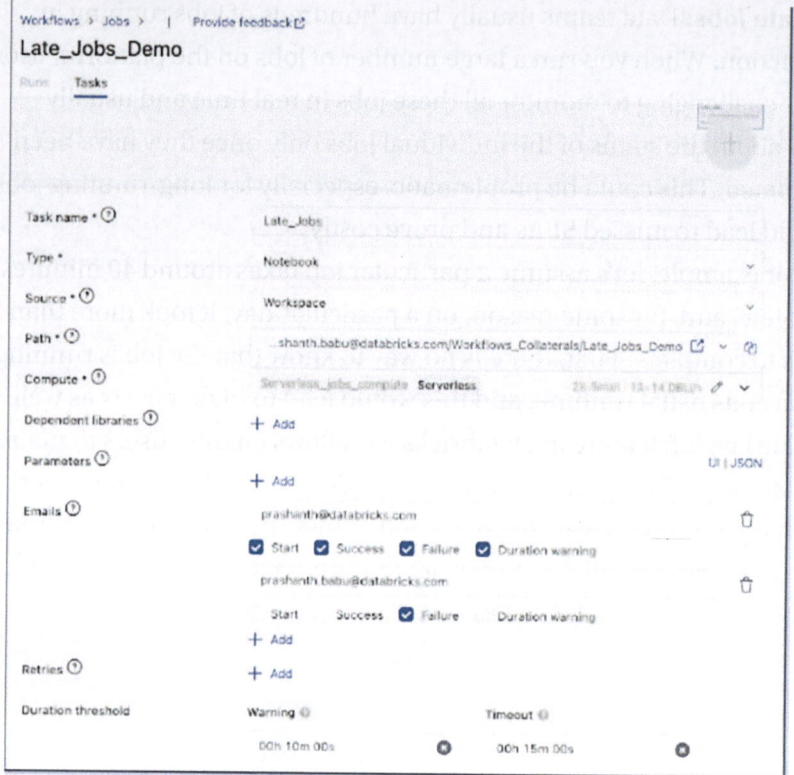

Figure 6-7. Late job run

So continuing with our example, the user can set the Warning option at 1 hr and Timeout option at 1.5 hrs. If the job takes more than 1 hr, a duration alert will be sent to the configured email(s). Further, if the job exceeds 2 hrs, the job will be stopped from executing. This gives users better control over their long running jobs.

Run Job Task Type - Modularize Jobs: Orchestration jobs can have multiple tasks with complex dependencies between them. More often than not, managing these complex jobs becomes challenging in terms of defining, testing, and troubleshooting. Modern software best practices usually emphasize modularizing complex code into reusable logical chunks.

Databricks Workflows Task has a task type "Run Job." This allows users to run a "child job" within a "parent job," which makes the overall workflow easier to comprehend and maintain. This effectively allows you to modularize your jobs, as you can now divide your DAGs into logical chunks or child jobs, which can be managed separately. Further, these modular child jobs can be reused in different parent workflows by parameterizing them. See Figure 6-8.

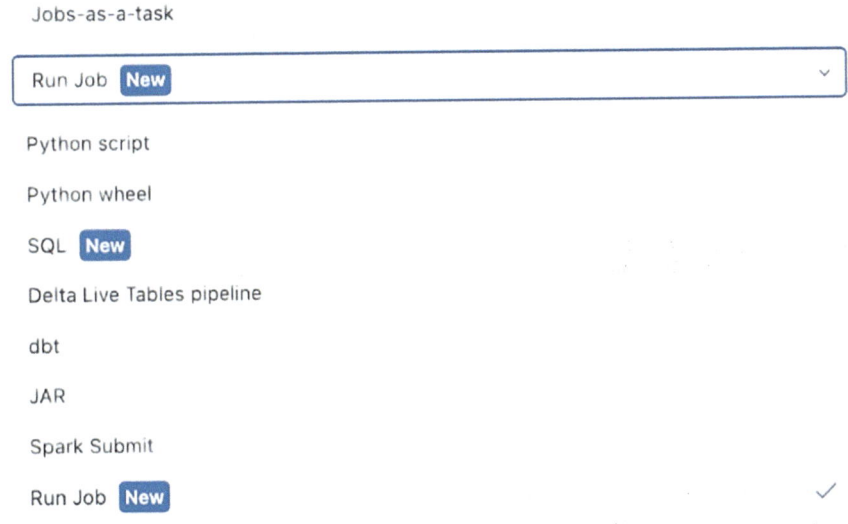

Figure 6-8. *Running job as a task*

Now, with Run Job as a task type, users can call the child jobs (previously defined) within the parent jobs, enabling them to create modular workflows.

In the next section, we will look into another aspect: monitoring Databricks workflows both at the job and task levels.

Monitoring Data Pipelines

Data engineers/admins need to have complete visibility of all the jobs
running on the orchestration platform to see the status of each job
and what jobs need troubleshooting in case they fail, thus having full
monitoring and observability capabilities. It will be challenging to create
custom dashboards to monitor your hundreds of jobs.

Databricks workflows give users a unified view of all job runs through
its Job Runs dashboard. With this, users not only can view all the jobs that
ran but also dive deep into individual runs for each job. Let's look into this
in a little more detail.

Job Run dashboard: The Job Run dashboard gives users a
comprehensive real-time view of all their jobs in a single workspace. These
are some of the most critical features in this dashboard:

1. **Finished Runs Chart:** This stacked bar chart
   depicts the number of job runs completed in the
   last 48 hours, with the option of redefining the
   time interval. The chart shows failed, skipped, and
   successful job runs.

2. **Jobs List** This table details all the job runs within the
   workspace. It is helpful as one can quickly assess the
   job runs for any job and navigate to a particular job
   run from this table if human intervention is needed
   in case the run fails.

3. **Top 5 Error Types:** This table lists the most frequent
   error types for all the jobs that ran within the
   selected timeframe. It helps identify a summary of
   the top error types across all workloads, enabling
   users to troubleshoot faster, take proactive
   measures, and minimize the negative impact on
   business operations downstream. See Figure 6-9.

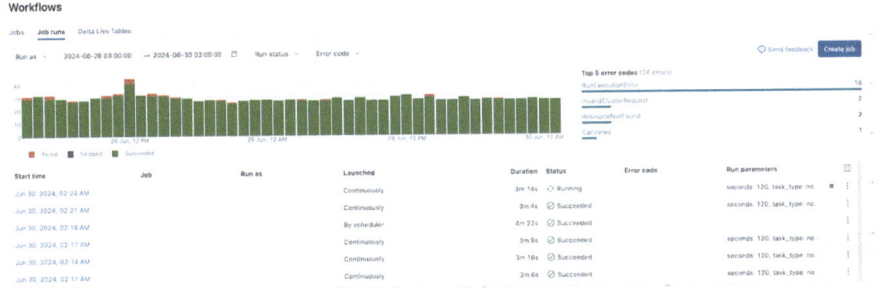

Figure 6-9. *Workflow monitoring dashboard*

Job Matrix View: Previously, we saw the job run dashboard, which gives an overview of all the jobs in real time. Now as a user one also needs to keep track of all the runs for an individual job. The Job Matrix View allows users to assess all the job runs and quickly see the health of each task within (see Figure 6-10).

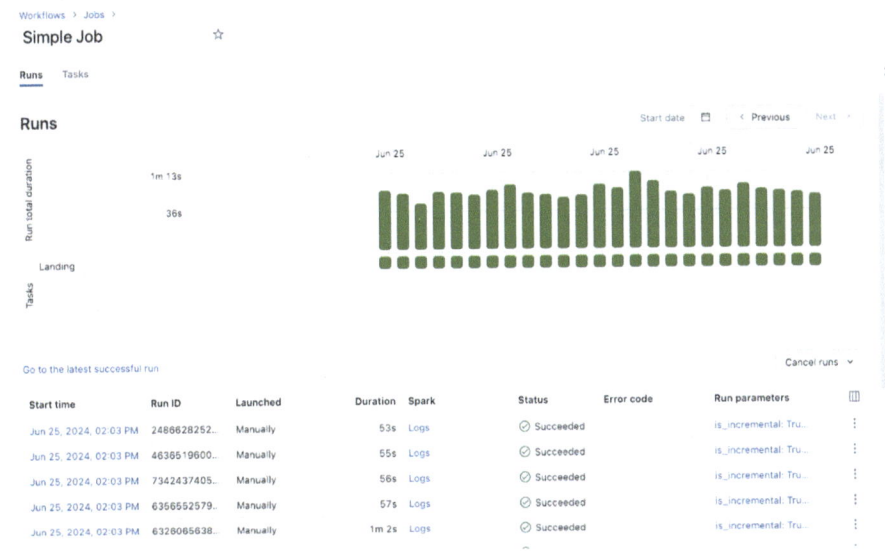

Figure 6-10. *Single job status*

Hovering over or clicking ae specific task allows us to identify the cause of the failure as the error message is displayed. Further, one can quickly go to a particular task and even to the underlying notebook. This is useful for seeing which step errored out and for troubleshooting quickly.

Conclusion

In this chapter, we explored how to orchestrate your data pipelines using Databricks workflows. Databricks jobs enable you to execute all your data processing and analysis tasks within a Databricks workspace. A job consists of one or multiple tasks that can be combined using dependencies. We then created a simple task and learned about the different parameters required for configuration at the task and job levels. We also learned about more advanced configurations, such as cluster reruns, where you can reuse a single cluster for all your tasks, and how to configure conditional dependencies for your tasks.

Finally, we examined the observability and monitoring aspects of Databricks workflows with the job-run dashboard and Job-View Matrix. In the next chapter, we will examine Delta Live Tables, which provides ETL capabilities on the Databricks platform.

CHAPTER 7

Data Engineering Part 2: Delta Live Tables

It is no secret that good, reliable data is the foundation of the lakehouse architecture. Organizations need clean, fresh, and reliable data to drive their analytics and data science projects, which in turn help them make decisions for key business initiatives.

However, most data engineers will agree that maintaining data quality and reliability at scale is quite complex and tedious. Apart from writing ETL transformations, they must spend much time on tasks like handling table dependencies, recovery, backfilling, retries, or error conditions. They must also manage the infrastructure, which turns simple ETL tasks into complex data pipelines.

In this chapter, we will introduce you to Delta Live Tables (DLT), which enable data engineers to concentrate on writing the transformation logic (the "what"), while Databricks manages the rest (the "how"). We will start with understanding what Delta Live Tables is and learn about concepts in declarative programming. Then we will look at some of the key features of DLT, including Change Data Capture (CDC), data quality and monitoring, enhanced autoscaling, and more.

© The Editor(s) (if applicable) and The Author(s),
under exclusive license to APress Media, LLC, part of Springer Nature 2024
N. Gupta and J. Yip, *Databricks Data Intelligence Platform*,
https://doi.org/10.1007/979-8-8688-0444-1_7

What Is Delta Live Tables?

Delta Live Tables makes it easy to build and manage reliable batch and streaming data pipelines that deliver high-quality data on the Databricks lakehouse platform. DLT uses a simple declarative approach using SQL and Python that helps data engineering teams simplify ETL development and management with pipeline development, automatic data testing, and visibility for monitoring and recovery. DLT also automates infrastructure management by handling cluster sizing, error handling, performance tuning, and orchestration. Therefore, using DLT data, engineers can now spend less time managing the tooling, focusing on data transformations, and getting value from data.

So, what is the difference between Delta tables and Delta Live Tables? Delta is a storage format, and the tables created on the underlying data are called *Delta tables*. Delta Live Tables is a declarative pipeline development that manages how data flows between Delta tables. See Figure 7-1.

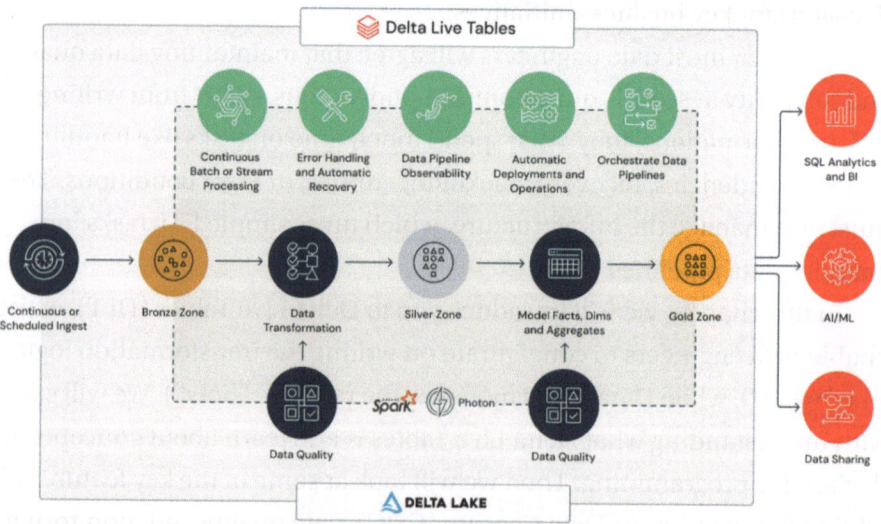

Figure 7-1. *Delta Live Tables overview*

Before we proceed, let's examine the different types of DLT datasets, namely, streaming tables, materialized views, and views.

> **Streaming tables:** A streaming table is a Delta table that supports incremental data processing. It is most suitable for ingestion workloads and pipelines that require data freshness and low latency. It is designed to read append-only data sources like Kafka, Kinesis, or Auto Loader.

> **Materialized views:** A materialized view (or live table) precomputes and stores results and keeps them fresh over time. It is refreshed according to the pipeline's update schedule and, more importantly, incrementally, thus reducing processing costs. Each time the pipeline updates, query results are recalculated to reflect changes in upstream datasets.

> **Views:** Views are temporary tables that should not be exposed outside of the DLT pipeline. They are just used like temp tables in standard SQL processing. Views are not published to public datasets.

Let's move on and see how we can build a simple DLT pipeline and explore some key features.

Data Ingestion Using DLT

The first step is to get data from DLT. This could be ingesting a number of raw files in a cloud storage folder or directly connecting to a streaming source like Kafka. It is important to note that data ingestion has to be reliable and scale efficiently. Under the hood, DLT ingests data using Auto Loader. We discussed Auto Loader in detail in Chapter 3. To recap, Auto

Loader incrementally processes new files as they land in the cloud storage. It can infer schema automatically and evolve schemas as the use cases require.

Listing 7-1 creates a Delta table called raw_txs and ingest JSON files from cloud storage into this table. Please note that DLT manages all Auto Loader configurations, like checkpointing, in the back end.

Listing 7-1. Creating a Streaming Live Table

```
CREATE STREAMING LIVE TABLE loan_bronze
AS SELECT * FROM cloud_files('/demos/dlt/loans/raw_
transactions', 'json', map("cloudFiles.inferColumnTypes",
"true"))
```

Listing 7-2 involved a batch data ingestion. Let's examine how to ingest from a Kafka source.

Listing 7-2. Data Ingestion from Kafka

```
@dlt.table
def sales():
  return (
    (spark.readStream
    .format("kafka")
    .option("subscribe", 'sales_trends')
    .option("kafka.bootstrap.servers", kafka_bootstrap_
    servers_tls)
    .option("kafka.security.protocol", "SSL")
    .option("startingOffsets", "earliest")
    .load()).select(col("key").cast("string").alias("eventId"),
    from_json(col("value").cast("string"), behavioral_input_
    schema).alias("json"))
  )
```

Before we build our silver layer, let's examine some important concepts, such as Change Data Capture and expectations.

Change Data Capture with DLT

One important use case during data ingestion, especially from databases and data warehouses, is capturing Change Data Capture (CDC) events into the data lake. CDC is a process of identifying any data changes, such as inserts, updates, or deletes, made to your data sources and moving those changes to the target.

Let's look at an example of how CDC can be implemented using Delta Live Tables. First, external tools, such as Debezium, Fivetran, Qlik Replicate, etc., can capture and record the history of data changes, say from external systems like databases, in logs; downstream applications consume these CDC logs. See Figure 7-2.

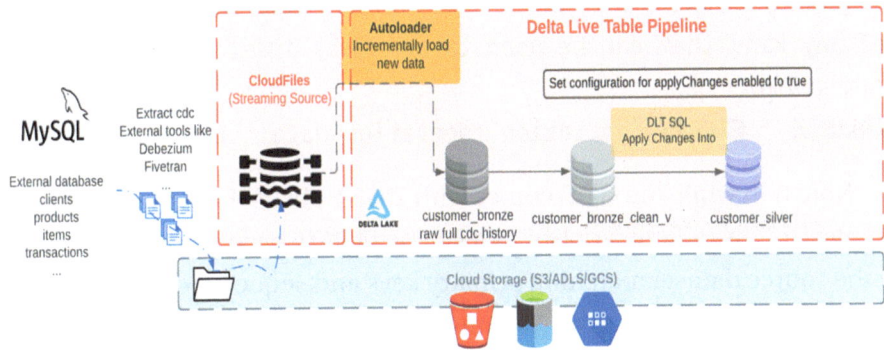

Figure 7-2. *Change Data Capture with DLT*

As a first step, we will move these logs into a cloud storage object or a message queue like Kafka. In the previous section, we discussed how to ingest data in the Delta Bronze Layer from either of these sources. Delta Live Tables allows you to apply changes from CDC seamlessly your tables, enabling incremental changes to flow through analytical workloads at scale easily. Let's quickly look into an example.

Before we begin to apply CDC, we need to ensure that the target table has most up-to-date data from the source table, as shown in Listing 7-3.

Listing 7-3. Creating a CDC Table

```
CREATE INCREMENTAL LIVE TABLE customers
COMMENT "Clean, materialized customers";
```

Once the table has been created, we use APPLY CHANGES to propagate the changes to the target table, as shown in Listing 7-4.

Listing 7-4. Updating the Live Table

```
APPLY CHANGES INTO live.customers
FROM stream(live.customers_cdc)
KEYS (id)
APPLY AS DELETE WHEN operation = "DELETE"
SEQUENCE BY operation_date --primary key, auto-incrementing ID
of any kind that can be used to identity order of events, or
timestamp
COLUMNS * EXCEPT (operation, operation_date, _rescued_data);
```

Note that while the CDC comes with INSERT, UPDATE, and DELETE events, DLT, by default, applies INSERT and UPDATE events from any record in the source dataset matching primary keys and sequenced by a field that identifies the order of events. You must use APPLY AS DELETE WHEN in SQL to handle DELETE events.

After CDC, we will move into another important feature supported by DLT—Slowly Changing Dimensions (SCD)—for both type 1 and type 2. In SCD Type 2, when the value of a record changes, a new line for the record is created and becomes the current record, while the older one is closed. In Type 1, there is only a simple append.

The following code explains how this can be easily achieved in DLT. To create a SCD2 table, all we have to do is leverage APPLY CHANGES with the extra option STORED AS {SCD TYPE 1 | SCD TYPE 2 [WITH {TIMESTAMP|VERSION}}], as shown in Listing 7-5.

Listing 7-5. Creating a Slowly Changing Dimension Type 2 Table

```
APPLY CHANGES INTO live.SCD2_customers
FROM stream(live.customers_cd)
  KEYS (id)
  APPLY AS DELETE WHEN operation = "DELETE"
  SEQUENCE BY operation_date
  COLUMNS * EXCEPT (operation, operation_date, _rescued_data)
  STORED AS SCD TYPE 2 ;
```

We now move into another important aspect of DLT called Expectations, which help maintain data quality throughout the DLT pipeline.

Delta Live Tables Expectations

One of the most important issues data engineers face while building data pipelines is ensuring proper data quality and establishing the trust of end users in the data they are using. Further, engineers often struggle to identify and resolve data quality issues once they discover them.

Delta Live Tables provides a data quality management feature called Expectations that helps users define data quality and integrity constraints within their DLT pipelines.

Expectations are optional clauses to which you constrain your DLT dataset declarations. They apply data quality checks on each record passing through a query into your table. See Listing 7-6.

Listing 7-6. Delta Live Tables Expectations

```
CREATE STREAMING LIVE TABLE loans_silver (
 CONSTRAINT `Payments should be this year` EXPECT (next_
 payment_date > date('2020-12-31')),
 CONSTRAINT `Balance should be positive` EXPECT (balance > 0
 AND arrears_balance > 0) ON VIOLATION DROP ROW,
 CONSTRAINT `Cost center must be specified` EXPECT (cost_
 center_code IS NOT NULL) ON VIOLATION FAIL UPDATE
)
AS SELECT * from loans_bronze
```

In the previous query, we have defined a few constraints on the
DLT table. An expectation typically consists of three parts: description,
invariant, and action when the condition fails. A description is a unique
identifier and allows you to track the metrics for the particular constraint.
An invariant returns a Boolean expression (True/False) based on the
defined condition. Finally, action defines what to do if the condition fails.

There are three actions you can apply to the failed records.

- **Warn:** In this action, the invalid records are written to
  the target tables, but failure is reported in as a metric
  for the dataset.

- **Drop:** The invalid records are dropped before the target
  table is written, and the number of records dropped is
  recorded.

- **Fail:** In this, the DLT pipeline is stopped, and the
  records have not been updated. Users need to check
  and update before manually restarting the pipeline.

Later in this chapter we will see how you can view data quality metrics
in the DLT monitoring UI. See Figure 7-3.

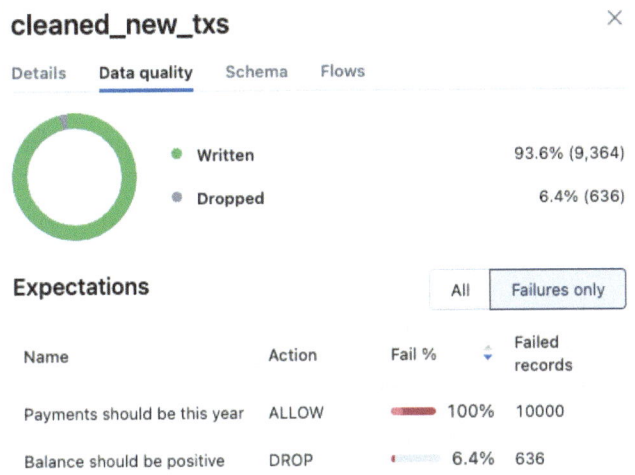

Figure 7-3. *Data quality metrics*

After creating the silver table, let's do some transformations and create a Gold table (see Listing 7-7), which will be the final in our medallion architecture.

Listing 7-7. Creating a Gold DLT Table

```
CREATE LIVE TABLE loans_gold
AS SELECT sum(revol_bal) AS bal, addr_state    AS location_code
FROM live.historical_txs GROUP BY addr_state
 UNION SELECT sum(balance) AS bal, country_code AS location_
 code FROM live.cleaned_new_txs GROUP BY country_code
```

We have defined the logic for our DLT so far. Let's create our DLT pipeline.

Creating a DLT Pipeline

After defining the logic for our bronze, silver, and gold tables, let's combine everything and create our first DLT pipeline. First, navigate to the DLT UI in the Dela Live Tables tab and click Create Pipeline, as shown in Figure 7-4. JSON mode is also available for quick parameter population, as shown in Figure 7-5.

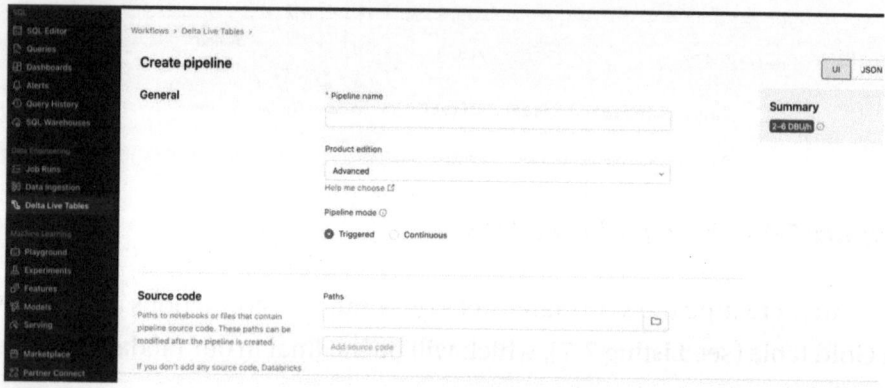

Figure 7-4. *Delta Live Tables user interface*

```json
{
    "id": "62dca0e2-cba2-4e75-ad15-b8f8f3b4a879",
    "pipeline_type": "WORKSPACE",
    "clusters": [
        {
            "label": "default",
            "autoscale": {
                "min_workers": 1,
                "max_workers": 5,
                "mode": "ENHANCED"
            }
        }
    ],
    "development": true,
    "continuous": false,
    "channel": "CURRENT",
    "photon": false,
    "libraries": [
        {
            "notebook": {
                "path": "/Users/nikhil.gupta@databricks.com/dlt-loans/01-DLT-Loan-pipeline-SQL"
            }
        }
    ],
    "name": "DLT_Demo",
    "edition": "ADVANCED",
    "storage": "dbfs:/pipelines/62dca0e2-cba2-4e75-ad15-b8f8f3b4a879",
    "data_sampling": false
}
```

Figure 7-5. *Sample DLT pipeline JSON file*

Here are the key parameters that must be provided:

- **Source code:** This is the path to the notebooks or files containing the pipeline code. The source code could be in multiple notebooks or files, and you can give file locations for all of them in this parameter.

- **Product edition:** DLT comes with four SKUs: Core, Pro, Advanced, and Serverless. The difference is the features they support. A comparison table is given here:

  https://www.databricks.com/product/pricing/delta-live

- **Pipeline mode:** We can run DLT pipelines in Triggered or Continuous mode. In Triggered mode, the pipelines update the data once and shut down until you run the pipeline manually or schedule the update. In Continuous mode, pipelines run continuously and ingest/process new data as it arrives.

- **Compute:** In this part, you define the compute resources for pipeline running. You can select a cluster policy if you want to use one for your cluster. Next is cluster mode, which has the options Enhanced Autoscaling, Legacy Autoscaling, or Fixed size. As a best practice, use Enhanced Autoscaling for your pipelines. We will discuss this later in the chapter. Finally, there is an option to select Photon Acceleration for your workloads.

You can run the pipeline in Development or Production mode to optimize pipeline execution. When the pipeline runs in Development mode (default), the cluster is reused in multiple runs to avoid the restarts. Also, the pipeline retries are disabled, so you can quickly fix any errors. In Production mode, the pipeline retries in case of operational issues like cluster failure.

Once you have defined the appropriate parameters for your pipeline, let's run it once and see the results. Figure 7-6 represents the high-fidelity lineage diagram for this DLT pipeline.

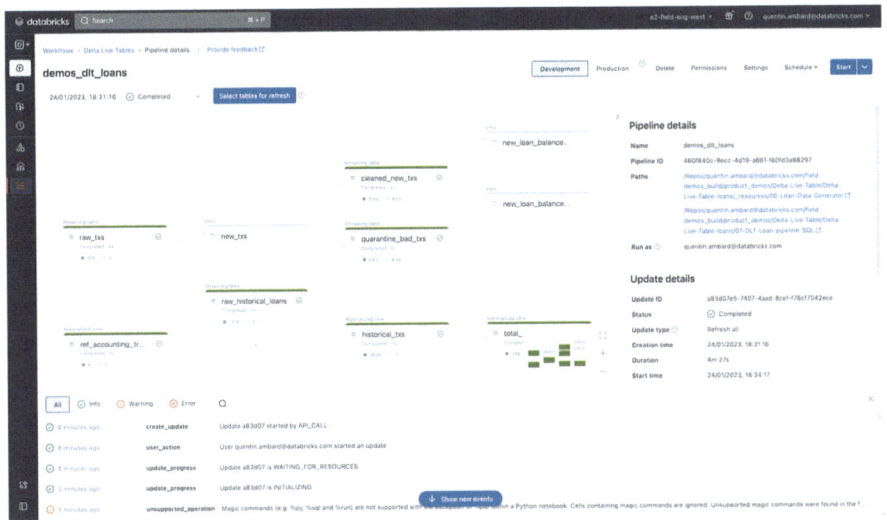

Figure 7-6. *DAG for a DLT pipeline*

It is important to note that if a particular table in the pipeline is out or you want to refresh a specific table without rerunning it, you can select Select Tables for Refresh and choose the tables you need to refresh or run again.

You can see table details, schema, and data quality metrics if you click on any table boxes in the DAG (Figure 7-6).

Next, we will move forward and examine other aspects of DLT, such as monitoring and logging, CI/CD, and enhanced autoscaling.

Logging and Monitoring

Each DLT pipeline emits all event logs to a predefined and unique storage location. The DLT event logs contain all information related to a pipeline, including audit logs, data quality checks, pipeline progress, and data lineage.

The logs are also visible in the DLT pipeline run UI page, where you can quickly investigate the errors. DLT provides a variety of error-handling capabilities, including retrying failed tasks, handling failed records, and detecting and fixing data quality issues, but in a nice GUI format. See Figure 7-7.

All	⊘ Info	① Warning	⊗ Error

⊘ 2 days ago	flow_progress	Flow 'new_loan_balances_by_country' has COMPLETED.
⊘ 2 days ago	flow_progress	Flow 'historical_txs' has COMPLETED.
⊘ 2 days ago	flow_progress	Flow 'total_loan_balances' is PLANNING.
⊘ 2 days ago	planning_information	Flow 'total_loan_balances' has been planned in DLT to be executed as COMPLETE_RECOMPUTE.
⊘ 2 days ago	flow_progress	Flow 'total_loan_balances' is STARTING.
⊘ 2 days ago	flow_progress	Flow 'total_loan_balances' is RUNNING.

Figure 7-7. *DLT task status*

These logs are exposed as Delta tables and used for monitoring, lineage, and data quality reporting using the BI tool of your choice. Figure 7-8 shows a sample dashboard that can be built on DBSQL.

Figure 7-8. *Dashboard for monitoring DLT job statuses*

Enhanced Autoscaling

Organizations are increasingly looking toward real-time and streaming workloads to provide them with the freshest possible data for their analytics and ML workloads, which can help them make decisions faster. However, streaming workloads have spiky and unpredictable data volumes, making it difficult for data engineers to avoid overprovisioning compute infrastructure, leading to higher costs.

The DLT Enhanced Autoscaling algorithm improves on the standard Databricks cluster autoscaling feature to handle streaming workloads more efficiently. It optimizes cluster utilization for streaming workloads to lower costs while ensuring your data pipeline has the resources to maintain consistent SLAs. See Figure 7-9.

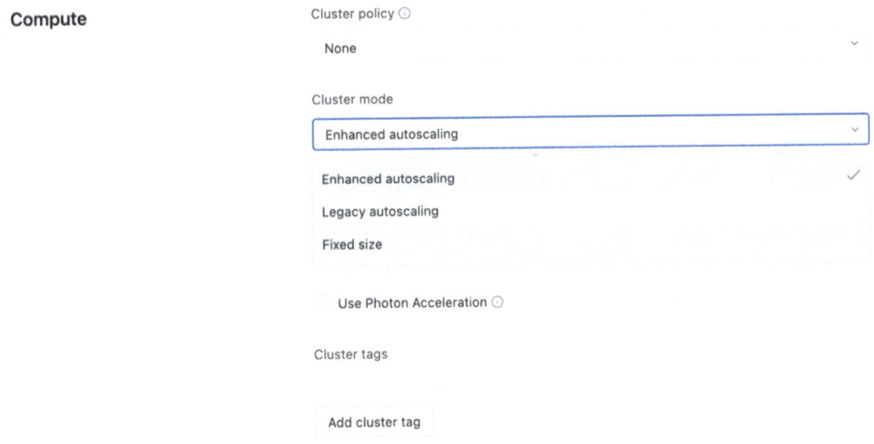

Figure 7-9. DLT *"Enhanced autoscaling"* option

The "Enhanced autoscaling" option maximizes resource utilization by shutting down nodes when utilization is low while guaranteeing that tasks are completed successfully. Further, when the workload increases, it only scales up to nodes that are needed, even if this is lower than the maximum number of nodes provisioned. DLT's "Enhanced autoscaling" option optimizes cluster utilization while minimizing overall end-to-end latency. DLT's enhanced autoscaling can be easily enabled on the pipeline during or after pipeline creation by setting Cluster Mode to "Enhanced autoscaling."

Runtime Channels

Traditional clusters require the maintenance of runtime versions, also known as Databricks Runtime (DBR). In Delta Live Tables, you can get the flexibility of both choosing the cluster VM type and having Databricks manage the runtime for you. The Channel drop-down is designed for this exact purpose. By default, the "current" channel uses the latest Databricks runtime, whereas the "preview" channel uses the upcoming runtime. See Figure 7-10.

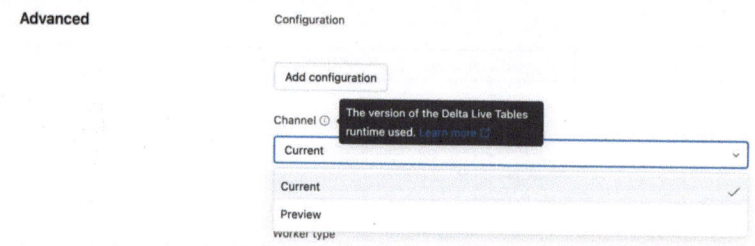

Figure 7-10. *DLT runtime channel*

Example: A Retail Sales Pipeline

Now we will example a retail sales pipeline. The source code is conveniently located at the Databricks' repo:

```
https://github.com/databricks/delta-live-tables-notebooks/blob/
main/sql/Retail%20Sales.sql
```

This example highlights these four features:

- Streaming pipeline

- Data validation

- Data lineage

- Validation dashboard

Streaming Pipeline

Listing 7-8 shows the raw sales order.

Listing 7-8. Raw Sales Order as Bronze Table

```
CREATE STREAMING LIVE TABLE sales_orders_raw
COMMENT "The raw sales orders, ingested from /databricks-
datasets."
TBLPROPERTIES ("myCompanyPipeline.quality" = "bronze")
AS

SELECT * FROM cloud_files("/databricks-datasets/retail-org/sales_
orders/", "json", map("cloudFiles.inferColumnTypes", "true"))
```

This raw pipeline is simply trying to stream the JSON files from the specified location. As a result, it is now simpler to build a streaming pipeline using DLT.

Data Validation

The next step is to perform data cleanup. The traditional ETL requires separate steps for error handling and data validation. As a result, this logic will be written in the SQL query, and other developers will try to decode the purpose. In DLT, there is a descriptive way to handle these records called Expectation, as shown in Listing 7-9.

Listing 7-9. Raw Sales Order as Bronze Table

```
CREATE STREAMING LIVE TABLE sales_orders_cleaned(
  CONSTRAINT valid_order_number EXPECT (order_number IS NOT
  NULL) ON VIOLATION DROP ROW
)
PARTITIONED BY (order_date)
COMMENT "The cleaned sales orders with valid order_number(s)
and partitioned by order_datetime."
TBLPROPERTIES ("myCompanyPipeline.quality" = "silver")
AS

SELECT f.customer_id, f.customer_name, f.number_of_line_items,
  TIMESTAMP(from_unixtime((cast(f.order_datetime as long)))) as
  order_datetime,
  DATE(from_unixtime((cast(f.order_datetime as long)))) as
  order_date,
  f.order_number, f.ordered_products, c.state, c.city, c.lon,
  c.lat, c.units_purchased, c.loyalty_segment
  FROM STREAM(LIVE.sales_orders_raw) f
  LEFT JOIN LIVE.customers c
      ON c.customer_id = f.customer_id
    AND c.customer_name = f.customer_name
```

Data Lineage

The flow chart in Figure 7-11 in the DLT job shows how the data moves from one place to another. Therefore, running it through another parsing tool to generate these diagrams is unnecessary.

Figure 7-11. *DLT lineage diagram*

Validation Dashboard

Each step automatically summarizes the expectations and data quality checks, saving time on the creation and upkeep of additional toolkits. Having these available automatically also reduces the time needed to evaluate the code to ensure data validation. See Figure 7-12 and Figure 7-13.

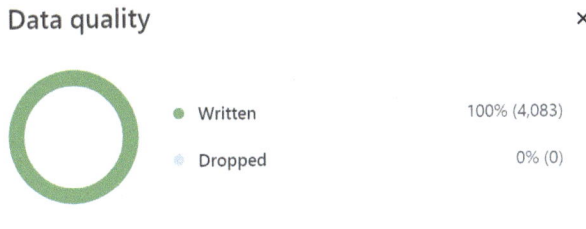

Figure 7-12. *DLT data quality dashboard*

Figure 7-13. *DLT Expectations*

Conclusion

Data teams are constantly on the go. However, with Databricks' Delta Live Tables, they can streamline reliable data pipelines and quickly find and manage enterprise data assets across various clouds and data platforms using Unity Catalog. Additionally, they can simplify the enterprise-wide governance of data assets, both structured and unstructured.

This chapter examined Delta Live Tables, which provides a declarative framework for developing, managing, and deploying ETL pipelines. DLT automatically manages your infrastructure, ensures high data quality and unifies batch and streaming workloads. We built a DLT pipeline and looked into important features like Change Data Capture, SCD Type 1 and 2 support, and DLT Expectations, which help maintain data quality. We also discussed various performance optimizations that DLT uses via enhanced autoscaling. Last but not least, the runtime version is managed for you automatically by default. There is no need to worry about managing the latest runtime, but cluster types are still available and are similar to interactive clusters.

CHAPTER 8

Data Warehousing with DBSQL

If you're a data analyst who primarily uses SQL to write queries and reports and create comprehensive dashboards for analysis using your favorite business intelligence (BI) tools, Databricks SQL (DBSQL) provides a comprehensive environment for running ad hoc queries and creating dashboards on data stored in your data lake.

Traditionally, SQL/BI use cases have most commonly been implemented by storing data in a data warehouse or a database, writing SQL queries in a SQL IDE, and, finally, using BI tools to build dashboards. However, with the lakehouse platform, you can handle all this without moving data to a different storage, like a data warehouse or a database.

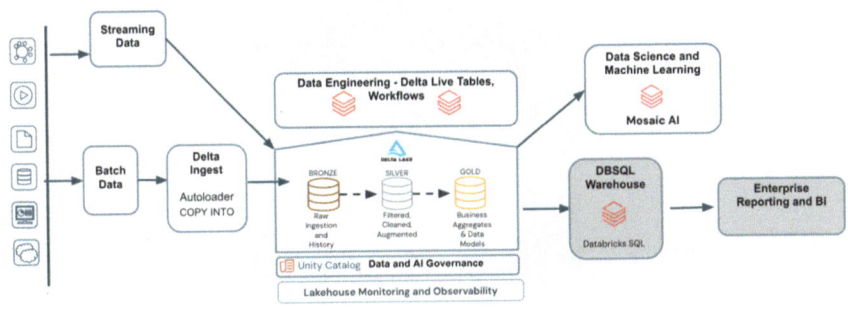

Figure 8-1. *Architecture diagram using Databricks SQL*

© The Editor(s) (if applicable) and The Author(s),
under exclusive license to APress Media, LLC, part of Springer Nature 2024
N. Gupta and J. Yip, *Databricks Data Intelligence Platform*,
https://doi.org/10.1007/979-8-8688-0444-1_8

In this chapter, we will learn how the Databricks platform provides the most complete end-to-end data warehousing solution for all analytics use cases.

Next, we will move on to understand various components and key features of Databricks SQL or DBSQL.

What Is Databricks SQL?

Databricks SQL is the collection of services bringing data warehousing capabilities and performance to your existing data lake through open formats and standard ANSI SQL. The DBSQL platform provides not only a SQL editor but also dashboarding tools that allow team members to collaborate with users directly in the Databricks workspace. Further, Databricks SQL integrates with a variety of BI tools via connectors or JDBC/ODBC so that analysts can author queries and dashboards using their favorite BI tools without adjusting to a new platform. See Figure 8-2.

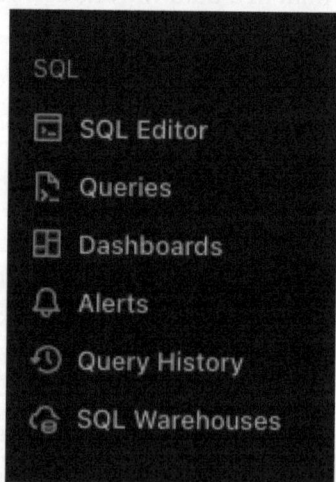

Figure 8-2. *SQL Persona section on Databricks sidebar*

In the following sections, we will look into a few key services in DBSQL.

SQL Warehouses

SQL warehouses are compute resources within DBSQL that run your SQL queries on data objects within Databricks SQL. Simply put, SQL warehouses provide processing capabilities in DBSQL similar to clusters in the data engineering part of the platform. There are three main types of SQL warehouses.

- **Classic:** This offers limited Databricks SQL functionality and basic performance features. Only use a classic SQL warehouse to run interactive queries for data exploration with entry-level performance and Databricks SQL features.

- **Pro:** This supports all the Databricks SQL functionality and delivers higher-performance features than Classic, including query federation, workflow integration, and data science and ML functions.

- **Serverless:** This is the most powerful and cost-effective option. The serverless SQL warehouse gives the most advanced performance features and supports all of the features available in the Pro type, along with instant and fully managed compute. Serverless compute spins up almost instantaneously with best-in-class price/performance.

Figure 8-3 shows how to set up a SQL warehouse.

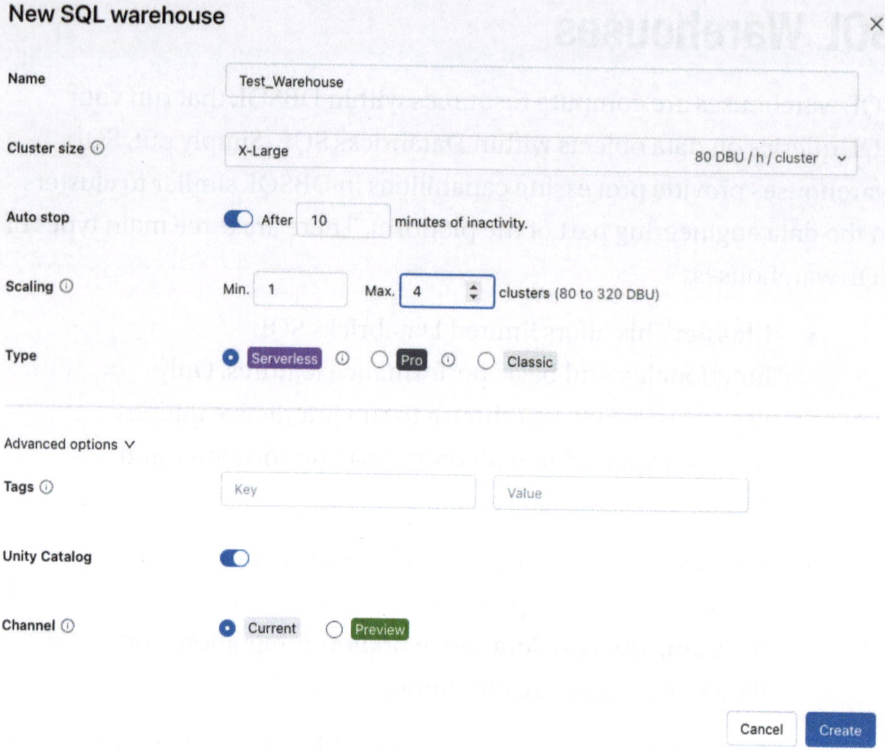

New SQL warehouse   ✕

Name	Test_Warehouse

Cluster size ⓘ	X-Large	80 DBU / h / cluster ⌄

Auto stop   🔵 After 10   minutes of inactivity.

Scaling ⓘ   Min. 1   Max. 4 ⇕ clusters (80 to 320 DBU)

Type   🔘 Serverless ⓘ ⚪ Pro ⓘ ⚪ Classic

Advanced options ⌄

Tags ⓘ | Key | | Value |

Unity Catalog   🔵

Channel ⓘ   🔘 Current   ⚪ Preview

                                            Cancel   Create

Figure 8-3. Setting up a SQL warehouse in Databricks

It is really simple to spin up a SQL warehouse. Once you click Create Warehouse, the first parameter to fill in is the warehouse's name. Next, you can also select the warehouse type from Classic, Pro, and Serverless.

Let's look into other important parameters to take into consideration.

- **Cluster Size:** SQL warehouses come in T-shirt sizes from X-Large to X-small. Please choose the size based on the latency and throughput. As a best practice, start from Medium and move up and down as per your needs.

- **Scaling:** A traditional cluster comes with one driver and a number of workers. When you autoscale a cluster, you are increasing the number of worker nodes. However, there is still only one driver node, and with high-frequency, low-latency workloads, it will become a bottleneck. With the SQL warehouses scaling feature, you determine the min and max number of clusters behind the endpoint, and it is these clusters (not workers) that can increase based on concurrency requirements.

Further, one of the key aspects of SQL warehouses that makes it really performant is Photon. Let's look into what Photon is and how we can enable it.

Photon

Photon is the next-generation ANSI-compliant vectorized query engine developed by Databricks to support workloads in DBSQL. It comes with hundreds of built-in optimizations, providing the best performance for all tools, query types, and real-world applications. This includes the AI-powered predictive I/O that eliminates performance tuning like indexing by intelligently prefetching data.

It's 100% compatible with Apache Spark APIs, which means you don't have to rewrite your existing code (SQL, Python, R, Scala) to benefit from its advantages.

While Photon is an optional feature in interactive clusters, it is activated by default for SQL warehouses. You can also enable Photon for All Purpose and Job Clusters options by toggling the switch on the Create Cluster page.

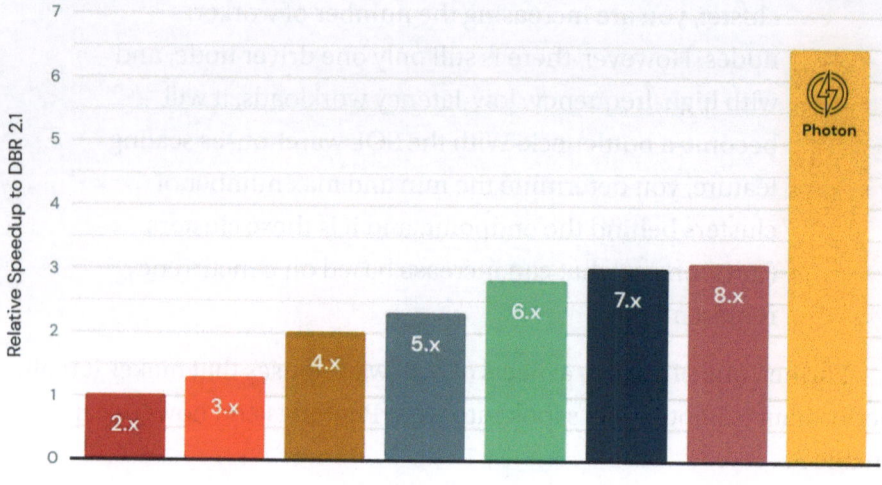

Figure 8-4. *TPC-DS 1TB performance by DBR version versus Photon (source.* `https://www.databricks.com/product/photon`*)*

Figure 8-4 shows the performance of Photon with regard to the Databricks runtime, showing that Photon is almost three times more performant than the DBR 8.x.

SQL Editor

The DBSQL UI provides a SQL editor (Figure 8-5) that you can use to author SQL queries using a familiar ANSI SQL syntax, browse available data, and create visualizations. You can also share your saved queries with other team members in the workspace. SQL Editor also supports functionalities such as autocomplete, autoformatting, auto-save, etc. Additionally, query updates can be scheduled to refresh automatically, as well as to issue alerts when meaningful changes occur in the data.

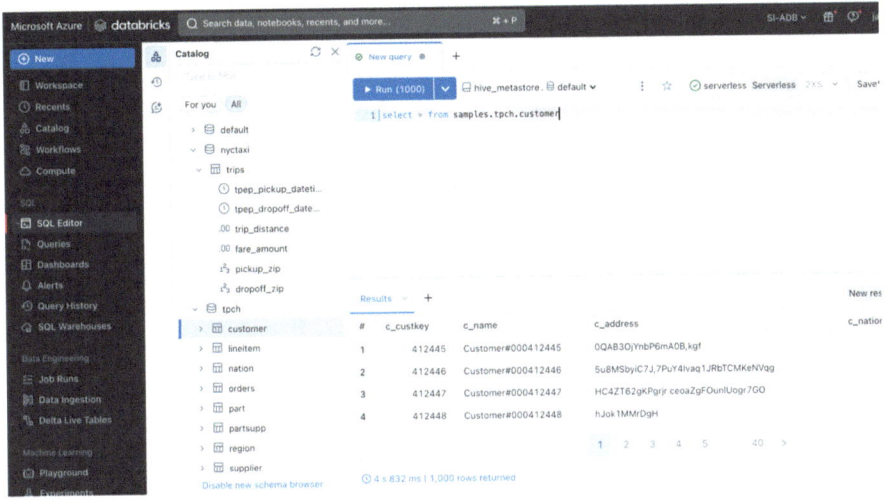

Figure 8-5. *SQL Editor in Databricks*

Introduction to AI/BI Dashboards

Databricks SQL enables data analysts to make sense of data through visualizations and drag-and-drop dashboards. Dashboards (now termed as *legacy dashboards*) within DBSQL allow users to combine both visualizations (via the built-in SQL Editor) and text boxes to give context to their data. Once built, dashboards can be easily shared with stakeholders, both within and outside the organization, via a web browser.

Databricks introduced a new generation of dashboards at DAIS'24 called AI/BI dashboards. AI/BI dashboards (formerly known as *lakeview dashboards*) allow analysts to quickly build highly interactive dashboards using natural language questions that analysts can ask. Further, these dashboards are integrated with the Databricks platform, which ensures fast performance at a high scale, while all security and governance policies are managed in Unity Catalog.

Let's look into how you can quickly build AI/BI dashboards in DBSQL. The dashboard has two tabs, a **Data tab** for searching for tables within the Unity catalog or writing queries that will serve as your Dataset(s), and a **Canvas tab** where your visualizations are created and assembled with the option to use natural language to generate tables and visualizations. Some of the capabilities included are sleek visualizations, cross-filtering, and periodic PDF snapshots via email.

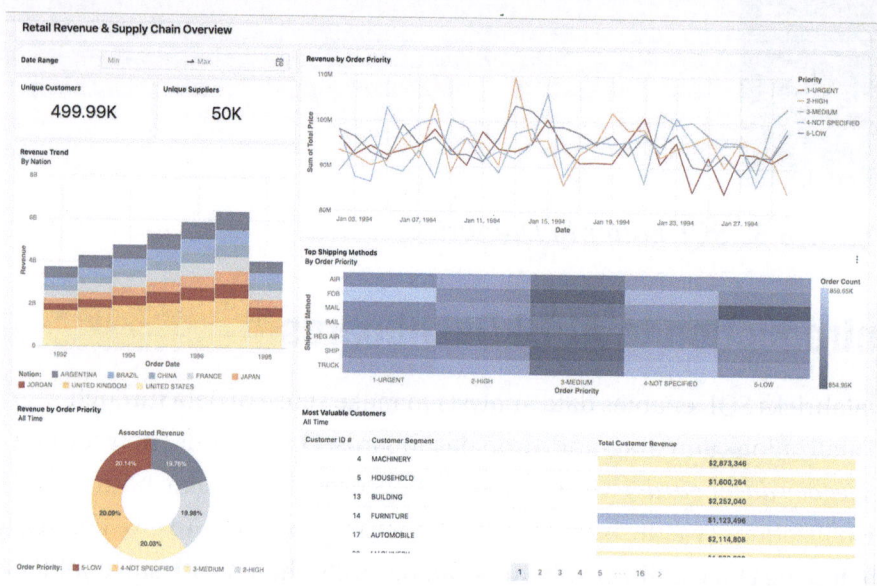

Figure 8-6. *AI/BI dashboards*

Finally, AI/BI dashboards allow users to publish their dashboards to the entire organization. This means that any authenticated user in your identity provider (IdP) can access the dashboard via a secure web link, even if they don't have Databricks workspace access.

Alerts

DBSQL allows users to set up alerts, which in turn send notifications if a particular condition is not met in the data. Take, for example, an inventory management table. One can set an alert on the table if the quantity of a particular product or SKU falls below a certain threshold. The notifications can either be delivered through email or to other platforms like Slack, Teams, etc., via webhooks.

In Chapter 9 , we will discuss using alerts to trigger model retraining when a threshold drops below certain levels (see Figure 8-7).

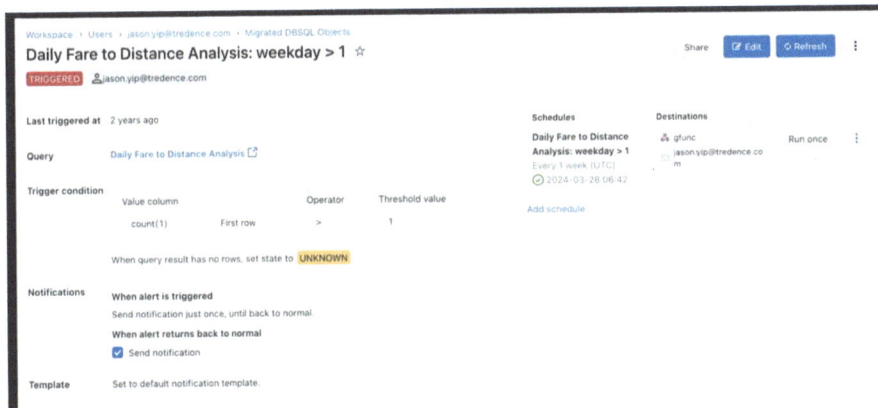

Figure 8-7. *Databricks alerts*

Query History and Profile

Query history in DBSQL gives you full visibility and details of query execution for all the queries executed on the SQL warehouses for the last 30 days. With a unified view, you not only can see the number of queries executed at a particular time but also quickly zoom into specific queries and debug issues, if any. See Figure 8-8.

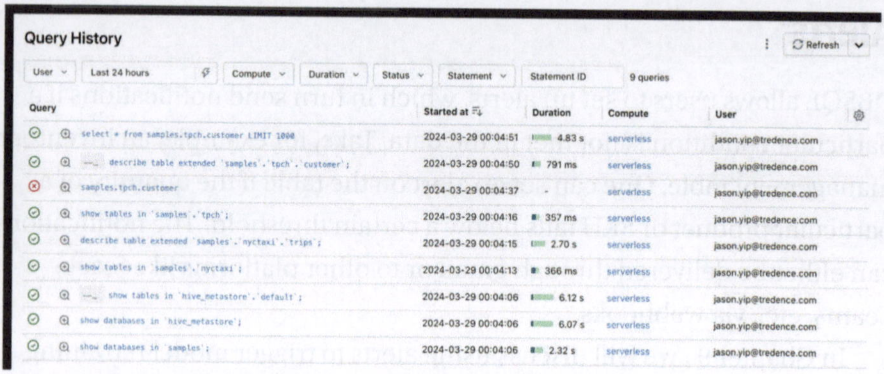

Figure 8-8. *DB SQL query history*

A query profile provides the ability to visualize the details of a query execution. The query profile helps you troubleshoot performance bottlenecks during the query's execution (see Figure 8-9). For example:

- You can visualize each query task and its related metrics, such as the time spent, number of rows processed, and memory consumption.

- You can identify the slowest part of a query execution at a glance and assess the impacts of modifications to the query.

- You can discover and fix common mistakes in SQL statements, such as exploding joins or full table scans.

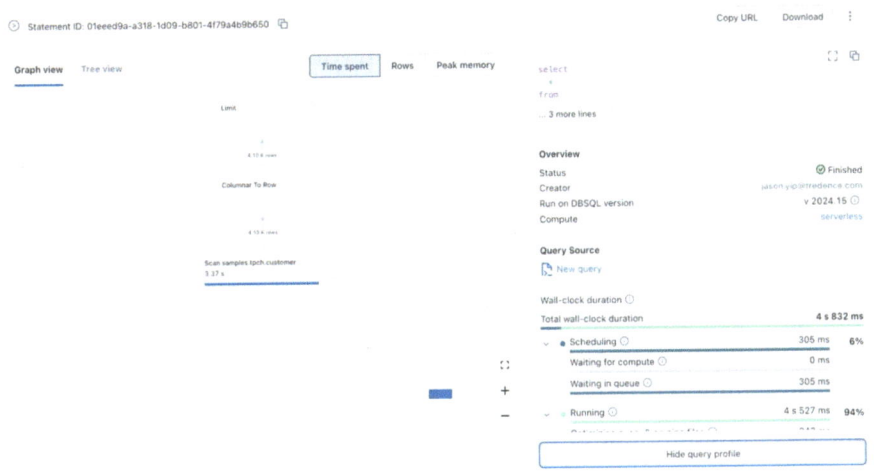

Figure 8-9. *Query profile*

After the overview of DBSQL, we will move on to do a deep dive into some important features of DBSQL.

Serverless Compute

Databricks Serverless is a paradigm in Databricks compute. Serverless is a fully managed service, eliminating the burden of capacity management, patching, upgrading, and performance optimization of the cluster. Additionally, Serverless simplifies the billing. In other words, you need to pay only once to Databricks for both Compute and Databricks costs. Although Serverless was initially introduced only for DBSQL workloads, it has then been expanded to other parts of the platform including Delta Live Tables, workflows, and notebooks as well.

As discussed previously, with any type of compute be it all-purpose clusters, jobs clusters, or even SQL warehouse (non-Serverless) the virtual machines are provided by the cloud provider for which you need to pay directly to them. This has two main effects - not only it takes three to four minutes for a cluster to come up or terminate but also the users have to manage these in terms of runtimes, machine types, cluster sizes etc. Secondly, there are usually two line items in your cloud bill - Databricks Costs ($DBU) and cloud VM costs. To look at the cost more in-depth, please refer to the chapter Databricks Pricing and Observability using System Tables.

With the introduction of Serverless, Databricks is basically "owning the compute." To put it simply, Databricks prepurchases these VMs from the respective cloud provider, and once you ask for a Serverless compute resource, it releases the specific number of VMs as per the request. Since this computer is fully managed by Databricks, it spins up or down in seconds rather than minutes. Furthermore, you only need to pay Databricks once for both Databricks costs and VM costs. Thus, Serverless compute brings a truly elastic environment that's instantly available and scales with your needs.

Constraints in DBSQL

Many data analysts have previous experience in relational databases, building entity-relational models using primary key/foreign key relationships. After normalization, they usually build multidimensional data models (referred to as *star schemas*) so that it is easy to understand and analyze data across these relational databases or data warehouses. Further, primary key/foreign keys help maintain data integrity and avoid errors during data processing and modification, thus helping maintain data quality.

Constraints on Databricks

Constraints in databases are rules that ensure data integrity and consistency by enforcing certain conditions or restrictions on the data stored in a table. Databricks supports standard SQL constraint management clauses, which can be divided into two categories:

- **Enforced constraints:** These are enforced on the tables/columns to ensure data quality and integrity: NOT NULL , UNIQUE, and CHECK.

- **Informational constraints:** These constraints are not enforced but explain the relationships between fields in the tables.

Enforced Constraints

Enforced column constraints are rules that apply to a single column in a table. Delta tables support the following column constraints:

- NOT NULL: This constraint ensures that a column must have a value for each row and cannot be null. This ensures data completeness and consistency.

- UNIQUE: This constraint ensures that all values in a column are unique and distinct. It prevents duplicate values in the table.

- CHECK: This constraint validates that a column's value meets a specific condition or a range of conditions, such as ensuring that a particular column is within a certain range, or a number is greater than a specific value. It helps ensure data accuracy and consistency. This constraint allows you to specify a Boolean expression that must evaluate as true for each row in the table. If the expression evaluates to false, an error is raised, and the statement is rolled back.

Let's look at an example of how to define these constraints on a table. The constraints can be set either while creating a new table or on an existing table. You can add constraints in a new table as follows:

```
CREATE TABLE T1 (
  id INT NOT NULL,
  quantity INT,
  date DATE,
  CONSTRAINT chk_quantity CHECK (quantity > 0)
);
```

To add a constraint to an existing table, you can use ALTER TABLE ADD CONSTRAINTS, and to drop such a constraint, you can use ALTER TABLE DROP CONSTRAINT.

```
ALTER TABLE T1 ADD CONSTRAINT dateWithinRange CHECK
(Date > '1900-01-01');
```

Informational Constraints: Primary Key Foreign Key

A Databricks lakehouse with Unity Catalog gives users the ability to build entity relationships that are simple to maintain and evolve. Also note that for now primary key and foreign key are informational only and they are not enforced. To leverage primary keys/foreign keys (PKs/FKs), your workspace should be UC-enabled with DBR version 11.1 and above.

Let's see how we can implement a primary key/foreign key relationship with an example. We can create two tables, P1 and F1. The P1 table has a primary constraint on the id column, and table F1 has a foreign key constraint on the p1_id column that refers to the id column in the P1 table.

```
CREATE TABLE P1 (
 id INT PRIMARY KEY,
 name STRING
)
USING delta;

CREATE TABLE F1 (
 f1_id INT,
 f1_date DATE,
 p1_id INT,
 FOREIGN KEY (p1_id) REFERENCES p1(id)
)
USING delta;
```

The "View relationships" button (Figure 8-10) in the Overview or Schema tab conveniently shows the relationship between tables (Figure 8-11).

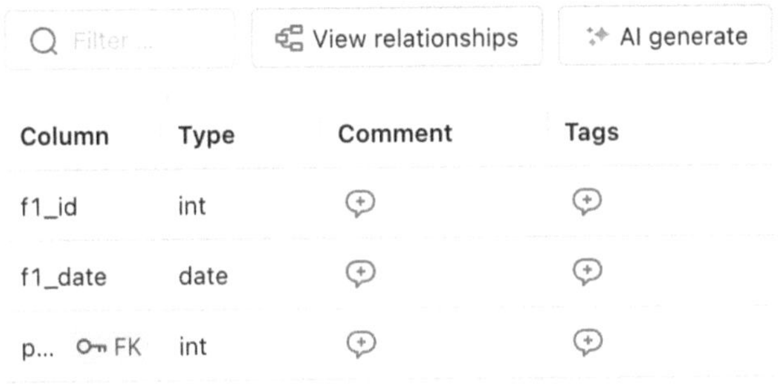

Column	Type	Comment	Tags
f1_id	int		
f1_date	date		
p... ⊶ FK	int		

Figure 8-10. *"View relationships" button*

Entity Relationship Diagram for unitygo.default.f1

Figure 8-11. ER diagram in Databricks

Next we move into streaming tables and materialized views. We touched on these two briefly in Chapter 9, but we will do a greater deep dive here.

Streaming Tables and Materialized Views

Some of the common challenges faced by data analysts while working in data warehouses include the inability to self-service ingest and fix data issues, the inability to have the most recent data for BI dashboards, and having to deal with slow BI dashboards because of the huge volume of underlying data.

Streaming tables and materialized views in DBSQL will allow SQL analysts to perform data engineering tasks and thus have real-time capabilities along with their existing workflows. It is important to note that both Streaming tables and materialized views require Unity Catalog and Serverless enabled in your workspace. In the next section, we will discuss these two features in detail.

Streaming Tables

A streaming table is a special type of table that enables ingestion in DBSQL. It is managed by Unity Catalog and supports append-only incremental and streaming data processing from various data sources. In reference to the medallion architecture, streaming tables are ideal for bringing data into the Bronze layer. Streaming tables enable continuous, scalable ingestion from any data source, including cloud storage, message buses (EventHub, Kafka), and more.

Streaming from a source requires the source to be append-only and never updated or deleted. To configure the streaming table to perform streaming ingestion of your source, you must specify the STREAM keyword. Say, for example, you have an S3/ADLS container, and a lot of new files are continuously arriving. You can create a streaming table by using the following syntax:

```
CREATE OR REFRESH STREAMING TABLE mystream
  AS SELECT * FROM STREAM read_files('s3://<bucket>/<path>/
  <folder>')
```

By default read_files processes all the files in the folder. To avoid this you can set the property includeExistingFiles option to false.

```
CREATE OR REFRESH STREAMING TABLE mystream
  AS SELECT * FROM STREAM read_files('s3://<bucket>/<path>/
  <folder>',, includeExistingFiles => false
)
```

Once the previous command is executed under the hood, a DLT pipeline is created for each streaming table. You can keep these tables updated and refreshed.

To load data from a system like Kafka, use the following command:

```
SELECT * FROM STREAM read_kafka(
  bootstrapServers => '<server:ip>',
  subscribe => '<topic>',
  startingOffsets => 'latest'
);
```

Materialized Views

A materialized view is a special type of view that precomputes and stores the results of a SQL query and automatically keeps them fresh over time.

A materialized view is a database object that stores a query's results as a physical table. Unlike regular virtual database views, which derive their data from the underlying tables, materialized views contain precomputed data that is incrementally updated on a schedule or on-demand. This precomputation of data allows for faster query response times and improved performance in certain scenarios.

Materialized views are especially useful in situations where complex queries or aggregations are performed frequently and the underlying data changes infrequently. By storing the precomputed results, the database can avoid the need to execute complex queries repeatedly, resulting in faster response times.

Create a Materialized View

Databricks SQL materialized view CREATE operations use a Databricks SQL warehouse to create and load data in the materialized view. Because creating a materialized view is a synchronous operation in the Databricks SQL warehouse, the CREATE MATERIALIZED VIEW command blocks until the materialized view is created and the initial data load finishes. A Delta

Live Tables pipeline is automatically created for every Databricks SQL materialized view. When the materialized view is refreshed, an update to the Delta Live Tables pipeline is started to process the refresh.

```
CREATE MATERIALIZED VIEW mv1
AS SELECT
  date, sum(sales) AS sum_of_sales
FROM
  table1
GROUP BY
  date;
```

Refresh a Materialized View

In Databricks SQL, you have the option to set up automatic refresh for a materialized view based on a predefined schedule. This schedule can be configured during the creation of the materialized view using the SCHEDULE clause or added later using the ALTER VIEW statement. Once a schedule is established, a Databricks job is automatically created to handle the updates.

```
REFRESH MATERIALIZED VIEW mv1;
```

Next, we move into another important feature: Lakehouse Federation, which allows you to query data stored in data sources without moving the data.

Lakehouse Federation

Lakehouse Federation gives the Databricks platform query federation capabilities. Query federation enables users and systems to run queries against multiple data sources without migrating all the data to one central location.

Most organizations have valuable data distributed across multiple data sources—databases, data warehouses, object storage systems, etc. This siloed data leads to incomplete data and insights, which hinders the ability to make informed decisions based on the full available data.

To query data across multiple data sources, users typically need to move or migrate their data to a central data location first, which usually takes time and effort. Lakehouse Federation addresses these critical pain points and makes it simple for organizations to expose, query, and govern siloed data systems as an extension of their lakehouse. The various systems include MySQL, PostgreSQL, Amazon Redshift, Snowflake, Azure SQL Database, Azure Synapse, BigQuery, and more from within Databricks without moving or copying the data, all within a simplified and unified experience.

Further, Unity Catalog's advanced security features, such as row and column-level access controls, discovery features like tags, and data lineage, are available across these external data sources, ensuring consistent governance.

To make a dataset available for read-only querying using Lakehouse Federation, you create the following (Figure 8-12):

- **A connection** that specifies a path and credentials for accessing an external database system

- A **foreign catalog** that mirrors a database in an external data system enabling you to perform read-only queries on that data system in your Databricks workspace, managing access using Unity Catalog

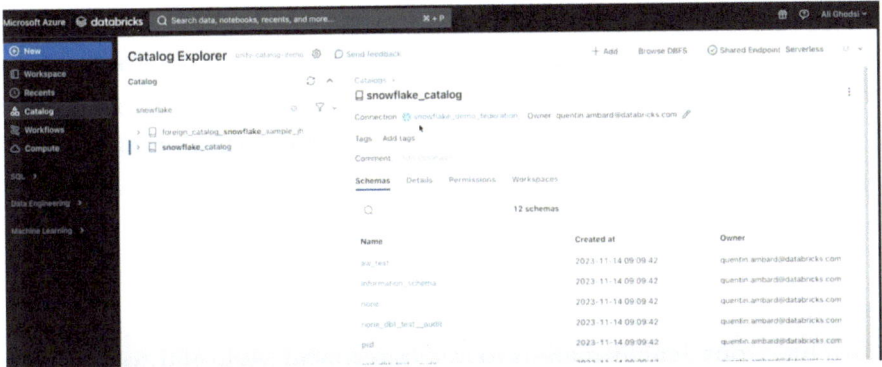

Figure 8-12. *Lakehouse Federation into Snowflake*

You can start to write queries against these tables in DBSQL and create visualizations to view the data.

As a best practice, Lakehouse Federation should not be used for real-time data processing, where latency is paramount, or complex data transformations, where vast amounts of data need to be ingested and processed.

AI Functions in DBSQL

In the age of large language models, there is an urgent need to combine AI output into a BI report so management can take action based on the results. However, this inferencing pipeline not only creates another layer of complexity but also requires seasoned data scientists and an ML Ops team to maintain, which can become costly.

Consume LLM Models in DBSQL

Now, there are multiple ways to consume these large language models within Databricks. The traditional way is to leverage the code provided on Huggingface. Though this approach is flexible, it will require integrating the sample code into an existing pipeline, which requires development work.

The next approach is to use the Model Serving API. Databricks has curated popular models and made them part of the platform. These are then exposed as the Foundation Model API. With the Foundation Model API, developers can access these carefully curated models out of the box without going through the deployment process and getting enhanced performance.

The third approach we will look into in detail here is AI functions in serverless SQL.

AI functions enable analysts to integrate any LLMs in SQL to enrich data and empower analysts to extract actionable insights

There are two types of AI functions provided by Databricks:

- Built-in functions backed by the Foundation Model APIs

- Custom functions backed by a Serverless serving endpoint

Built-in functions invoke a state-of-the-art generative AI model to perform tasks such as sentiment analysis, classification, and translation. Let's examine some common built-in functions.

- **ai_analyze_sentiment:** Given text, output sentiment of the text like positive, negative, neutral, mixed.

- **ai_classify:** Ask the LLM to do classification. A good use case is to ask an LLM to determine if the text contains PII, which is to ask it if the text ["contains PII", "no PII"].

- **ai_extract:** Ask the LLM to extract any entities. Similar to regex patterns but you no longer need to write a regex. You only need to tell the function what you want to extract. For example, "Place" will allow you to extract a place name.

- **ai_gen:** Prompting at scale. Given a list of questions, ask the LLM to output a list of answers, given in table format.

For a list of AI functions, please visit the Databricks website: `https://docs.databricks.com/en/large-language-models/ai-functions.html`.

`Let's` put the AI function in action. Consider this Kaggle Amazon review dataset.

`We can download` it to Databricks and create a Delta table. Using DB SQL's built-in AI functions, we can extract the sentiment from the text, successfully connecting AI with BI. See Figure 8-13.

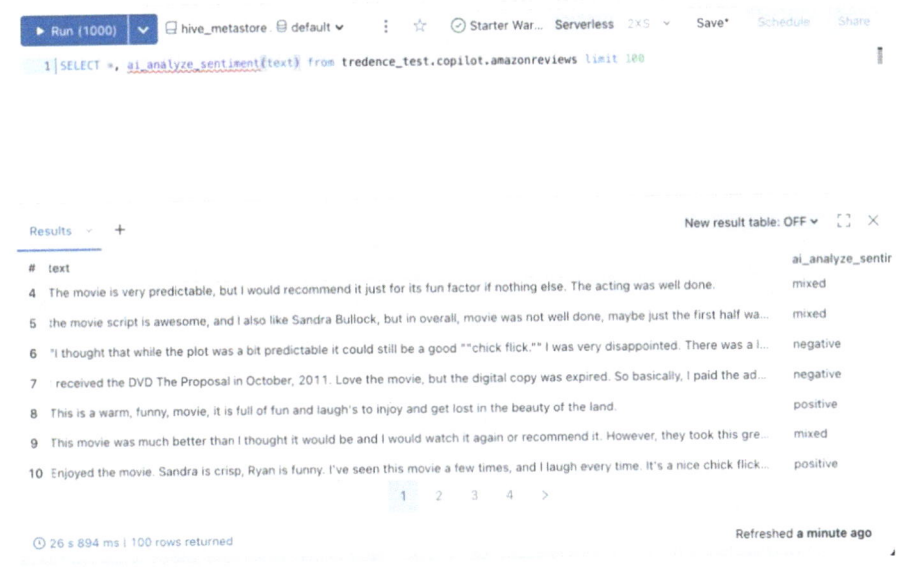

Figure 8-13. *Sentiment analysis with DB SQL*

Finally, after persisting the results in Unity Catalog, we can publish the inferred dataset to Power BI or Tableau.

Custom Functions Backed by a Serverless Serving Endpoint

The ai_query() function allows you to serve your ML and LLM models using Databricks Model Serving and query them using SQL. To do so, this function invokes an existing Databricks Model Serving endpoint and parses and returns its response. You can use ai_query() to query endpoints that serve custom models hosted by a model-serving endpoint, foundation models made available using Foundation Model APIs, and external models, which are third-party models hosted outside of Databricks.

Let's look into an example that queries the model behind the sentiment-analysis endpoint with the text dataset and specifies the request's return type.

```
SELECT text, ai_query(
    "sentiment-analysis",
    text,
    returnType => "STRUCT<label:STRING, score:DOUBLE>"
  ) AS predict
FROM
  catalog.schema.customer_reviews
```

In the next part of the chapter, we will examine how you can connect your BI tools through DBSQL.

Integrate BI Tools with Databricks

Organizations usually deploy transformational and BI tools such as PowerBI, Tableau, Looker, etc., for enterprise-wide dashboards and reporting needs. Moreover, many data analysts have been proficiently using these tools for quite some time. Databricks provides validated integrations with your BI tool of choice, allowing you to connect to your data using SQL warehouses or clusters. As a recommended practice, analysts get the best experience when they connect their BI tools to optimized gold tables via SQL warehouses.

In this section, we will focus on connecting PowerBI to Azure Databricks. There are two main ways to connect PowerBI to Azure Databricks. The first is to publish to PowerBI Online from Databricks. The second popular method is to connect Power BI Desktop to Databricks. Let's explore both these methods.

Publish to PowerBI Online from Databricks

This allows users to publish tables from Databricks Catalog Explorer UI directly to PowerBI workspaces. In short, this is a one-click publish of UC datasets to PowerBI workspaces. This method supports both DirectQuery and Import modes. Moreover, you can publish entire schemas with table relationships (PK/FK). Some of the requirements are that the data must be on Unity Catalog, the compute must be UC enabled, users must have a premium PowerBI License, and users must enable "Users can edit data models in Power BI service (preview)" under the Workspace settings and Data model settings. In Figure 8-14, we can go to the Catalog tab and select either the full schema or a particular table. Next, in the drop-down, select "Use with BI tools."

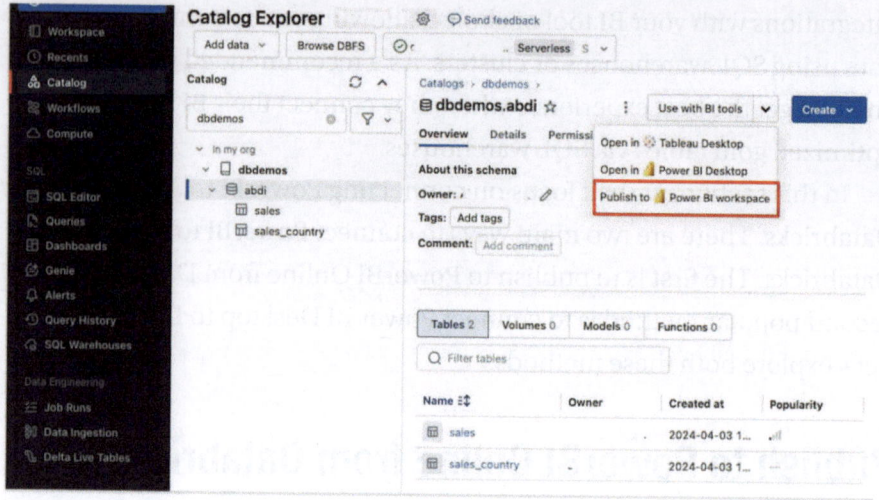

Figure 8-14. *Publishing to PowerBI Online*

You can select Publish to Power BI. This will ask you to authenticate with your Microsoft account via Entra ID. Once the authentication is completed, the user can select the PowerBI workspace and Dataset Mode (Direct Query or Import Mode). Thereafter, click Publish to PowerBI, and you can start to query this dataset in your PowerBI workspace.

Connect Power BI Desktop to Databricks

Users can also connect their PowerBI Desktop to Delta Lake Tables via the SQL warehouse for a full modeling experience in PowerBI. Further, there are three mainly used storage modes that PowerBI offers for tables. First is Import mode, wherein all the data is loaded in PowerBI's in-memory cache. Second is DirectQuery mode, wherein the data remains in the source system and the metadata is stored in PowerBI. Finally, a newer feature is Hybrid mode, which combines the Import and Direct Query modes by using partitions. The user can select the mode they want to use depending on the use case.

Let's now look at how to connect Databricks SQL warehouse with PowerBI. In Partner Connect, once you click PowerBI and choose the Databricks Compute resource that you want to connect, it downloads the connection file. You can open that file with Power BI Desktop, and your connection will be automatically configured. After selecting the connectivity mode, you can start querying the tables.

Conclusion

Databricks SQL gives complete warehousing capabilities on the lakehouse platform and provides features to data analysts for various BI use cases. SQL warehouses, especially Serverless, provide an enhanced compute to process SQL queries and provide a connection to various BI tools.

Some of the key features discussed in the chapter include Lakehouse Federation, AI functions, materialized views, streaming tables, and constraints, which include a primary key/foreign key relationship. Finally, we saw how easily you integrate you BI tool of choice with the Databricks platform with PowerBI as a case study.

CHAPTER 9

Machine Learning Operations Using Databricks

Databricks not only provides exceptional data processing capabilities but also offers a wealth of opportunities to develop machine learning use cases.

Databricks' machine learning capabilities have evolved significantly over the years. Since 2021, various user personas have been actively engaging with the platform. These personas include:

- **Data scientists:** They unlock the power of algorithms and models.

- **Data engineers:** They craft robust pipelines for seamless data flow.

- **Machine learning engineers:** They skillfully orchestrate model deployment.

In this chapter, we will examine the different components in Databricks that support machine learning, including model development, deployment, inferencing, and monitoring. You will be able to learn how to deploy an ML model to Databricks. These concepts are critical in the later

© The Editor(s) (if applicable) and The Author(s),
under exclusive license to APress Media, LLC, part of Springer Nature 2024
N. Gupta and J. Yip, *Databricks Data Intelligence Platform*,
https://doi.org/10.1007/979-8-8688-0444-1_9

chapters when we discuss GenAI as a lot of the components will be reused as we advance to GenAI. If you are already familiar with the end-to-end ML lifecycle, this chapter can serve as a refreshment and prepare you for the concepts to come in later chapters.

Machine Learning with Databricks

While SQL and data engineering are the top portion of the menu, machine learning is also an integral part of Databricks. In 2021, Databricks was named a leader in the Gartner Magic Quadrant for data science and machine learning platforms.[1] See Figure 9-1.

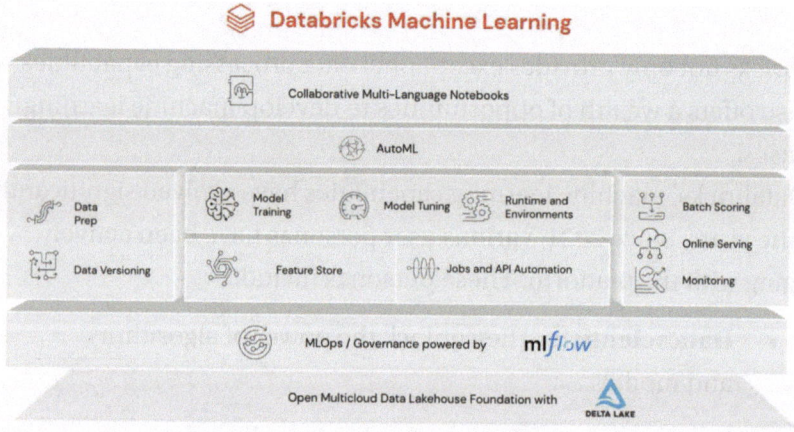

Figure 9-1. *Databricks machine learning stack*

Figure 9-2 shows five components in the ML platform that we will focus on.

[1] https://databricks.com/blog/2021/03/04/databricks-named-a-leader-in-2021-gartner-magic-quadrant-for-data-science-and-machine-learning-platforms.html

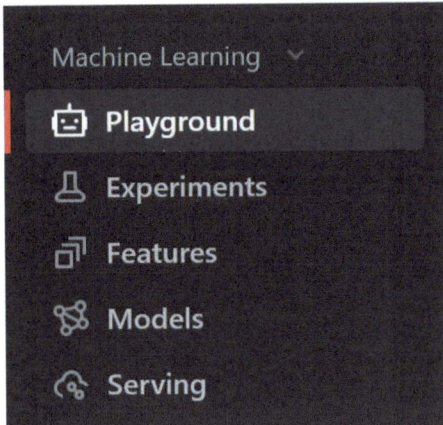

Figure 9-2. *Machine learning persona in Databricks UI*

Experiments

Experiments are individual pages to track ML training runs. It provides an overview of everything related to your training configuration and results as well as lineage to your dataset and ML model. You can use MLflow to log these values into the experiment page, therefore providing a one-stop shop for all the trails you run without losing out the configuration for the champion model, aka the best-performing model. See Figure 9-3.

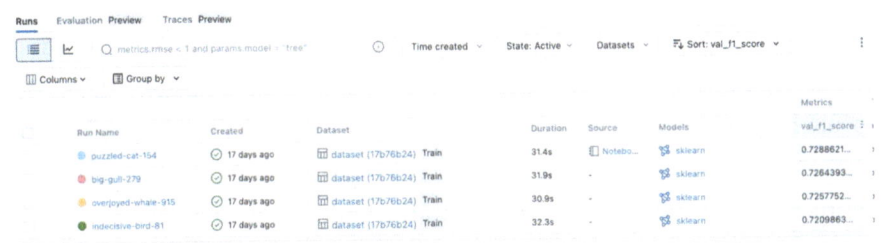

Figure 9-3. *Experiments page for one of the models*

The best way to start with ML is to leverage the Glass Box AutoML provided out of the box by Databricks.

It's worth noting that to use Unity Catalog with AutoML, the cluster access mode must be **Single User**, and you must be the designated single user of the cluster. Even administrators will not be able to run AutoML on behalf of a user. The selected "table" must also be a "table." Materialized views are not accessible via Single User clusters. See Figure 9-4.

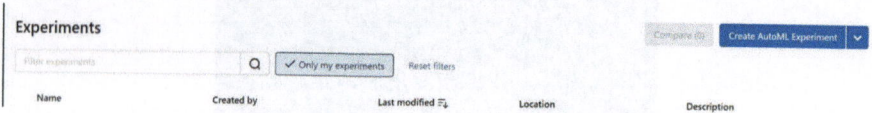

Figure 9-4. *Creating an AutoML experiment*

What Is the Glass Box Approach to Automated Machine Learning?

Machine learning is a highly iterative task. Data scientists spend a lot of time trying out different algorithms and tuning hyperparameters to find the best-performing model. However, these repetitive tasks can be automated using AutoML.

Unfortunately, most platforms, such as Azure Machine Learning, are black boxes, while capable of picking the best model, because the code to train the model is not provided. Hence, it is difficult to replicate the best-performing model and make further enhancements. Databricks' Glass Box approach provides all the source codes that generated all the models, not just the best-performing ones but all the models evaluated, allowing data scientists to customize the models with the source code provided.

Machine Learning Lifecycle: MLOps

A typical MLOps lifecycle contains the following stages:

1. Data prep

2. Model building

3. Model deployment

We can conveniently use Databricks to do everything. Furthermore, to harness the distributed nature of Spark, we can also use libraries like Horovord and Petastorm to scale out the model training. See Figure 9-5.

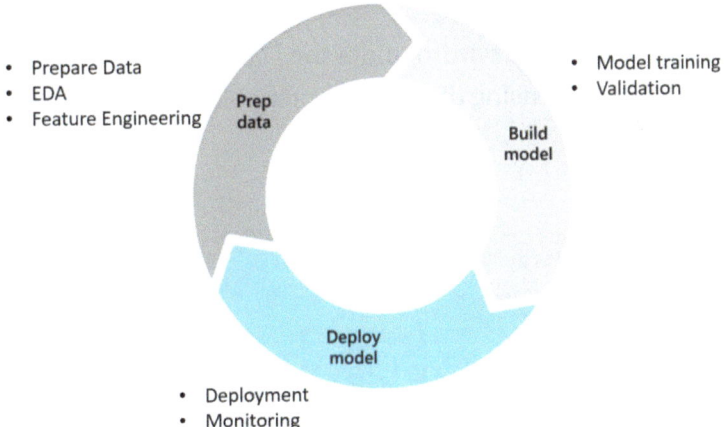

Figure 9-5. *Machine learning life cycle*

In the following sections, we will take an example dataset and go through every step in Figure 9-5. There will be further clarifications, but rest assured that every step will be covered.

We will demonstrate AutoML using a classification problem on a flight dataset, which can be found in Kaggle. This problem aims to predict whether a flight will be delayed or canceled based on historical flight data. Of course, in reality, the real reason for a flight delay or cancellation can be caused by a lot of factors beyond the flight itself, like weather or staff

185

shortage. This exercise does not demonstrate how to build a state-of-the-art on-time flight predictor. It simply uses a dataset to illustrate the workflow of ML Ops using Databricks.

ML Example: Predicting Flight Delays with Databrick's AutoML

Prepare Data

First, we need to upload the data to Databricks. This can be done very easily with JDBC or simply by uploading the CSV file. Then, we can create tables from there.

The Create New Table wizard under the Data tab can be used to upload data and create a table using the UI or a notebook. See Figure 9-6.

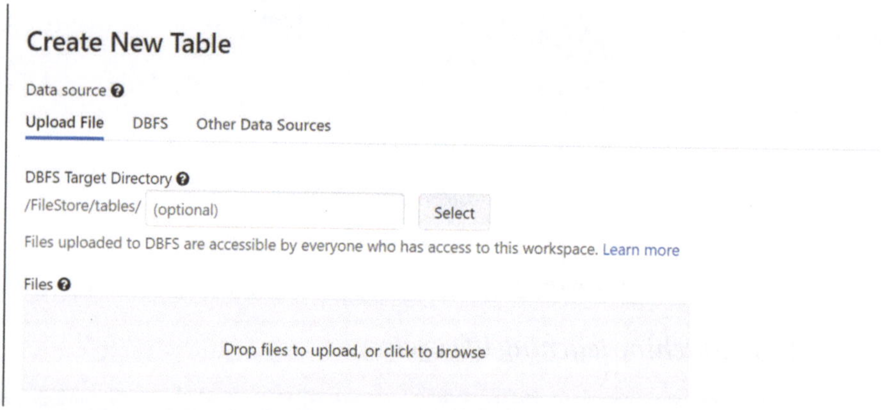

Figure 9-6. *Creating a new table using the Databricks UI*

Exploratory Data Analysis

Databricks has integrated Pandas Profiling for Exploratory Data Analysis (EDA). Pandas Profiling is an open-source library that precomputes some statistics that data scientists usually want to know and saves these into properly formatted HTML. Once we hook up our data in the AutoML interface, it will also generate an EDA report from Pandas Profiling. See Figure 9-7.

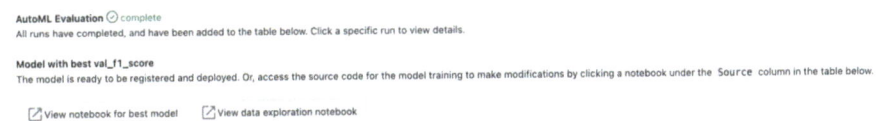

Figure 9-7.   *The "View data exploration notebook" button can be found on the experiment page*

By clicking "View data exploration notebook," we can also examine how Databricks uses Pandas profiling to perform EDA. See Figure 9-8.

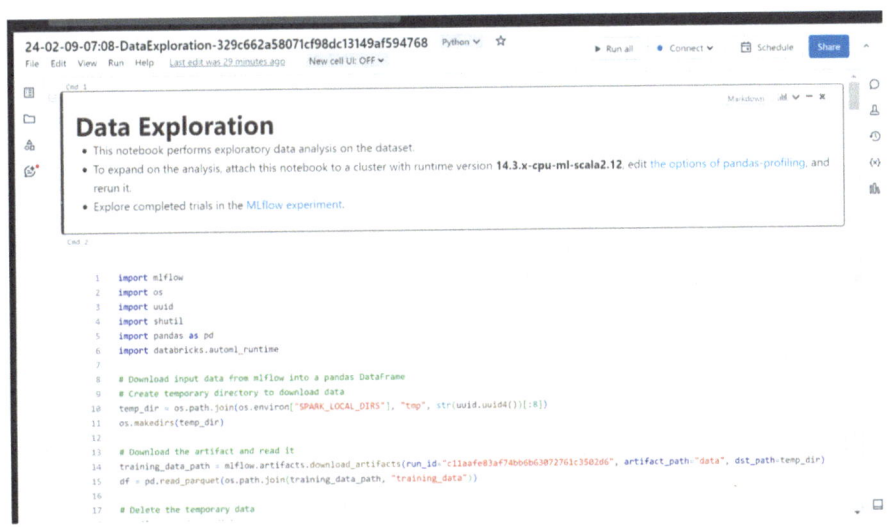

Figure 9-8.   *Data exploration notebook*

Feature Engineering

In most cases, we need to transform raw data into something useful that the model can use for better predictions. For example, in our flight delay example, we can compute the percentage of delayed flights by airport or airline. We can then save these into a feature store, so other team members can reuse them and understand how they were built (see Figure 9-9). Features can also be joined with raw data in AutoML to increase the accuracy of the prediction.

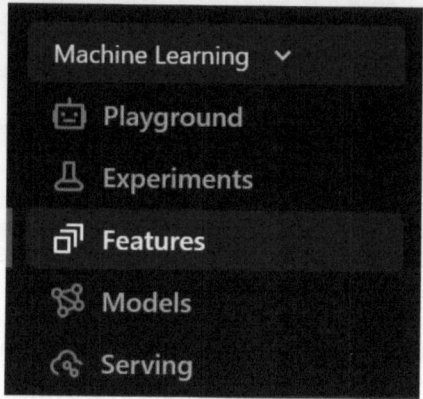

Figure 9-9. *Features tab in Databricks*

A feature table can be registered either by a dataframe or a table.

Data Exploration at Scale

Let's look at data exploration at scale.

Pandas Profiling

Data scientists often need to understand the data distribution to decide whether the data is useful, whether imputation is required, or, in extreme cases, whether to exclude specific columns from the model training. For example, if the column contains all nulls or empty values, it will not contribute anything to the machine learning model as it cannot learn anything.

Pandas Profiling has always been done using the Python pandas library. The limitation of a Pandas library is that it can run on only one machine despite a cluster of machines available. The shortcoming is that the memory will be limited if the process can run on only one machine. If the dataset cannot fit into the memory, Pandas profiling cannot generate a report. This often leads to sampling of the data such as `df.sample(fraction=0.5)`. This often leads to a misrepresentation of the data distribution. For example, if the sampled rows don't contain any nulls, then it will lead to data scientists believing that the column does not contain nulls.

As of Databricks runtime 14.3 LTS, Databricks will still try to sample the dataset for profiling and Auto ML. However, as of 2023, YData released Spark support for the popular profiling library, which keeps the same interface but takes a Spark Dataframe instead. The following is an example:

`https://ydata.ai/resources/ydata-profiling-the-great-debut-of-pandas-profiling-into-the-big-data-landscape`

The Pandas version requires reading Parquet files as Pandas, which is not optimal because Delta format contains transactions of Parquet. See Listing 9-1.

Listing 9-1. Pandas Profiling Sample Usage

```
import pandas as pd
import databricks.automl_runtime

training_data_path = mlflow.artifacts.download_artifacts(run_id
="a0922defd3b542acb2b4bb0956aeb0bf", artifact_path="data", dst_
path=temp_dir)
df = pd.read_parquet(os.path.join(training_data_path,
"training_data"))

from ydata_profiling import ProfileReport
df_profile = ProfileReport(df,
```

```
                            correlations={
                                "auto": {"calculate": True},
                                "pearson": {"calculate": True},
                                "spearman": {"calculate": True},
                                "kendall": {"calculate": True},
                                "phi_k": {"calculate": True},
                                "cramers": {"calculate": True},
                            }, title="Profiling Report", progress_
                            bar=False, infer_dtypes=False)
profile_html = df_profile.to_html()
displayHTML(profile_html)
```

Spark version—reading the table in Databricks—can preserve the integrity of Delta format, as shown in Listing 9-2.

Listing 9-2. YData's Spark Support

```
df = spark.table("kaggle.flight_featured_detla")
from ydata_profiling import ProfileReport
df_profile = ProfileReport(df,
                        correlations={
                            "auto": {"calculate": True},
                            "pearson": {"calculate": True},
                            "spearman": {"calculate": True},
                            "kendall": {"calculate": True},
                            "phi_k": {"calculate": True},
                            "cramers": {"calculate": True},
                        }, title="Profiling Report",
                        progress_bar=False,
                        infer_dtypes=False)
```

BREAKING CHANGES

The pandas-profiling package naming was changed to ydata-profiling.

Data Summarization Using dbutils

Fortunately, despite not being as comprehensive as Pandas profiling, most notably missing correlation matrices, starting in Databricks Runtime 9.0, there's a summarize feature available under dbutils, which is using Spark as a compute for the statistics that data scientists requires.

Listing 9-3 shows example usage.

Listing 9-3. Summarization Function in dbutils

```
df = spark.table("kaggle.flight_featured_detla")
dbutils.data.summarize(df)
```

Feature Store

A feature store is a centralized repository that enables data scientists to find and share features and also ensures that the same code used to compute the feature values is used for model training and inference (see Figure 9-10).

Similar to Unity Catalog, consider this a place where the data scientists will look for their features. A feature is usually a calculation that has been tested and agreed upon among the team and provides value. Not only is it useful to the team, but it can potentially be shared among different models to save time in the discovery process.

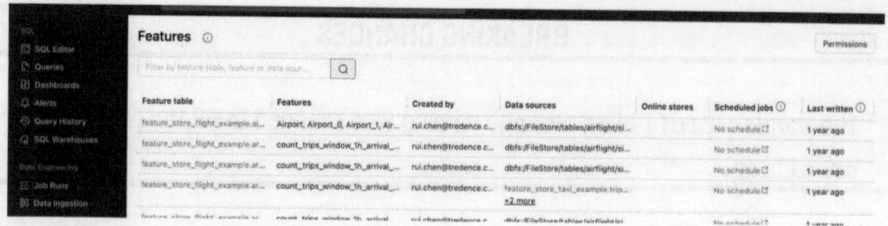

Figure 9-10. *Databricks feature store*

Why Use Databricks Feature Store?

Databricks Feature Store is fully integrated with other components of Databricks. Along with Unity Catalog, it provides powerful lineage tracking for all the features from source data to the model. The following are the advantages of this out-of-the-box feature store:

- **Discoverability**. The Feature Store UI, accessible from the Databricks workspace, lets you browse and search for existing features.

- **Lineage**. When you create a feature table in Databricks, the data sources used to create the feature table are saved and accessible. For each feature in a feature table, you can also access the models, notebooks, jobs, and endpoints that use the feature.

- **Integration with model scoring and serving**. When you use features from Feature Store to train a model, the model is packaged with feature metadata. When you use the model for batch scoring or online inference, it automatically retrieves features from Feature Store. The caller does not need to know about them or include logic to look up or join features to score new data. This makes model deployment and updates much easier.

- **Point-in-time lookups**. Feature Store supports time series and event-based use cases that require point-in-time correctness. For deep dive into time-series tables, please refer to the following documentation:

  ```
  https://docs.databricks.com/en/machine-learning/
  feature-store/time-series.html#how-time-series-
  feature-tables-work
  ```

To create a feature table in the feature store, it is possible either with a delta table or with a Spark dataframe. Bear in mind that if your table requires a primary key, you can use the syntax in Listing 9-4 to add a primary key.

Listing 9-4. Syntax to Add a Primary Key

```
ALTER TABLE <full_table_name> ADD CONSTRAINT <pk_name> PRIMARY
KEY(pk_col1, pk_col2, ...)
```

The main class to be used is called FeatureEngineeringClient, as shown in Listing 9-5.

Listing 9-5. FeatureEngineeringClient Class

```
from databricks.feature_engineering import
FeatureEngineeringClient
fe = FeatureEngineeringClient()
```

To register a feature table with an existing delta table, see Listing 9-6.

Listing 9-6. Registering a Delta Table as a Feature Table

```
fe.register_table(
  delta_table='kaggle.flight_featured_detla',
  primary_keys='flight_id',
  description='Flight features'
)
```

To create a feature table with a dataframe, see Listing 9-7.

Listing 9-7. Writing a Feature Table Using Dataframe

```
fe.write_table(
  name='kaggle.flight_featured_detla',
  df = flight_features_df,
  mode = 'overwrite'
)
```

Refer to the Feature Engineering Python API for comprehensive usage information:

https://api-docs.databricks.com/python/feature-engineering/latest/index.html

Finally, we can either look up a feature with the Python API or leverage AutoML, and then we can join features in the feature store easily. In the next section, we will discuss how to create an AutoML experiment (see Figure 9-11).

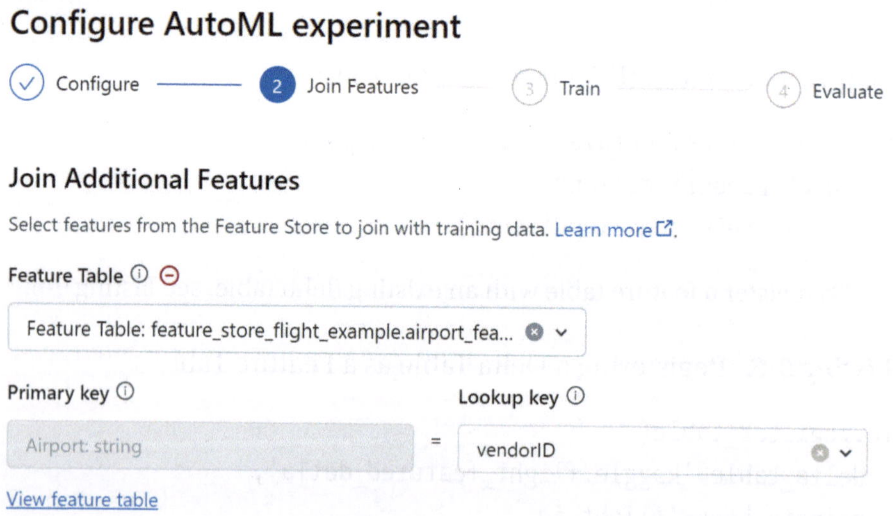

Figure 9-11. *Joining a feature table in AutoML experiment*

Model Building

Let's talk about model building.

Model Training

As mentioned, the next most time-consuming task after careful feature engineering is to train and tune your model and sometimes carefully select its algorithm or architecture to achieve the best accuracy while making predictions. With Databrick's AutoML, we can seamlessly select the data from the Data tab, allowing it to perform hundreds of selections automatically and saving data scientists hours of effort to build these from scratch.

Of course, the best way to learn how to build a model within Databricks is through Databricks. That makes the Glass Box AutoML an attractive approach to start with.

While AutoML does not solve all the machine learning problems in the world, it does, however, provide a framework that can solve some very typical machine learning problems, saving data scientists time to gain insight into the quality of the models that can be built with the dataset.

The following ML problem types are supported by Databricks AutoML (Figure 9-12):

- *Classification*

  Classification allows you to assign each observation to one of a discrete set of classes, such as good credit risk or bad credit risk.

- *Regression*

  Regression allows you to predict a continuous numeric value for each observation, such as annual income.

- ***Forecasting***

  Time-series forecast allows you to predict a future value
  based on a hierarchy, for example a future store sale in
  each city of each state in the United States.

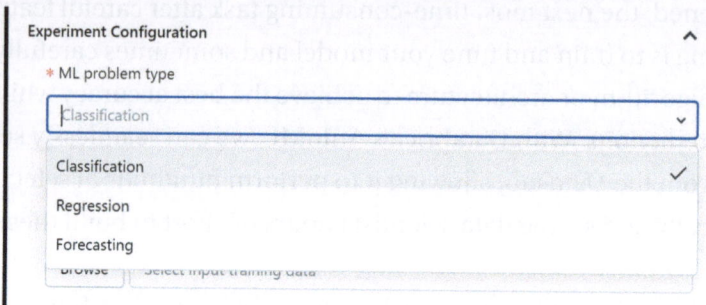

Figure 9-12. *ML problem types in AutoML*

As shown in Table 9-1, the interface contains only a few drop-downs
and is fully integrated with the feature tables and the Delta tables created
and persisted on the Data tab. The next step is to choose a prediction
target. Finally, we can also choose how to handle *imputation*, a process
of handling nulls in the dataset. Auto ML will then handle the rest of the
model selection, hyperparameters tuning, and presenting the results along
with the notebook.

Table 9-1 illustrates how many different algorithms Databricks
will try in each ML problem type. In our example, we are using
Classification models.

Table 9-1. *Databrick Algorithms*

Classification Models	Regression Models	Forecasting Models
Decision trees	Decision trees	Prophet
Random forests	Random forests	Auto-ARIMA (available in Databricks Runtime 10.3 ML and above)
Logistic regression	Linear regression with stochastic gradient descent	
XGBoost	XGBoost	
LightGBM	LightGBM	

Next, we can configure our experiment by choosing the training dataset and the target variable (see Figure 9-13). We can also configure how we want to treat null values, which is often referred to as *imputation*. Please ensure that you have selected a machine learning runtime-enabled cluster for the experiment, which is often an oversight.

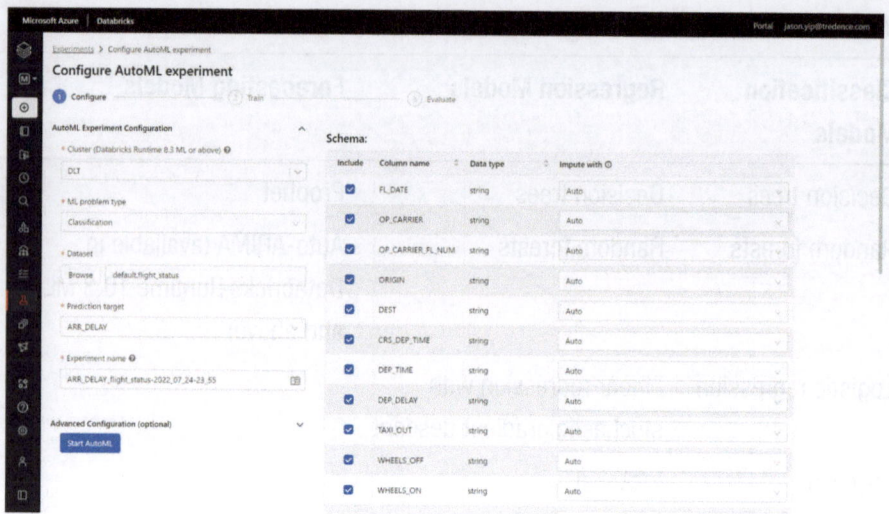

Figure 9-13. *Databrick's AutoML interface*

Finally, we can run the training and wait for the magic to happen. Please note that by no means does Databricks AutoML try to produce a production-grade model because it aims to simplify the process of parameter search, and most importantly, if the dataset is too large, it will try to take a sample of the dataset (Figure 9-14). If there is a need to train a huge dataset, we can consider using distributed training. The details are beyond the scope of this book, but the documentation can be found on Databricks' website:

https://docs.databricks.com/en/machine-learning/train-model/distributed-training/index.html

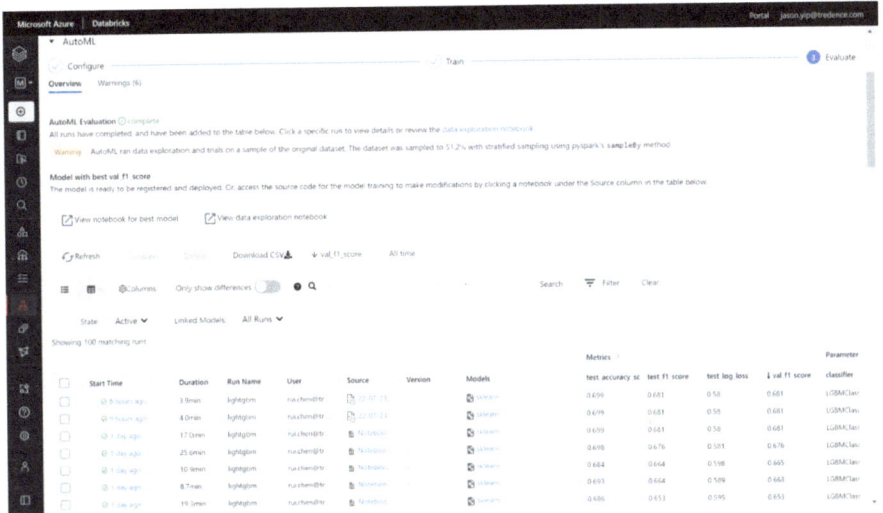

Figure 9-14. *AutoML will automatically select the top 100 most relevant results*

Validation

Before we deploy the model to production, we need to validate the model with test dataset. We would usually split the data into a training set and a test set with ratios like 70/30 split or 80/20 split. We can do random split, or splitting based on a business key, aka stratified split. Some more advanced approaches can include training/validation/test split with the validation is used for hyperparameters tuning, and until after the final model is decided, we can evaluate the model with a test set. We can split into 60/20/20 for train/val/test in these scenarios. AutoML uses the latter approach for splitting.

In Databricks Runtime 10.1 ML and above, we can specify a time column for splitting for classification and regression problems. This provides flexibility when some problems are highly based on chronological order.

Deploy Model

Let's talk about the deploy model.

Deployment

Once you have verified the champion model's parameters and metrics and you are ready to take the next step to deploy this model for further review with stakeholders, you can register the model right from the experiment. For best practices, you should always register your mode in Unity Catalog, as shown in Figure 9-15 and Figure 9-16.

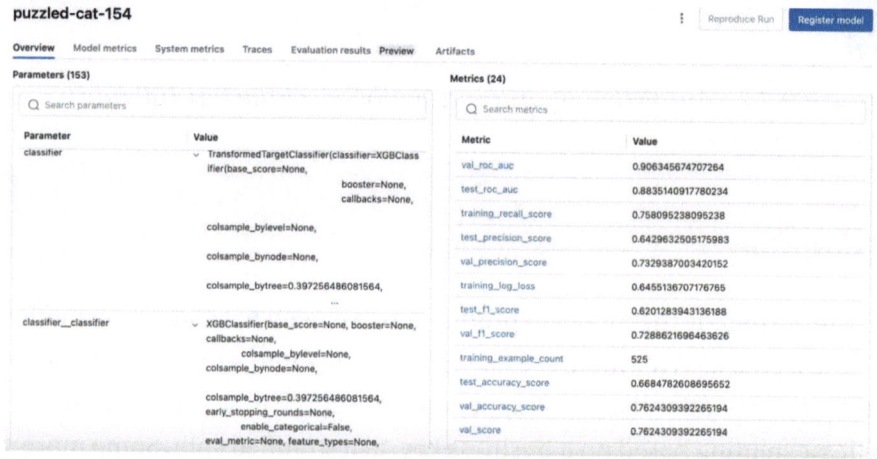

Figure 9-15. *Model overview page in an experiment*

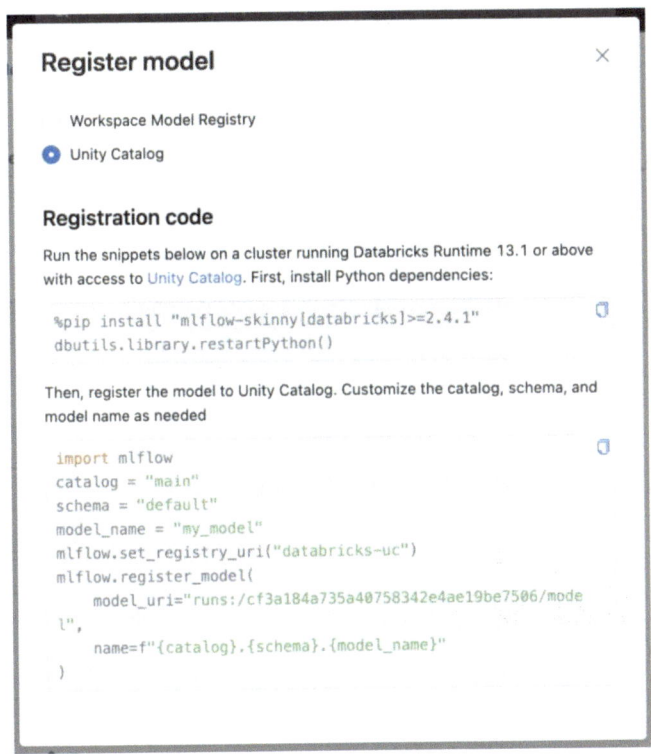

Figure 9-16. *Model registration*

Databricks model registry provides a portal to manage our models' versions, tag them, and create a model-serving endpoint right from the model. All these models are registered using MLflow. See Figure 9-17 and Figure 9-18.

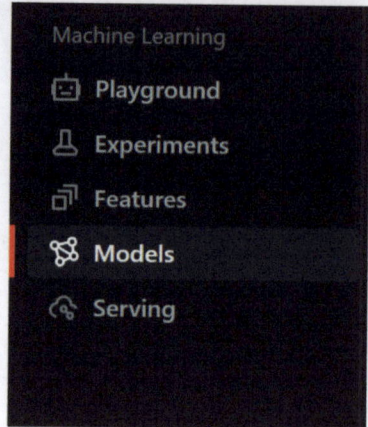

Figure 9-17. *Models tab in Databricks UI*

Figure 9-18. *Registered models in Databricks*

Model Serving/Inferencing

On the Models tab, we can also create an API endpoint, aka serve the model, for external consumption (Figure 9-19). This will generate a REST API endpoint, allowing the model to be easily accessed externally using Python or other programming languages.

Figure 9-19. *Model Serving and inferencing*

Unity Catalog tightly integrates with Lakehouse Monitoring, so it is recommended that an inference table be set up for output (Figure 9-20).

Figure 9-20. *Set up model inference tables*

Once a serving endpoint is created, we can query it. The "Query endpoint" button on the top right of the page will show us the exact commands to call the API (Figure 9-21). This API will generate inference results and save into the table shown, in our case, `unitygo.default.flight_delay_output_payload`. Databricks will always append `_payload` for the inference table.

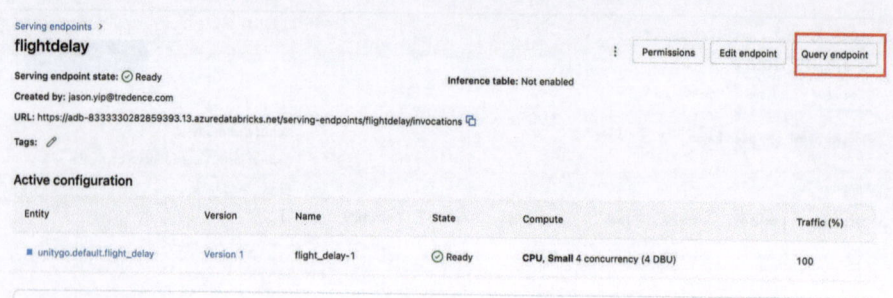

Figure 9-21. *Query endpoint button in a served endpoint*

To summarize, here is how the workflow looks:

1. Enable inference tables on your endpoint, as in Figure 9-20.

2. Schedule a workflow to process the JSON payloads using the code in the Query Endpoint button, as in Figure 9-21.

3. (Optional) Join the unpacked requests and responses with ground-truth labels to allow model quality metrics to be calculated.

4. Create a monitor over the resulting Delta table and refresh the metrics by using Lakehouse Monitoring, which we will discuss next.

Monitoring

ML models are never built once and run forever. So, we do need to retrain our models. The question is when we need to retrain. Some people decide to train the model daily, but a more reactive approach is to detect data drift and trigger a retrain when it happens. Simply put, data drift is the change in input data that causes the model's performance to degrade over time. This can be caused by missing data in the pipeline, for example. With Unity

Catalog, Databricks provides out-of-the-box Lakehouse Monitoring, which includes drift detection, that leverages the SQL workspace to build a drift monitoring dashboard. Additionally, triggers can be set up to retrain a model when drift happens.

Next we will dive into Lakehouse Monitoring, a powerful monitoring product that comes with no cost to Databricks customers.

Lakehouse Monitoring

Your table must be Unity Catalog enabled to use Lakehouse Monitoring. Otherwise, the monitoring option will not be visible. While everything in Databricks can be configured with code, the easiest way to get started is via the user interface. We can navigate to the Quality tab in any UC-enabled table and set up a monitor. Please note that only one type of monitoring can be set up in any given table. See Figure 9-22 and Figure 9-23.

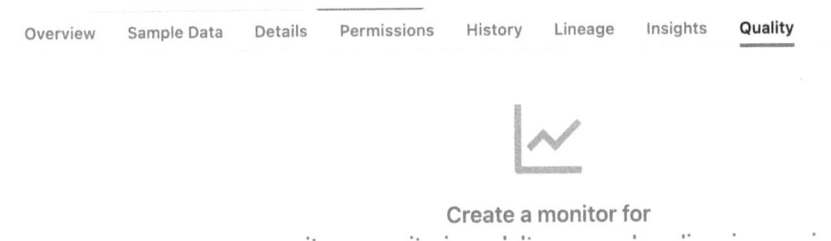

Figure 9-22. Setting up Lakehouse Monitoring

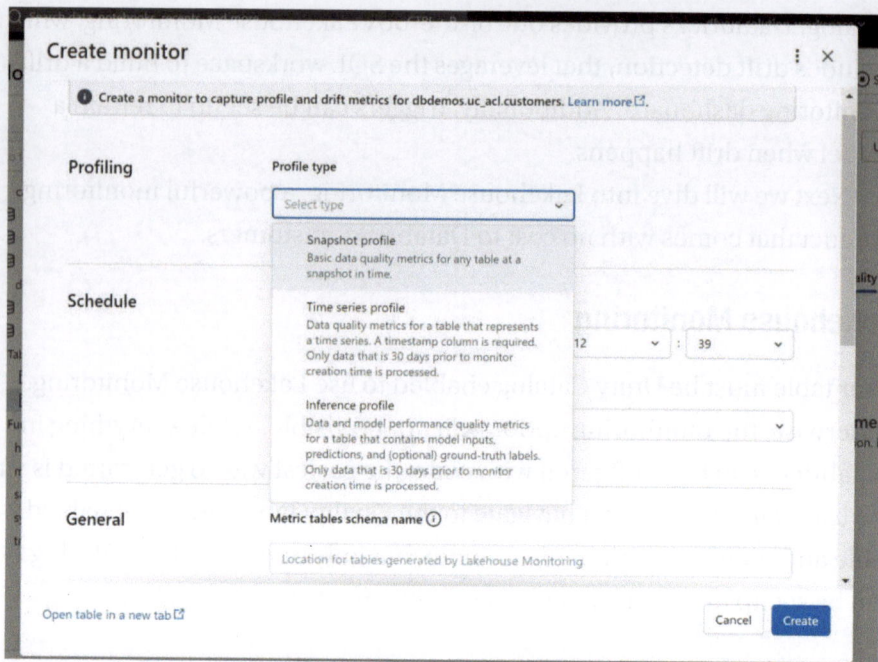

Figure 9-23. *Creating a monitor using Lakehouse Monitoring*

Here are the options available when setting up a monitor in Lakehouse
Monitoring:

> **Snapshot profile:** Designed for basic quality metrics
> for any table at a snapshot in time. While Delta
> Live Tables (DLT) comes with basic expectations,
> a snapshot profile is similar to the data profiling
> function in dbutils for exploratory data analysis
> over time.

> **Time series profile:** This isn't talking about a time
> series forecast but rather determining data drift
> occurrence given a timestamp column. The drift
> analysis metrics are shown in Table 9-2.

Table 9-2. *AI Model Monitoring Metrics*

Column Name	Type	Description
chi_squared_ test	struct<statistic: double, pvalue: double>	Chi-square test for drift in distribution.
ks_test	struct<statistic: double, pvalue: double>	KS test for drift in distribution. Calculated for numeric columns only.
tv_distance	double	Total variation distance for drift in distribution.
l_infinity_ distance	double	L-infinity distance for drift in distribution.
js_distance	double	Jensen–Shannon distance for drift in distribution. Calculated for categorical columns only.
wasserstein_ distance	double	Drift between two numeric distributions using the Wasserstein distance metric.
population_ stability_ index	double	Metric for comparing the drift between two numeric distributions using the population stability index metric.

Inference profile: This is designed to measure classification and regression influence results like precision and recall and R2 scores. Currently, data scientists need to maintain the codebase to calculate these numbers, and each team will try to use a different library depending on what model they are using. Databricks' approach is model agnostic and is calculated completely based on an inference table. On top of that, it includes fairness and biases, which is a baby step forward to Responsible AI. The metrics include *predictive parity*, *predictive equality*, *equal opportunity*, and *statistical parity*. See Figure 9-24.

Figure 9-24. *Microsoft's* `Responsible AI` *standards*

All of these profilers can detect PII in the data using AI, and sensitive columns will be tags as PII in the catalog.

Why Profiling?

When building an ML model, it is critical to understand the statistics in every stage of the process. For example, if a column contains lots of nulls, it will not be suitable to be included in the model as it will not provide a lot of values, and it will further degrade the training and inference performance.

Data drift is a concept that determines whether model retraining is required. Most of the time ML models are not trained once and run forever. For example, ChatGPT usually comes with a knowledge cutoff time. That's because, like humans, ML models also need to get up-to-date with their knowledge to make better decisions or predictions. There are many different ways to measure drift, and calculating these drift metrics at scale is often very challenging.

Finally, by monitoring the model metrics like F1 and R2 over time, we will also know when a model retrain or refactor is required. In the case of retraining with more data that does not improve the model metrics, it is time to refactor the model to provide more high-quality data, which goes back to the need to monitor the data drift and understand the statistical distribution of the raw data. See Figure 9-25.

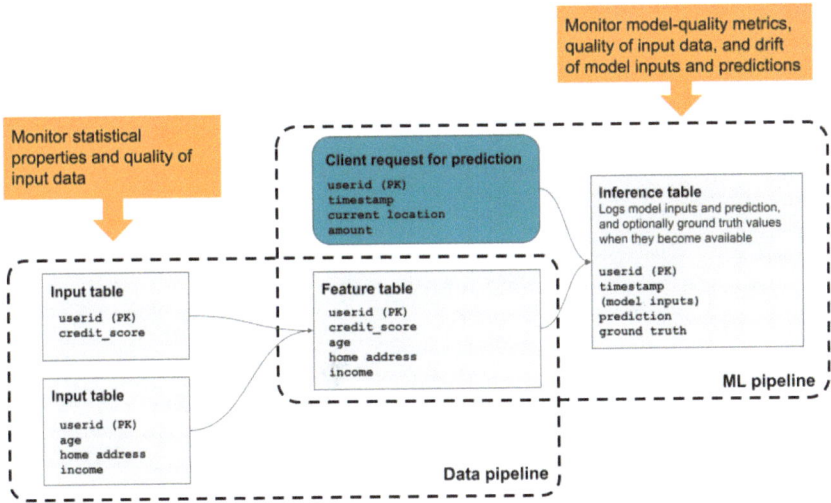

Figure 9-25. *Lakehouse Monitoring table schema*

Lakehouse Monitoring provides a one-stop interface for setting up all these statistics and storing them in different tables so they can be reused. Dashboards are also created for ease of visualization and quick insights, saving teams numerous hours of research and development effort. See Figure 9-26 and Figure 9-27.

To learn more about the monitoring metrics tables, please visit the documentation here:

https://docs.databricks.com/en/lakehouse-monitoring/monitor-output.html

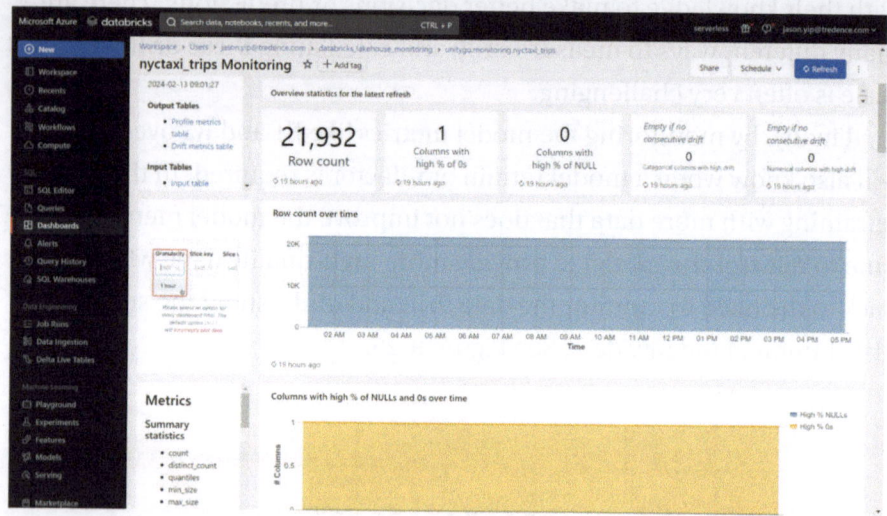

Figure 9-26. *Lakehouse Monitoring report #1*

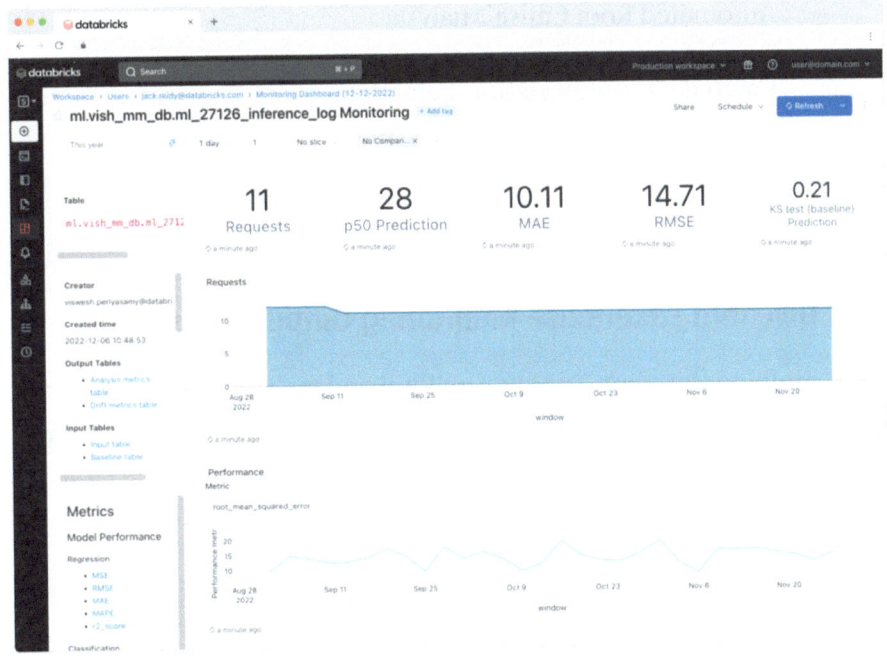

Figure 9-27. *Lakehouse Monitoring report #2*

Databricks Lakehouse Monitoring is a framework that enables a self-serve data platform with proactive issue management.

- Auto-Generated Reports

  Share quality updates organization-wide with auto-generated dashboards, and use ready-made metrics and analytics tools for easy issue exploration in your data products.

- Unified Monitoring

  Monitor the quality of all data products with a single tool, regardless of the framework or platform used to build them. Merge quality and business metrics effortlessly in your lakehouse to gauge your data products' impact.

- Automated Root Cause Analysis

  Catch data product issues before they reach consumers with cost-effective "insurance." Boost efficiency with smart automation in your data and AI pipelines, avoiding unnecessary retraining.

Deep Dive into Lakehouse Monitoring Output Tables

YData profiling (formerly Pandas profiling) and the summarize command provide **invaluable insights** for data scientists to analyze datasets with Spark compute, yet it doesn't give you access to the raw data in a table format. The importance of getting this data into a reusable format includes:

- *Setting up custom alerts:* When including basic statistics in a data drift report, we can trigger an alert to re-train a model.

- *Creating reports beyond the given interface:* Rarely does any team not have an existing dashboard, so integrating the analysis into an existing dashboard, like Power BI and Tableau, is an important part of the team process.

- *Comparing statistics between two different tables (with Spark):* Often we want an efficient way to compare the differences between two different tables. While YData profiling provides capabilities in Pandas to compare two different datasets, it currently does not support using Spark. On the other hand, the dbutils command does not allow comparing two different datasets.

- *Keeping track of historical differences:* Understanding the trend of the data allows the team to understand if there is missing data due to source issues or incomplete data pull. For example, if there is a job that was rerun but produced fewer rows than it used to have or the standard deviation compared to the last month dropped drastically compared to historical runs, it would be important to understand the root cause in case it impacted the model.

Figure 9-28 illustrates input tables versus generated tables as well as the relationship between generated tables and the dashboard.

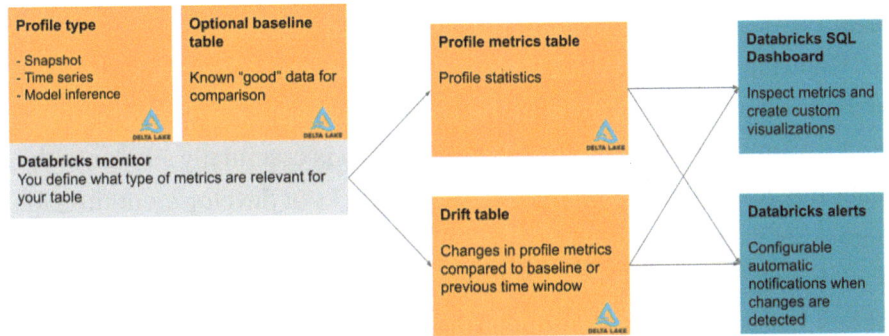

Figure 9-28. *Lakehouse Monitoring tables relationships*

Consider if we wanted to monitor the table `nyctaxi_trips,` a baseline table can also be specified optionally to measure drift. Lakehouse Monitoring will generate two new tables automatically:

- `nyctaxi_trips_drift_metrics`
- `nyctaxi_trips_profile_metrics`

Figure 9-29 illustrates the output table in Unity Catalog.

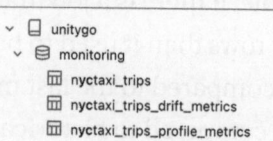

Figure 9-29. *Lakehouse Monitoring tables that were being generated*

These tables contain a lot of information. To examine it closely, please refer to the documentation from Databricks website:

https://docs.databricks.com/en/lakehouse-monitoring/monitor-output.html#column-schemas-for-generated-tables

The usage of these tables is also fully transparent. Databricks generates all the queries and uses them in the Databricks Dashboard—everything can be found in the workspace. This approach is essentially similar to the Glass Box Auto ML approach—it will save weeks of development time for teams that want to kick off an ML monitoring project.

The help you get started, Databricks has created sample notebooks, including the datasets and models for Lakehouse Monitoring:

https://docs.databricks.com/en/lakehouse-monitoring/create-monitor-api.html#example-notebooks

Figure 9-30 shows examples of queries and dashboard generated by Lakehouse Monitoring.

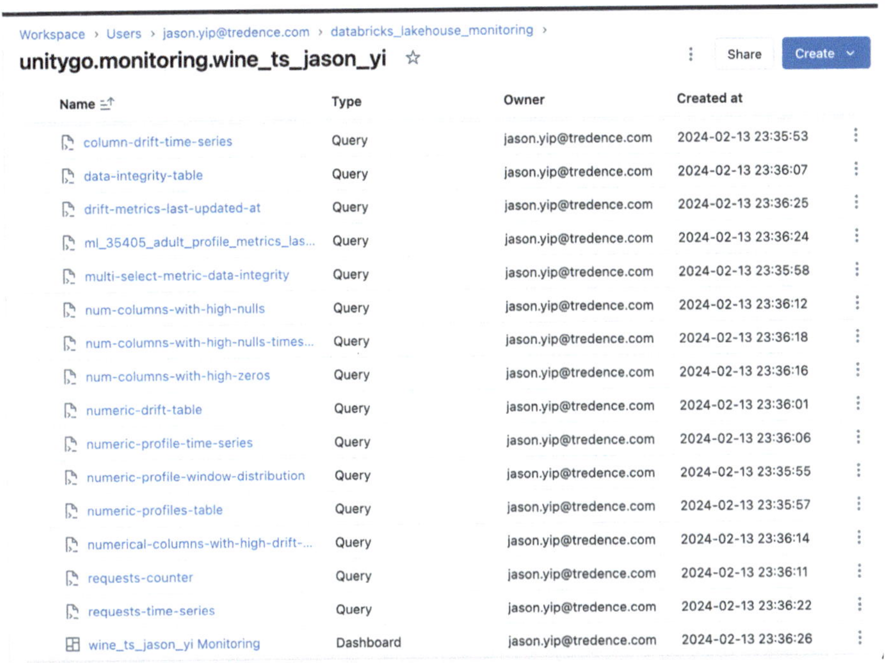

Workspace › Users › jason.yip@tredence.com › databricks_lakehouse_monitoring ›

unitygo.monitoring.wine_ts_jason_yi ☆

Name	Type	Owner	Created at	
column-drift-time-series	Query	jason.yip@tredence.com	2024-02-13 23:35:53	⋮
data-integrity-table	Query	jason.yip@tredence.com	2024-02-13 23:36:07	⋮
drift-metrics-last-updated-at	Query	jason.yip@tredence.com	2024-02-13 23:36:25	⋮
ml_35405_adult_profile_metrics_las...	Query	jason.yip@tredence.com	2024-02-13 23:36:24	⋮
multi-select-metric-data-integrity	Query	jason.yip@tredence.com	2024-02-13 23:35:58	⋮
num-columns-with-high-nulls	Query	jason.yip@tredence.com	2024-02-13 23:36:12	⋮
num-columns-with-high-nulls-times...	Query	jason.yip@tredence.com	2024-02-13 23:36:18	⋮
num-columns-with-high-zeros	Query	jason.yip@tredence.com	2024-02-13 23:36:16	⋮
numeric-drift-table	Query	jason.yip@tredence.com	2024-02-13 23:36:01	⋮
numeric-profile-time-series	Query	jason.yip@tredence.com	2024-02-13 23:36:06	⋮
numeric-profile-window-distribution	Query	jason.yip@tredence.com	2024-02-13 23:35:55	⋮
numeric-profiles-table	Query	jason.yip@tredence.com	2024-02-13 23:35:57	⋮
numerical-columns-with-high-drift-...	Query	jason.yip@tredence.com	2024-02-13 23:36:14	⋮
requests-counter	Query	jason.yip@tredence.com	2024-02-13 23:36:11	⋮
requests-time-series	Query	jason.yip@tredence.com	2024-02-13 23:36:22	⋮
wine_ts_jason_yi Monitoring	Dashboard	jason.yip@tredence.com	2024-02-13 23:36:26	⋮

Figure 9-30. *Objects created by Lakehouse Monitoring; everything is open source*

MLOps Best Practices

Building an ML model is rarely a single-person effort. Even if a single data scientist is working on model building, they will require collaboration with other people. Often, best practices must be shared across the team so there is no difference to the ML pipeline on every model. If best practices are followed, working across the team will save time without repeatedly learning the code base on every single model. Hence, learning from our experience, Databricks has open-sourced the internal best practices to develop an ML model, called the MLOps stack.

```
https://github.com/databricks/mlops-stacks
```

The architecture diagram in Figure 9-31 represents the process from development to deployment for an ML model. There are three components provided in the repo:

- **ML Code:** Example ML project structure (training and batch inference, etc.), with unit-tested Python modules and notebooks

- **ML Resources as Code:** ML pipeline resources (training and batch inference jobs, etc.) defined through Databricks CLI bundles

- **CI/CD (GitHub Actions or Azure DevOps):** GitHub Actions or Azure DevOps workflows to test and deploy ML code and resources

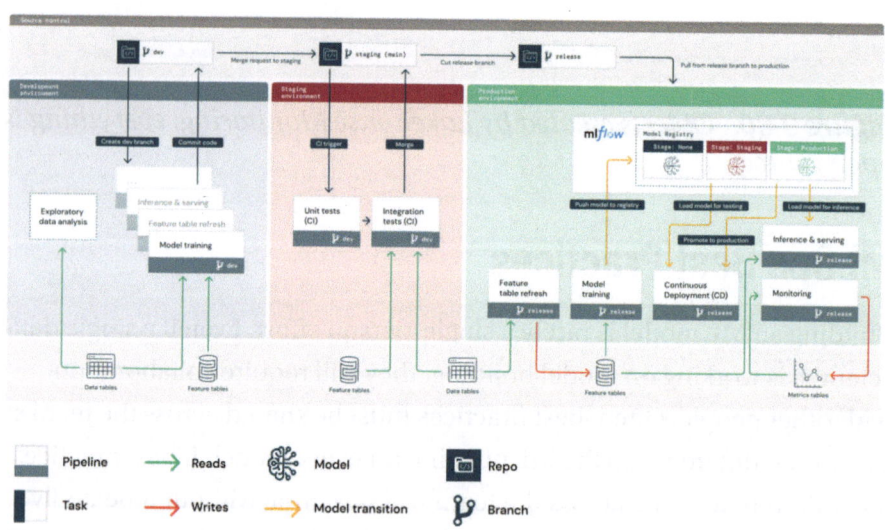

Figure 9-31. *Databricks' MLOps stack*

By leveraging Databricks' best practices, teams can focus on generating business values from their ML model rather than dealing with infrastructure setup. The three items can be grouped into two major parts, which is CI/CD and ML Project. We will focus on the ML Project portion in this chapter and look at the CI/CD portion later in Chapter 15. See Figure 9-32.

```
Select if both CI/CD and the Project should be set up, or only one of them.
You can always set up the other later by running initialization again:
    CICD_and_Project
  ▸ Project_Only
    CICD_Only
```

Figure 9-32. *Databricks' MLOps stack*

Once we have initialized the MLOps stack with `databricks bundle init mlops-stacks` and choose the Project Only option, the CLI will create template folder structures along with an NYC ML example in the folders so we can follow the code. Figure 9-33 shows what the folder structure look like, but as the project continues to evolve. The exact project details can be found in the `README.md` file under the main project folder.

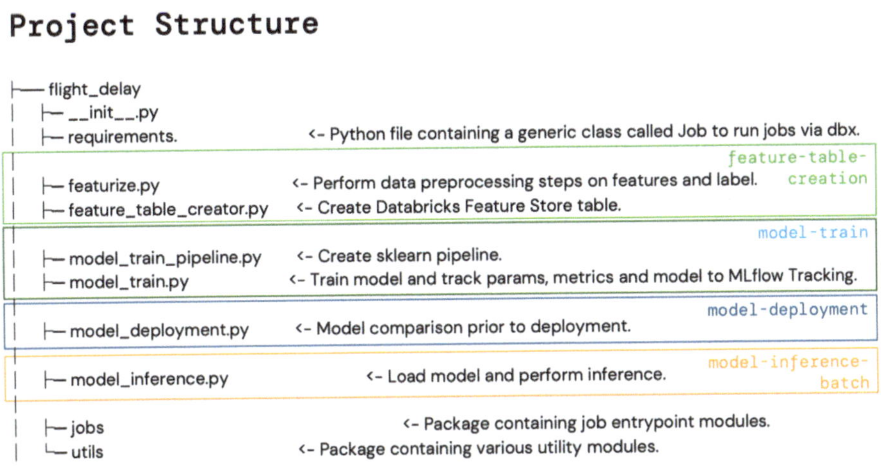

Project Structure

```
├── flight_delay
│   ├── __init__.py
│   ├── requirements.          <- Python file containing a generic class called Job to run jobs via dbx.
│                                                                    feature-table-
│   ├── featurize.py           <- Perform data preprocessing steps on features and label.   creation
│   ├── feature_table_creator.py  <- Create Databricks Feature Store table.
│                                                                         model-train
│   ├── model_train_pipeline.py  <- Create sklearn pipeline.
│   ├── model_train.py         <- Train model and track params, metrics and model to MLflow Tracking.
│                                                                    model-deployment
│   ├── model_deployment.py    <- Model comparison prior to deployment.
│                                                                 model-inference-
│   ├── model_inference.py         <- Load model and perform inference.          batch
│   ├── jobs                          <- Package containing job entrypoint modules.
│   └── utils                  <- Package containing various utility modules.
```

Figure 9-33. *Suggested project structure*

With the provided sample notebooks, teams can use source control to develop their ML model and deploy via the Git actions provided, saving time to re-developing reusable standards and focusing on producing business values.

Conclusion

In this chapter, we spent a lot of time taking an ML model from development to production because data scientists need to spend time coding up the model and then tuning the models. In between they also need to work closely with data engineers and ML engineers to ensure they are getting the latest data and deploying the latest model for testing.

When using Databricks' Glass Box Auto ML approach along with other toolsets like feature stores, data scientists can now speed up their process of getting to a baseline model, which is an important milestone to evaluate the effectiveness of the input data. Then they can leverage the code generated to build a production model while seamlessly collaborating with data engineers and ML engineers using the intuitive interfaces.

Databricks has recently taken the ML model building to the next level. While many tools out there can manage the MLOps life cycle, Databricks is the only platform that allows a team of data experts to work together seamlessly without having to jump through multiple hoops of toolsets.

CHAPTER 10

Generative AI with Databricks

Ever since ChatGPT was released to the public, there has been no shortage of interest in chatbots or generative artificial intelligence (GenAI). But what exactly is GenAI, and how does Databricks come into the picture? And how it can help organizations deploy their own chatbot or develop their own GenAI applications? In this chapter, we will first learn the concepts around GenAI. Then we will discuss how Databricks and the newly acquired company Mosaic ML will work together and transform the industry once more. This chapter lays some background regarding the journey of GenAI and introduces the Databricks offering in the GenAI space.

What Is Generative AI?

According to Gartner:

> *"Generative AI can learn from existing artifacts to generate new, realistic artifacts (at scale) that reflect the characteristics of the training data but don't repeat it. It can produce a variety of novel content, such as images, video, music, speech, text, software code and product designs."*

Generative AI uses several techniques that continue to evolve. Foremost are AI foundation models, which are trained on a broad set of unlabeled data that can be used for different tasks, with additional fine-tuning. Complex math and enormous computing power are required to create these trained models, but they are, in essence, prediction algorithms.

Today, generative AI most commonly creates content in response to natural language requests—it doesn't require knowledge of or entering code—but the enterprise use cases are numerous and include innovations in drug and chip design and material science development.

Figure 10-1 explains how generative AI and ChatGPT are different. From a very high level, generative AI is a technique that tries to generate some new content by learning from vast amounts of similar content. For example, when trying to generate English language content, it could have trained on all the text on Wikipedia to start with. However, that's just an understatement. With that said, not something every household would have access to the resources required to train these models, despite the number of models increasing by the day. As a result, similar to transfer learning, a lot of data scientists would use so-called foundation models to enhance the AI with some internal knowledge.

The enhancement process can be done via retrieval augmented generation (RAG) or fine-tuning. There is a fundamental difference between the two. RAG is trying to optimize the data, whereas fine-tuning is trying to optimize the model. We will discuss both these in greater detail later in the chapter.

Finally, if resources are available and the goal is to train a fully domain specific model without bias, Mosaic ML's training platform will help you do that albeit at a much reduced cost. The results of these are large language models (LLMs). ChatGPT is an application created on top of the LLMs to serve as a chatbot, providing an intuitive interface for the general public to use. But then on the other hand, in the case of GenAI, we might think that the larger the model the better, but in fact this is not the case. The world is still learning how to optimize the data for optimized throughput of the tasks required. An excellent example is Databricks' bespoke LLM model for auto-documentation generation, which costs about $1,000 to train. Figure 10-1 is a quick reference to these standard terms.

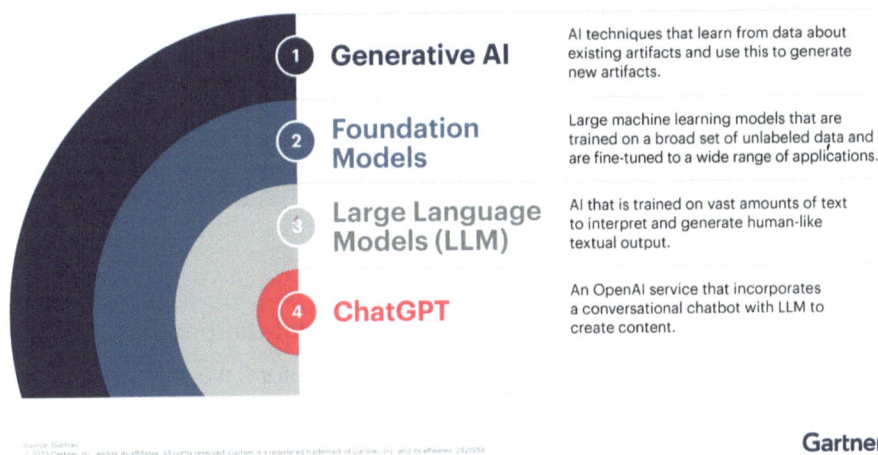

Figure 10-1. *Differences between AI models (source:* `https://www.gartner.com/en/insights/generative-ai-for-business`*)*

Databricks Generative AI

Databricks provides a lot of tools for you to take control in model training all the way to governing the model. Figure 10-2 illustrates all the capabilities that Databricks provides for organizations to use and build their next GenAI use case.

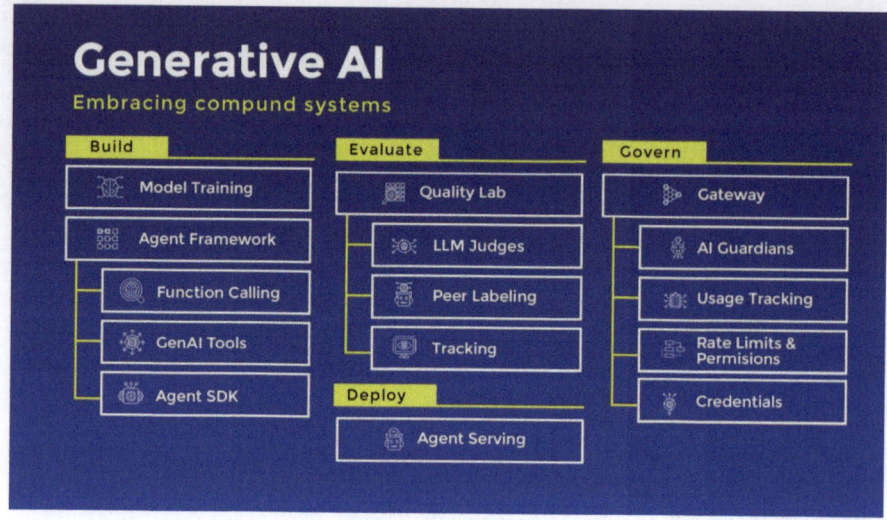

Figure 10-2. Databricks' generative AI offerings

In this chapter, we will examine the basic out-of-the-box features that Databricks provides, and then in subsequent chapters, we will discuss how you can leverage all the advanced tools to enhance your GenAI offering.

While we will discuss the details in later chapters, here are the high-level functionalities of each stack:

- **Build:** This includes the Mosaic AI stack to refine an LLM either through RAG or fine-tuning.

- **Evaluate:** This is part of the AI Agent framework to allow evaluation with metrics as well as getting peer feedback.

- **Deploy:** There is a one-line command to deploy a nonproduction app for a user acceptance test.

- **Govern:** There is an extension of MLflow to manage internal and external LLM APIs.

The GenAI Journey

With GenAI dominating the world now, many organizations start by allowing their employees to play around with these models. To combine the data from within the organization, some of them with more budget want to train a model of their own from scratch. While others keep the model for use within their organization, some also decide to open up their model to the world and become foundation models.

Figures 10-3 and 10-4 present two different views of this journey. Figure 10-3 represents the journey or maturity an organization can get with GenAI. From the left, we have prompt engineering, then retrieval augmented generation, fine-tuning, and, lastly, pre-training.

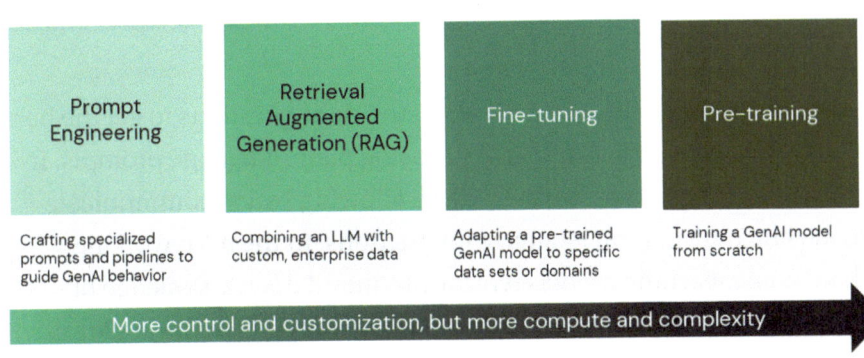

Figure 10-3. *The GenAI journey*

While these building blocks are available, not every organization will need to reach the last step of pre-training due to the computation, aka cost, as well as complexity, aka expert knowledge, required to reach the next level. Figure 10-4 illustrates this idea in another way. Note that pre-training is the most time-consuming and complex process. We will walk you through the journey, but before you decide to move to the next step, it is best to consider the trade-off between time and cost as well as whether experts are available to validate the results.

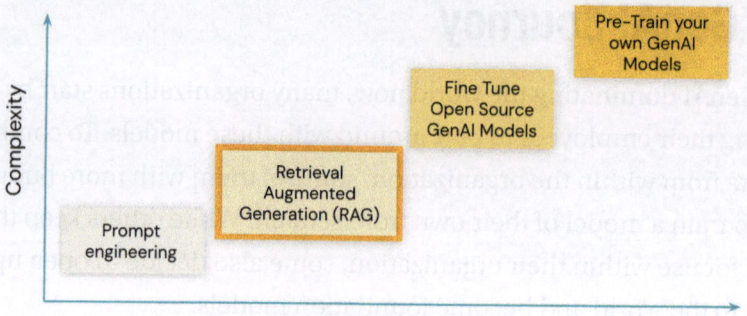

Figure 10-4. *Alternate aspect of the GenAI journey with time and complexity involved*

Prompt Engineering

Prompt engineering is the art of asking the right questions to get the best output from a **language model (LLM)** using plain-language prompts. It enables direct interaction with the LLM, allowing you to communicate with it using only natural language instructions. In the past, working with machine learning models typically required deep knowledge of datasets, statistics, and modeling techniques. However, today, LLMs can be "programmed" in English and other languages, making them more accessible to a broader audience.

Here are some key points about prompt engineering:

- **Best Practices for Prompting:**

  - **Clear communication:** Clearly communicate what content or information is most important.

  - **Structured prompts:** Structure your prompts by starting with the role or context, followed by input data and then the instruction.

- **Varied examples:** Use specific and varied examples to help the model narrow its focus and generate accurate results.

- **Constraints:** Use constraints to limit the scope of the model's output and prevent factual inaccuracies.

- **Break down complex tasks:** Divide complex tasks into a sequence of simpler prompts.

- **Self-evaluation:** Instruct the model to evaluate or check its own responses before producing them.

- **Creativity:** Be creative! The more open-minded and creative you are, the better your results will be.

- **Types of Prompts:**

  - **Direct prompting (zero-shot):** The simplest type of prompt that provides only an instruction without examples.

    - **Example:** "Can you give me a list of ideas for blog posts for tourists visiting New York City for the first time?"

  - **Role prompting:** Assign a role to the model and ask it to understand your goals and objectives before designing a prompt.

    - **Example:** "You are a mighty and powerful prompt-generating robot. Design a prompt for the best outcome based on the context and data provided."

  - **Chain-of-thought prompting:** Break down complex tasks into a sequence of simpler prompts.

Remember, being a great prompt engineer doesn't require coding experience. Creativity and persistence will benefit you greatly on your journey in this evolving field of LLMs and prompt engineering[1]2.

Mosaic AI Playground

Riding on the wave of generative AI, many corporations have released foundation large language models. However, without background in programming, it is difficult to use these models. Databricks has optimized a few models and curated in the "Playground" section for experimentation, providing a standard interface to interact with these models. It is also an interactive environment where users can simultaneously experiment and "chat" with various large language models and compare results.

The curated models include the following:

– **Llama2 70B Chat**

    Llama-2-70B-Chat is a state-of-the-art 70B parameter language model with a context length of 4,096 tokens, trained by Meta. It excels at interactive applications that require strong reasoning capabilities, including summarization, question-answering, and chat application

– **Mixtral-8x7B Instruct**

    Mixtral-8x7B Instruct is a high-quality sparse mixture of experts model (SMoE) trained by Mistral AI. Mixtral-8x7B Instruct can be used for a variety of tasks such as question-answering, summarization, and extraction.

- **MPT 30B Instruct**

  MPT-7B-8K-Instruct is a 6.7B parameter model trained by MosaicML for long-form instruction following, especially question-answering on and summarization of longer documents. The model is pre-trained for 1.5T tokens on a mixture of datasets, and fine-tuned on a dataset derived from the Databricks Dolly-15k and the Anthropic Helpful and Harmless (HH-RLHF) datasets. The model name you see in the product is mpt-7b-instruct, but the model specifically being used is the newer version of the model.

- **MPT 7B Instruct**

  MPT-30B-Instruct is a 30B parameter model for instruction following trained by MosaicML. The model is pre-trained for 1T tokens on a mixture of English text and code, and then further instruction fine-tuned on a dataset derived from Databricks Dolly-15k, Anthropic Helpful and Harmless (HH-RLHF), CompetitionMath, DuoRC, CoT GSM8k, QASPER, QuALITY, SummScreen, and Spider datasets.

  ```
  https://learn.microsoft.com/en-us/azure/
  databricks/machine-learning/foundation-
  models/supported-models
  ```

Use Cases

Out of the box, Databricks provides four different use cases in the playground, but large language models are not limited to these four use cases. Users can chat with these models and get responses from the prompt window.

Sentiment Analysis

You will be provided with a tweet, and your task is to classify its sentiment as positive, neutral, or negative.

Example:

Input: The new Spiderman movie was meh

Output: Sentiment: Neutral

Unstructured Text Parsing

You will be provided with unstructured data, and your task is to parse it into JSON format.

Example:

Input: Advanced Micro Devices, Inc. is an American multinational semiconductor company based in Santa Clara, California, that develops computer processors and related technologies for business and consumer markets. The company was founded in 1969 by Jerry Sanders and a group of other technology professionals

Output: { "name": "Advanced Micro Devices, Inc.", "location": "Santa Clara, California", "founders": ["Jerry Sanders"], "founding_year": 1969, "products": ["computer processors", "related technologies"], "markets": ["business", "consumer"] }

Summarization

You will be provided with a document and asked to summarize it.

Example:

Input: Copy activity supports Azure Databricks Delta Lake connector to copy data from any supported source data store to Azure Databricks delta lake table, and from delta lake table to any supported sink data store. It leverages your Databricks cluster to perform the data movement, see details in Prerequisites section.

Output:

1. The copy activity in a data integration tool supports the use of the Azure Databricks Delta Lake connector.

2. This feature enables data copying from any supported source data store to an Azure Databricks delta lake table, and from the delta lake table to any supported sink data store.

3. The copy activity utilizes a user's Databricks cluster to perform data movement.

4. Users should refer to the Prerequisites section for further details and requirements.

Document Q&A

You will be provided with a document and asked a question about it.

Example:

Document:

Farm Utility Exemption Notice

The full text can be found on this website due to its length:

`https://www.dfa.arkansas.gov/images/uploads/exciseTaxOffice/`
`et1441.pdf`

Question: Is a greenhouse a qualifying agricultural structure?

Answer: Yes, a greenhouse is a qualifying agricultural structure according to the document. It is listed under C of the definition of qualifying agricultural structures as a greenhouse used for commercial production.

While the playground allows developers to play around with the model use cases, behind the scenes it is using Databricks' serverless technology to power the LLM API. You can use the UI, the Foundation Models APIs, the Python SDK, the MLflow Deployments SDK, or the REST API to query supported models. The APIs are compatible with OpenAI, so you can even use the OpenAI client for querying.

Retrieval Augmented Generation

While ChatGPT democratized LLM-based chatbots for consumer use, companies need to deploy personalized models that meet needs like the following:

- Privacy requirements on sensitive information

- Preventing hallucination

- Specialized content, not available on the Internet

- Specific behavior for customer tasks

- Control over speed and cost

- Deploy models on private infrastructure for security reasons

To accomplish this, organizations often need to provide internal documents to ground the model with truth. This process requires converting context into something called **embeddings**. Embeddings are mathematical representations (vectors) of the semantic content of data, typically text or image data. Depending on the use case, there are many ways to generate embeddings. But in the case of GenAI, embeddings are generated by a large language model, for example, BAAI's BGE-Large-EN (https://huggingface.co/BAAI/bge-large-en). They are a key component of many GenAI applications that depend on finding documents or images that are similar to each other.

Figure 10-5 illustrates how we can convert a knowledge graph to embeddings, where the nodes can be viewed as internal documents and the edges can be viewed as references. You can see each component of the input graph is converted to a numeric calculation, and these numbers are helpful for machine learning or GenAI tasks.

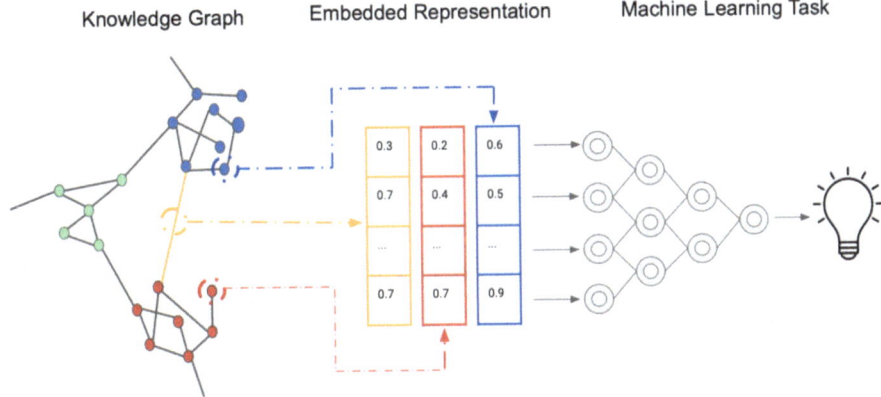

Figure 10-5. *Representation of embedding from a knowledge graph (source: https://en.wikipedia.org/wiki/Knowledge_graph_embedding)*

Figure 10-6 gives a sample workflow using LangChain to connect the document embeddings to Databricks' vector index and sync with Databricks' vector database. An application can then be built over this architecture. Later in this chapter, we will introduce **Mosaic AI Agent Framework**, an offering by Databricks to deploy the LLM application for evaluation with ML Flow LLM Judges or expert users.

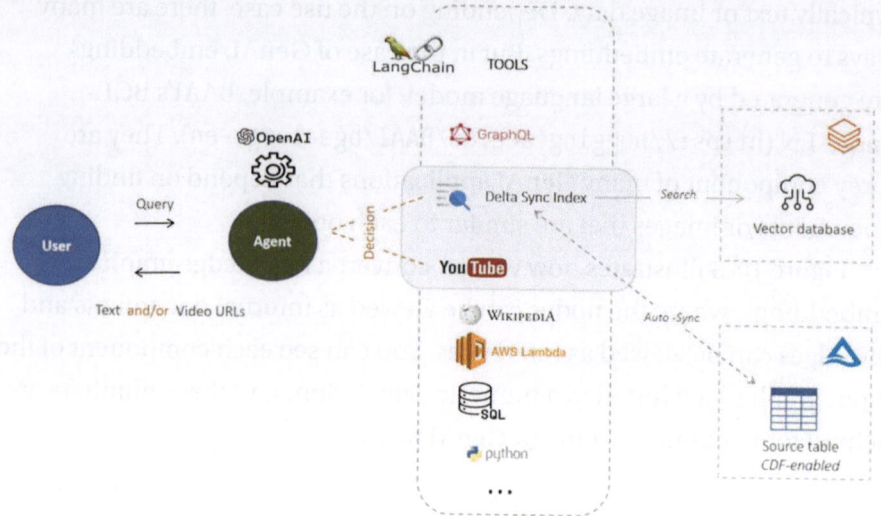

Figure 10-6. *AI Agent workflow with Dataricks vector database (source:* `https://medium.com/@tsiciliani/using-ai-agents-with-databricks-vector-search-8b688d7ed41a`)

As the data volume increases, it has become increasingly hard to optimize the performance of large data applications. To solve the performance issues, Databricks has released a suite of tools so developers can focus on developing the pipeline to achieve higher quality rather than worrying about performance tuning and maintaining infrastructure. These tools include the following:

- **Fully** `managed foundation models` **providing pay-per-token base LLMs.**

    The first step of our LLM workflow is to generate embeddings, either based on text or binaries. The Databricks Foundation Model API provides performance guarantees for some foundation models for different use cases. In the case of embedding, BGE Large (English) is provided with an API interface,

so developers can calculate the embedding at scale.
Table 10-1 lists some of the foundation models that
come out of the box from Databricks. They can be used
as an API endpoint without acquiring any compute,
simplifying the deployment requirements.

Table 10-1. *Databricks Foundation Model API*

Model	Task type	Endpoint
DBRX Instruct	Chat	`databricks-dbrx-instruct`
Meta-Llama-3-70B-Instruct	Chat	`databricks-meta-llama-3-70b-instruct`
Meta-Llama-2-70B-Chat	Chat	`databricks-llama-2-70b-chat`
Mixtral-8x7B Instruct	Chat	`databricks-mixtral-8x7b-instruct`
MPT 7B Instruct	Completion	`databricks-mpt-7b-instruct`
MPT 30B Instruct	Completion	`databricks-mpt-30b-instruct`
GTE Large (English)	Embedding	`databricks-gte-large-en`
BGE Large (English)	Embedding	`databricks-bge-large-en`

- **A** `vector search` **service to power semantic search
  on existing tables in your lakehouse.**

  A vector database is a specialized database to store
  embeddings. To ensure the performance is guaranteed,
  a vector index will be created for a specific column.
  Databricks Vector DB will either calculate the
  embeddings for you if it is a text column in a delta
  table or will sync the embeddings to an index if it is
  generated by binaries when the values are stored in a
  delta table or API can be used to sync the index if no

table is provided. Either way, embeddings will need to be present in a highly performant and scalable format to perform **similarity search**.

In summary, Databricks provides multiple types of vector search indexes (see Figure 10-7).

- **Managed embeddings:** These provide a text column and endpoint name, and Databricks synchronizes the index with your Delta table.

- **Self-managed embeddings:** You compute the embeddings and save them as a field of your Delta table; Databricks will then synchronize the index.

- **Direct index:** When you want to use and update the index without having a Delta table.

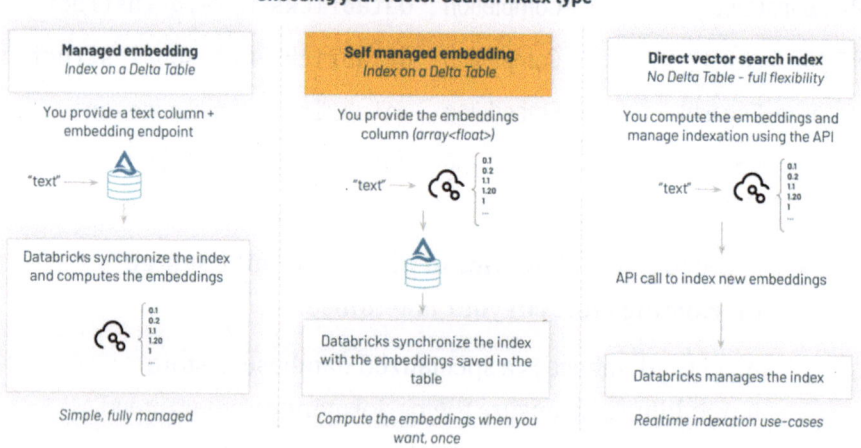

Figure 10-7. *Databricks vector search index types*

Similarity Search: The Magic Behind the Scenes

In the previous section, we discussed that vector search is based on the similarity algorithm. The text is first encoded into some vector form, and the similarity between the question and answer will be measured to give the best answer. Cosine similarity (Figure 10-8) or dot product is an algebraic operation that takes two equal-length sequences of numbers (usually coordinate vectors) and returns a single number. When the result is 0, it is completely different; whereas when the result is 1, it is identical. Vector DB uses this function to search for relevant documents to answer a specific question.

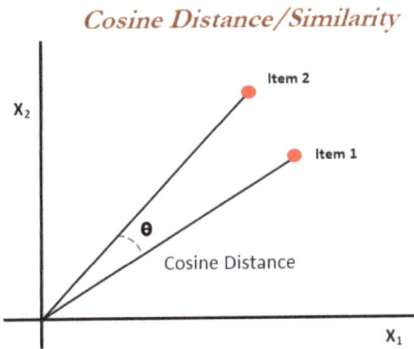

Figure 10-8. *Cosine similarity*

Mosaic AI Vector Search does not use cosine similarity. However, according to Databricks, Mosaic AI Vector Search uses the Hierarchical Navigable Small World (HNSW) algorithm for its approximate nearest neighbor searches and the L2 distance metric to measure embedding vector similarity. If you want to use cosine similarity, you need to normalize your datapoint embeddings before feeding them into vector search.

In other words, given a set of similar vectors, using either L2 distance or cosine similarity, we can use the HNSW to refine the search to ensure Databricks gains efficiency in terms of finding the most relevant chunk of document at scale. L2 and Cosine similarity are both an acceptable solutions for the search. Figure 10-9 shows HNSW.

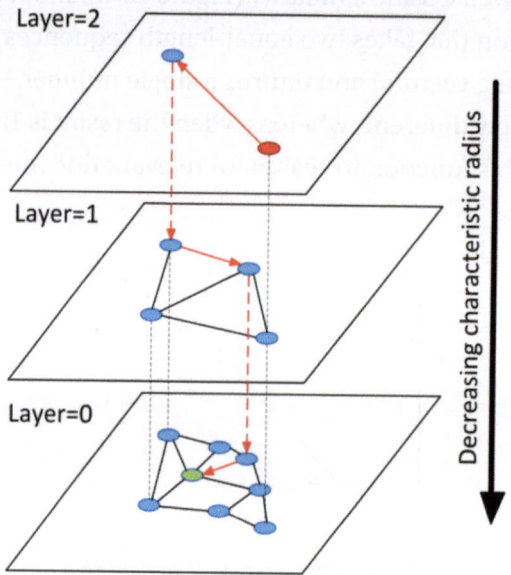

Figure 10-9. *Illustration of the hierarchical NSW idea*

After learning the basic concepts and building blocks of RAG we will look at an example of how to create an end-to-end RAG application

A Practical Example for RAG: Using Structured Data

Let's start by looking at the raw components Databricks provides to accelerate the development of a RAG application. These components are essential for any RAG application. Many companies are worried about vendor lock-in, but Databricks can be used as a serving platform. In the next chapter, we will discuss how to use Databricks to create an end-to-end RAG application.

The typical steps to create an end-to-end RAG app are as follows:

1. Create a feature serving endpoint. This step is required only if the data is being consumed outside of Databricks.

2. Calculate embedding and sync it into a vector database.

3. Create a LangChainTool that uses the endpoint to look up relevant data and log it with MLflow.

4. Evaluate the model using MLflow or human feedback.

Step 1: Feature and Function Serving

In case developers want to create the app outside of Databricks but still want to utilize the data within Databricks lakehouse platform, Databricks has made it easy to make data available via an API endpoint and automatically scale up and down as demand changes for the data, eliminating the needs to extract the data outside of Databricks.

There are two ways to serve these features. One is to expose the data, either via a delta table or via a function, to an API endpoint, and another is to sync the features into an external feature store.

- Online tables can be created easily via the UI or API, as shown in Figure 10-10.

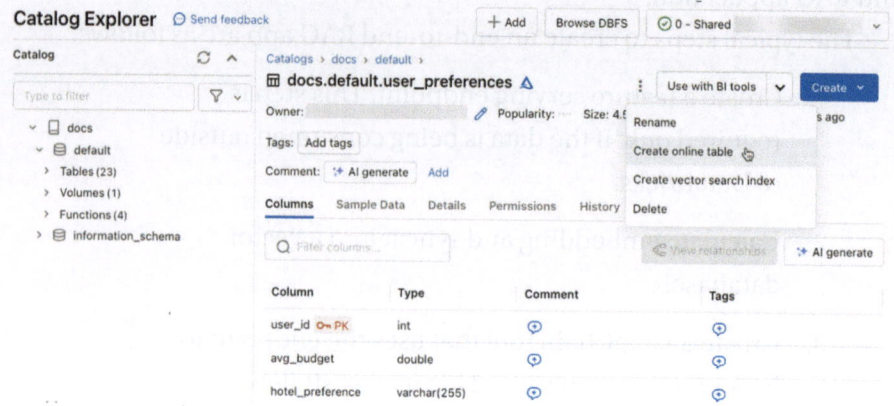

Figure 10-10. *Databricks online table*

- Sync features to external feature stores, as shown in Figure 10-11.

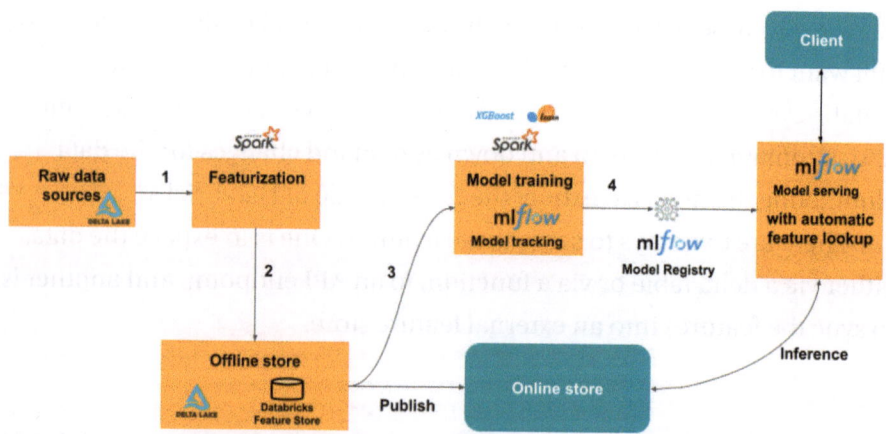

Figure 10-11. *Online external*

Table 10-2 shows which external feature stores Databricks supports for specific features.

Table 10-2. *External Feature Stores That Databricks Supports*

Online Store Provider	Publish with Feature Engineering in Unity Catalog	Publish with Workspace Feature Store	Feature Lookup in Legacy MLflow Model Serving	Feature Lookup in Model Serving
Amazon DynamoDB	X	X (Feature Store client v0.3.8 and above)	X	X
Amazon Aurora (MySQL-compatible)		X	X	
Amazon RDS MySQL		X	X	
Azure Cosmos DB	X	X (Feature Store client v0.5.0 and above)	X	X
Azure MySQL (Single Server)		X	X	
Azure SQL Server		X		

Listing 10-1 shows the sample code to publish to an online feature store.

Listing 10-1. Syncing Features to Amazon Dynamo DB

```
import datetime
from databricks.feature_engineering.online_store_spec import
AmazonDynamoDBSpec
# or databricks.feature_store.online_store_spec for Workspace
Feature Store

# do not pass `write_secret_prefix` if you intend to use the
instance profile attached to the cluster.
online_store = AmazonDynamoDBSpec(
  region='<region>',
  read_secret_prefix='<read-scope>/<prefix>',
  write_secret_prefix='<write-scope>/<prefix>'
)

fe.publish_table( # or fs.publish_table for Workspace
Feature Store
  name='ml.recommender_system.customer_features',
  online_store=online_store,
  filter_condition=f"_dt = '{str(datetime.date.today())}'",
  mode='merge'
)
```

Step 2: Calculate Embedding and Sync to a Vector Database

The Databricks Foundation Model API (FMAPI) can be used outside of Databricks. As discussed, Mosaic AI vector search will calculate the embeddings for you automatically. But it is also possible to use the embedding endpoint outside of Databricks. Listing 10-2 is the code to calculate the embedding using FMAPI.

Listing 10-2. Using Databricks FMAPI to Calculate Text
Embeddings

```
def calculate_embedding(text):
    embedding_endpoint_name = "databricks-bge-large-en"
    url = f"https://{mlflow.utils.databricks_utils.get_browser_
    hostname()}/serving-endpoints/{embedding_endpoint_name}/
    invocations"
    databricks_token = mlflow.utils.databricks_utils.get_
    databricks_host_creds().token

    headers = {'Authorization': f'Bearer {databricks_token}',
    'Content-Type': 'application/json'}

    data = {
        "input": text
    }
    data_json = json.dumps(data, allow_nan=True)

    print(f"\nCalling Embedding Endpoint: {embedding_endpoint_
    name}\n")

    response = requests.request(method='POST', headers=headers,
    url=url, data=data_json)
    if response.status_code != 200:
        raise Exception(f'Request failed with status {response.
        status_code}, {response.text}')

    return response.json()['data'][0]['embedding']
```

Step 3: Create a LangChainTool to Perform Various Tasks

It is typical to use a LangChainTool to perform the tasks, aka `from langchain.agents import initialize_agent`. But it can also be anything you want to do with LangChain or LlamaIndex. It is not a restriction but rather a suggestion because MLflow supports logging only with LangChain, OpenAI and Huggingface for now. The code in Listing 10-3 will log your LangChain model as artifacts.

Listing 10-3. Logging Model with LangChain Flavor

```
mlflow.langchain.log_model()
```

Step 4: MLflow LLM Evaluation

Similar to traditional machine learning models, LLMs also need to be evaluated to ensure the output is accurate. However, because the output can be nondeterministic and very often there is no single ground truth to compare against, ML Flow has provided a few ways to evaluate an LLM model, and the team is continuously working to update the functionalities.

1. **Use Default Metrics for Predefined Model Types with mlflow.evaluate()**

   MLflow comes with a few predefined model types; with each model type, it leverages some open-source libraries to compute the metrics. The types are described as follows:

- **question-answering:** `model_type="question-answering"`:

  - **exact-match:** Measures the exact match between the predicted answer and the true answer

  - `toxicity` [1] : detects if the answer contains toxic or harmful content

  - `ari_grade_level`[2] and `flesch_kincaid_grade_level` [2] : evaluate the readability of the answer based on its complexity and grade level

- **text-summarization:** `model_type="text-summarization"`:

  - ROUGE[3]: Measures the similarity between the predicted summary and the true summary

  - `toxicity`[1]: Detects if the answer contains toxic or harmful content

  - `ari_grade_level`[2] and `flesch_kincaid_grade_level`[2] : Evaluate the readability of the answer based on its complexity and grade level

- **text models:** `model_type="text"`:

  - `toxicity`[1]: Detects if the answer contains toxic or harmful content

  - `ari_grade_level`[2] and `flesch_kincaid_grade_level`[2]: Evaluate the readability of the

Listing 10-4 shows the code to run an evaluation.

Listing 10-4. Running an Evaluation of a Dataset

```
results = mlflow.evaluate(
    basic_qa_model.model_uri,
    eval_df,
    targets="ground_truth",  # specify which column
    corresponds to the expected output
    model_type="question-answering",  # model type
    indicates which metrics are relevant for this task
    evaluators="default",
)
```

2. **LLM Metrics:** Unlike traditional machine learning, where there is a formula for each metric, LLM metrics are evaluation criteria provided to a powerful LLM, by default GPT 4, to evaluate an answer either against ground truth or providing prompts. The following are the provided interfaces:

- **Answer_similarity:** Give a score on how similar the answer with respect to the ground truth.

- **Answer_correctness:** Give a score on the correctness of the answer with respect to the ground truth.

- **Answer_relevance:** Determine how relevant the answer is with respect to the ground truth.

- **Relevance:** Given both ground truth and context (for example, history of Databricks) to determine how relevant of the answer with respect to the ground truth.

- **Faithfulness:** It evaluates only with the provided context with the output to determine if the claim can be inherited from the context.

It is understandable that the concept is challenging to understand at first glance. However, reviewing default prompts from MLflow will help answer some of the doubts you might have in your mind:

```
https://github.com/mlflow/mlflow/blob/
master/mlflow/metrics/genai/prompts/v1.py
```

MLflow provides these examples by default to ensure we give enough hints to the model. It is recommended that you give examples as input, but you can also evolve from the default ones.

Figure 10-12 shows example output of the similarity metric.

answer_similarity/v1/score	answer_similarity/v1/justification
4	The output provided by the model aligns well w...
2	The output provided by the model does correctl...
5	The output provided by the model aligns very c...
4	The output provided by the model aligns well w...

Figure 10-12. *Example output of MLflow metrics*

3. **Evaluation data:** This is the data your model is
    evaluated by. It can be a Pandas dataframe, a Python
    list, a numpy array, or an `mlflow.data.dataset.`
    `Dataset()` instance. This dataset usually contains
    input dataset and ground truth labels, as shown in
    Listing 10-5.

Listing 10-5. Example Input Dataset for Ground Truth Data

```
{
    "inputs": ["What is MLflow?",],
    "ground_truth": [ "MLflow is an open-source platform for
    managing the end-to-end machine learning lifecycle. It was
    developed by Databricks, a company that specializes in big
    data and machine learning solutions. MLflow is designed to
    address the challenges that data scientists and machine
    learning engineers face when developing, training, and
    deploying machine learning models.",],
}
```

We have demonstrated the ability to use Databricks as a serving
endpoint as well as the open-sourced version of MLflow to do an
RAG application and evaluate its performance. To simplify all these
operations, we can easily use AI Agent framework and everything shown
in Figure 10-3. Without first understanding the core pieces of operations, it
will be easy to think that Databricks is a lock-in platform, but in fact, it is an
open platform. All the tools are built upon the basic components discussed
earlier.

After looking at the RAG Applications, we will look into the Fine-
Tuning API.

Mosaic AI Fine-Tuning API

In the world of LLMs, the cost of training and the hardware requirements increase as the stage moves from prompt engineering all the way to pre-training. Not only that but the technical knowledge required also increases. Table 10-3 illustrates the skills requirements as well as hardware requirements for each stage.

Table 10-3. *Role and Hardware Requirements for Each Step of the GenAI Journey*

	Prompting	RAG	Fine-Tuning	Pre-Training
Role	English	Data Engineers	Data Scientists	Research Scientists
Hardware	CPU	CPU	GPU	GPU clusters

According to Open AI, fine-tuning lets you get more out of the models available by providing:

- Higher-quality results than prompting
- Ability to train on more examples than can fit in a prompt
- Token savings due to shorter prompts
- Lower-latency requests

Referencing Table 10-3, understanding the resources and skills requirement as well as the training dataset, one should consider tweaking the prompt before getting into fine-tuning. It is necessary to gather more ground truth data for the model so the fine-tuned model can provide a more accurate response to a specific topic.

As the name suggests, fine-tuning is a process to get to some specific knowledge faster, but it comes with a cost unless it is really needed. For example, the model cannot answer some specialized medical questions that often involve a lot of nuances only training medical professionals would know how to answer. Then it can be a good use case for fine-tuning.

Open AI has published a detailed guide on prompt engineering, and we can try these with Mosaic ML Playground:

`https://platform.openai.com/docs/guides/prompt-engineering`

Fine-Tuning Example

Databricks has integrated Mosaic ML's fine-tuning API into the platform. The details of the fine-tuning API can be found at the MosaicML website:

`https://docs.mosaicml.com/projects/mcli/en/latest/finetuning/finetuning.html`

The advantage of integrating MosaicML with Databricks is that now the fine-tuned model will be supported by the Databricks platform with Model Serving and Model Registry, it will also be able to take advantage of the managed MLflow feature. Everything is integrated into a single environment.

Despite the warning, if you are really familiar with the process and also have a good dataset available, fine-tuning can achieve amazing results with a low cost:

`https://www.databricks.com/blog/creating-bespoke-llm-ai-generated-documentation`

Pre-Training

Pre-training is the most costly and would require the most effort to accomplish (`https://www.databricks.com/blog/ai2-olmo-is-here`). Because everything will be created from scratch, one must create a model like traditional deep learning; only it will require perhaps billions of times more data and much more commodity hardware, which is not something a small-to-medium enterprise would want to do.

A Case Study of AI2's OLMo, a Truly Open-Source Large Language Model

The Open Language Model (OLMo) is a collaboration between Databricks and Allen Institute for AI, and we will examine the requirements to re-create this model (`https://arxiv.org/pdf/2402.00838.pdf`). See Figure 10-13.

> **Dataset:** In traditional deep learning, the sample size required per category is about a few thousand. By comparison, the Dolma dataset is an open dataset of **3 trillion** tokens from a diverse mix of web content, academic publications, code, books, and encyclopedic materials.

Source	Doc Type	UTF-8 bytes (GB)	Documents (millions)	GPT-NeoX tokens (billions)
Common Crawl	web pages	9,022	3,370	2,006
The Stack	code	1,043	210	342
C4	web pages	790	364	174
Reddit	social media	339	377	80
peS2o	STEM papers	268	38.8	57
Project Gutenberg	books	20.4	0.056	5.2
Wikipedia, Wikibooks	encyclopedic	16.2	6.2	3.7
Total		**11,519**	**4,367**	**2,668**

Figure 10-13. *Composition of the data used in the model training*

Model training: Because the data volume is so huge, it can no longer fit in one GPU, it is required to distribute across multiple GPUs. In section 3.1 of the AI2 paper, it discusses the distributed framework in detail.

Model architecture: A proper model architecture must be implemented for the model. It is not prebuilt like foundation models. Section 2.1 of the AI2 paper discusses such architecture for 1B, 7B as well as 65B parameters.

Hardware: This might be the most expensive and most difficult part to achieve. Not to mention there are currently very limited availability of high-end GPUs on the market, they are reserved for researchers who would deliver ultimate value to the company.

In the OMLo model, it uses MosaicML with 27 nodes on the cluster, where each node consists of 8x NVIDIA A100 GPUs with 40GB of memory and 800Gbps interconnect. In total, 216 GPUs will be required to pretrain this model. Unless someone who really understands the ins and outs of LLMs as well as there is high ROI on these projects, organizations usually stop their GenAI journey at fine-tuning.

Gen AI Pricing

While the pricing of the GenAI infra is usually use per hour and can be found at the following Databricks websites. For information about DBU hours, please refer to Chapter 16.

Model Serving:

```
https://www.databricks.com/product/pricing/
model-serving
```

```
https://www.databricks.com/product/pricing/
foundation-model-serving
```

Vector search:

```
https://www.databricks.com/product/pricing/
vector-search
```

Model training:

```
https://www.databricks.com/product/pricing/
mosaic-training
```

There is one concept that is not fully explained, which is pay per token. The price is per 1 million tokens. Please note that one token does not directly translate to one English word or certain bytes. For example, ASCII is 1 byte, and Unicode ranges from 1 byte to 4 bytes. The concept in LLMs is similar, but it is not so straightforward.

What Are Tokens and Tokenizers?

The very short version is to split text into smaller chunks for the model to consume because with any model there is a capability to take in some text at once. Tokenizers are used to split some text into subwords, aka tokens. To learn more about tokenizers, please refer to this blog post from Huggingface: `https://huggingface.co/docs/transformers/main/ tokenizer_summary`.

More important, the real question is, how do we estimate how many tokens my input text will generate? To answer this question, we need first to understand what tokenizer each model is using; see Table 10-4.

Table 10-4. *Tokenizer Used in Some Popular Large Language Models*

Model	Tokenizer
DBRX	GPT-4
	(https://www.databricks.com/blog/introducing-dbrx-new-state-art-open-llm)
Llama 2	Bytepair encoding
	(https://ai.meta.com/research/publications/llama-2-open-foundation-and-fine-tuned-chat-models/ Section 2.2 Tokenizer)
Mistral	Byte-fallback BPE tokenizer
	(https://huggingface.co/docs/transformers/main/model_doc/mistral)
MPT	EleutherAI/gpt-neox-20b
	(https://huggingface.co/mosaicml/mpt-30b)

Understanding the tokenizer is just the first step. While you can very easily load a tokenizer with one line of code, as shown in Listing 10-6, we also need to have a way to estimate the number of tokens we need for our task in order for cost estimation.

Listing 10-6. Getting the Tokenizer from the Model

```
from transformers import AutoTokenizer, AutoModelForCausalLM

tokenizer = AutoTokenizer.from_pretrained("name-of-tokenizer")
```

The good news is that there are some python or JavaScript tools that we can utilize to estimate the number of tokens:

Open AI: `https://cookbook.openai.com/examples/how_to_count_tokens_with_tiktoken`

Llama: `https://github.com/belladoreai/llama-tokenizer-js/`

Mistral: `https://github.com/imoneoi/mistral-tokenizer/`

With these tools, we can easily calculate the number of tokens in input text to estimate the cost of using the GenAI services. Of course they can also be found in the system tables, but that will be after the job finishes running. Please refer to Chapter 16 for more information.

Conclusion

Navigating the LLM world is very challenging, and addressing those blockers with the right solution is something that requires careful consideration, especially from an ML lifecycle perspective. The Databricks GenAI stack provides a powerful solution for accelerating machine learning and AI capabilities at competitive pricing points. Databricks provides flexibility and customization options that traditional ML platforms lack or provide at a higher price. With the GenAI capabilities in Databricks, organizations can focus on creating value, whether it is managing data, tracking experiments, packaging code, or deploying models into Unity Catalog, thereby streamlining the entire LLM life cycle with governance.

CHAPTER 11

Large Language Model Operations

We discussed machine learning operations (MLOps) in an earlier chapter. In this chapter, we will discuss a similar topic called large language model (LLM) operations. This chapter has certain similarities with the chapter on generative AI (GenAI), but we will mainly focus on the operations part of machine learning and the benefits of it as a practice. We will also dive deep into using different techniques and libraries in the industry to perform these operations, which Databricks also supports.

MLOps and LLMOps are related but distinct concepts in artificial intelligence (AI) and machine learning (ML). Here's a brief overview of each.

Machine Learning Operations

MLOps aims to streamline the machine learning life cycle by combining machine learning practices and DevOps. In the previous chapter, we discussed the MLOps stack from Databricks, which combines ML templates and DevOps templates that are ready to deploy. This section will revisit the roles and responsibilities as well as the end goal for MLOps.

© The Editor(s) (if applicable) and The Author(s),
under exclusive license to APress Media, LLC, part of Springer Nature 2024
N. Gupta and J. Yip, *Databricks Data Intelligence Platform*,
https://doi.org/10.1007/979-8-8688-0444-1_11

- Collaboration and communication between data scientists and ML engineers

- Using CI/CD workflow (GitHub actions, Azure DevOps) to automate the ML life cycle, including feature engineering, model training, and deployment along with infrastructure as code

- Version control and management of ML models and data (features)

- Continuous integration and delivery (CI/CD) of ML models, including Model Serving

- Monitoring of ML model performance, including data drift, model drift, concept drift, etc., using Lakehouse Monitoring

Large Language Model Operations

It is only a natural transition with the historical singular focus on LLMs that large language model operations (LLMOps) ensures we are doing the right things when handling the huge amount of data and model outputs. LLMs are complex AI models that require significant computational resources, data, and expertise to develop, deploy, and maintain. Everything from the cost to the curation of data and ensuring few mistakes are made is crucial to the project's success. LLMOps builds upon MLOps principles and adds additional considerations, such as the following:

- Scalability and performance optimization for large models and datasets, mainly for cost reasons

- Specialized software and hardware requirements (e.g., GPUs, MosaicML)

- Advanced techniques for model pruning, knowledge distillation, and inference optimization (e.g. Mixture-of-Expert)

- Application of the model, such as chatbots, search engines, and recommendation systems

In summary, building on the foundation of MLOps, LLMOps is a suite of specialized tools focused on handling the unique challenges of LLMs including prompting, RAG, fine-tuning, and pre-training. Figure 11-1 outlines the flow and toolings required for LLMOps.

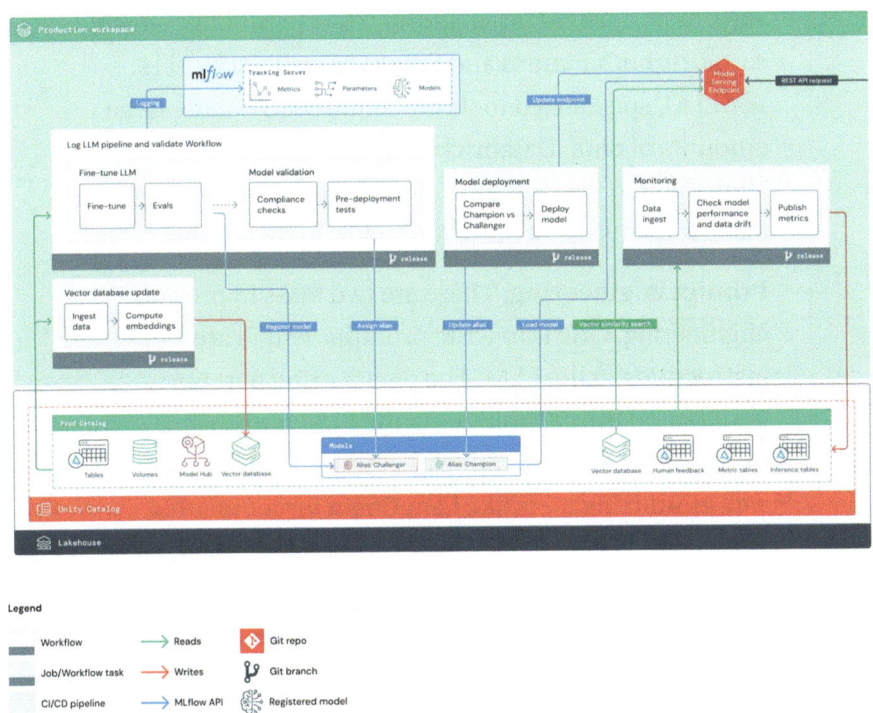

Figure 11-1. *LLMOps*

Components of LLMOps

First, let's review the components required to make LLMOps successful. While some of these components are similar to MLOps, as the volume of data increases, the process has become data-centric instead of model-centric. In other words, in MLOps, the goal was to test as many models as possible, but in LLMOps, a lot of work will go into ensuring the data is high-quality.

- **Exploratory data analysis (EDA):** As discussed in the previous chapter, Databricks offers a few different ways to perform EDA, including YData profiling (which supports both Pandas and Spark) as well as `dbutils`. For LLM, specialized tools are required to handle large amounts of data. Databricks acquired Llacai, which allows you to visualize and clean up data easily. We will discuss the usage of these tools later in this chapter.

- **Prompt engineering:** There are two tasks in prompt engineering. One is to write prompts, which are instructions to the LLM. The engineering part is to understand the capabilities of the LLM and ensure the prompts are generating meaningful outputs, as well as utilizing tools like LangChain for templating and creating a chain of thoughts process. Although prompt engineering is the first step, prompt quality, aka providing clear instructions and detailed steps for the question or an evaluation process, will greatly help the LLM to provide a good answer. And as the context length increases, people are starting to put very long pages of instructions for their prompt.

- **Retrieval augmented generation (RAG):** Prompt
  engineering is a way to instruct the model to search for
  the knowledge you want. But like humans, LLMs are
  also limited to the knowledge that's exposed to them
  during training time. Things like company-specific
  information not publicly available on the Internet will
  not be available in the models. Hence, the process
  of RAG is to update the model by providing extra
  information such as PDFs or PowerPoints. This is the
  process of RAG.

Figure 11-2 shows a typical RAG workflow.

***Figure 11-2.** Typical workflow of RAG*

- Vector database

  Vectors or embeddings are an essential part of the
  RAG process. They are numerical representations
  of the data. To answer a question, similarity search
  is often used. There are a few terms we need to
  consider:

  - **Vector index:** A specialized data structure
    optimized to facilitate similarity search within a
    collection of vector embeddings. It is read-only and
    needs to be rebuilt when content changes.

- **Vector library:** A tool to manage vector embeddings and conduct similarity searches. They predominantly:

  - Operate on in-memory indexes.

  - Focus solely on vector embeddings, often requiring a secondary storage mechanism for the actual data objects.

- **Vector database:**

  - Store both the vector embeddings and the actual data objects, permitting combined vector searches with advanced filtering.

  - Offer full CRUD (create, read, update, delete) operations, allowing dynamic adjustments without rebuilding the entire index.

  - Are generally better suited for production-grade deployments due to their robustness and flexibility.

Databricks offers vector search backed by the serverless architecture and provides the vector index service, combined with Unity Catalog and Delta tables, which can be served as a database. It offers the following advantages:

- Auto-syncs with the source Delta table

- Columns in Delta table are filters

- Unity Catalog governance and lineage

- Integrated with Model Serving for embedding generation

Figure 11-3 shows a sample workflow for an RAG application, which combines feature and function serving.

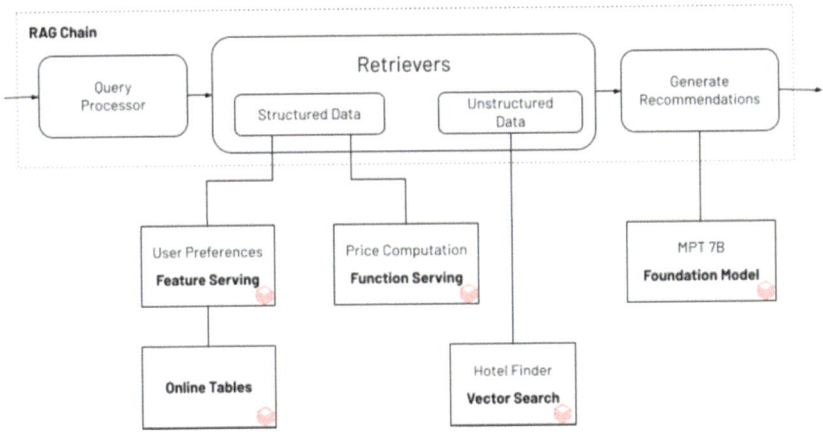

Figure 11-3. *Workflow of RAG in Databricks*

- **Model fine-tuning:** If we use the previous analogy, prompting is similar to seeking help from a consultant and providing clear instructions to them, assuming they have existing knowledge. RAG is the process of giving them extra documents or reference materials to enhance their knowledge. Sometimes, these are not enough because of a brand new domain. For example, an LLM might not be trained with highly specialized medical knowledge. That's when we need to provide a lot more datasets but still a relatively small amount compared to pre-training, for the fine-tuning process. Instead of prompting and RAG using the existing model, the fine-tuning process will create a new LLM from the base models. For example, DBRX base versus DBRX Instruct are two different models.

- **Model pre-training:** Model pre-training is a process of creating an LLM from scratch. In Chapter 13, we will discuss how Databricks trained DBRX using the platform available to customers. The difference between "pre-training" and "model training" in traditional machine learning is that the base model that's pre-trained often needs to be enhanced or fine-tuned. The output of this process is a base model, e.g., DBRX base.

- **Model evaluation with human feedback:** As opposed to using standard metrics like F1 and R2 scores, evaluating LLMs is more challenging and constantly evolving, primarily because LLMs often demonstrate uneven capabilities across different tasks. An LLM might excel in one benchmark, but slight variations in the prompt or problem can drastically affect its performance. Just think about it: not everyone gets the same result from Google. That's why in addition to new evaluation suites coming out every now and then, usually when a new company releases a new LLM, they will release a new suite.

However, even machine learning models require human feedback, and LLM is even more so. That's why Databricks MLflow comes with an interface for human evaluation. And application developers should purposely develop a feedback mechanism to collect user feedback.

- **Model packaging and deployment:** Similar to the MLOps pipeline, LLMOps also consists of various components, such as the mode API, RAG pipeline, and prompt engineering templates.

- **Model Serving and inference:** After we have developed and deployed a great new model, the last step is to serve it as an API endpoint and start generating outputs, aka inferencing. Databricks Model Serving and the MLflow deployment server (formerly AI gateway) can be used to standardize model API interfaces for real-time inference, and Spark can be leveraged to do offline distributed inference.

Deep Dive into Each Process

In Chapter 12, we discuss how these processes can be applied. However, there are often details beyond the chat window that we need to pay attention to when transitioning from an LLM user to an LLM application developer. This section is designed to get you started with these concepts.

Prompt Engineering

As discussed in the previous section, good prompt engineering involves giving clear instructions to a consultant to execute your task. When you give instructions, you might start with a few sentences, but you rarely end with that and expect high-quality outcomes. The key lies in providing clear and concise instructions.

Prompt Templates

While ancient knowledge is transferred by word of mouth, modern knowledge can be placed into a template. The purpose of templating is to allow best practices to be captured in a repeatable form so everyone can take advantage of it. A simple example is "What is the {input_model} model?" where "model" can be any machine learning model. We can consider using LangChain for this purpose, as shown in Listing 11-1.

Listing 11-1. Prompt Template with LangChain

```python
import os
import openai
from openai import OpenAI
from langchain import PromptTemplate

client = OpenAI(
    api_key="databricks-api-token",
    base_url="https://adb-xxxxxxxxx.xx.azuredatabricks.net/
    serving-endpoints"
)

template = PromptTemplate(
    input_variables=["input_model"],
    template="What is {input_model} model?"
)

prompt=template.format(input_model="Mixture of Expert")

response = client.chat.completions.create(
    messages = [{"role": "user", "content": prompt}],

    model="databricks-dbrx-instruct",
    max_tokens=256
)

generated_text = response.choices[0].message.content

# Use the generated text in your Databricks workflow
print(generated_text)
```

Please note that because we are hitting a Foundation Model API as an endpoint, we don't need a Databricks cluster to run the previous code. We can easily execute the previous code in a Python notebook on our local machine, as shown in Listing 11-2.

Listing 11-2. Actual Output from Python Code

A Mixture of Expert (MoE) model is a type of machine learning model that is composed of several "expert" models, each of which specializes in handling a particular subset of the data. These expert models are weighted and combined together using a "gating" mechanism, which determines how much each expert should contribute to the final prediction. This allows the MoE model to effectively handle a wide variety of data and make more accurate predictions.

Chain of Thoughts

Clear communication means creating a good prompt template, and creating step-by-step instructions will greatly help the LLM to provide a high-quality answer. This process is called *chain of thoughts* (https://arxiv.org/abs/2201.11903), as shown in Listing 11-3.

Listing 11-3. Chain of Thought Template

```
Think step by step and explain your reasoning:
{input}
Step 1: {question_1}
{answer_1}
Step 2: {question_2}
{answer_2}
Step 3: {question_3}
{answer_3}

Final Answer: {final_answer}
```

Providing step-by-step instructions, as shown in https://arxiv.org/abs/2201.11903, has been proven to increase an LLM's performance.

Retrieval Augmented Generation

As mentioned, to update an LLM with new knowledge, we need to conduct a similarity search with the text converted to vector form. Databricks provides all the components discussed earlier, aka vector index, vector library, and vector database.

A vector index can be conveniently created on the user interface on any table that is Unity Catalog enabled. As illustrated in Figure 11-4, an index can be created by going into the table interface itself by clicking the Create button.

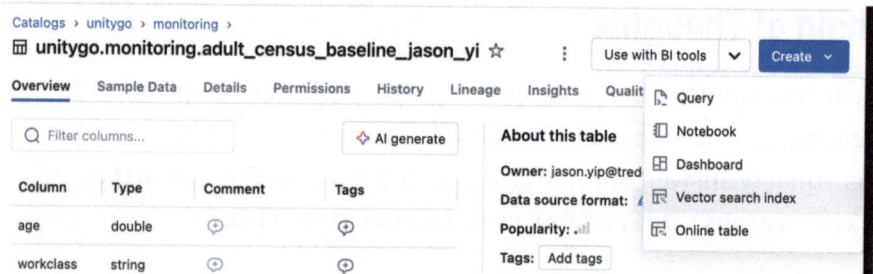

Figure 11-4. *Creating a vector search index*

The index creation has different options, as shown in Figure 11-5.

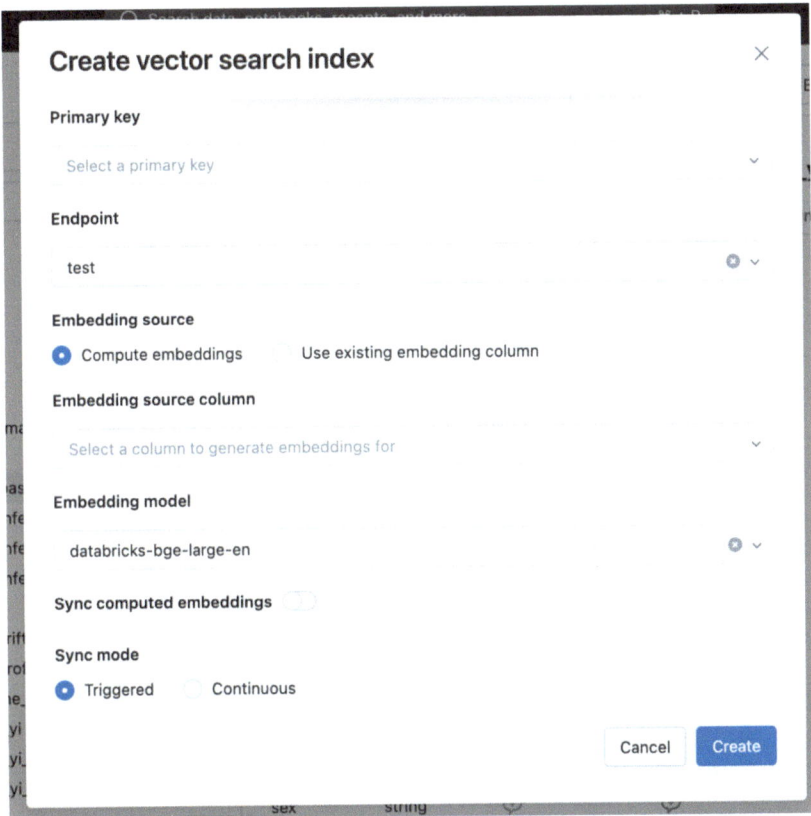

Figure 11-5. *Creating a vector search index UI*

- The primary key can be used to identify unique embedding entries so it will not create duplicates.

- The endpoint is the vector search endpoint, which is serverless compute, that can be used to compute the embedding or perform similarity search.

- "Embedding source column" can be used to generate embedding based on a text-based column, but for binary (e.g., PDF or images), embeddings can be stored

267

in a column in a Delta table. The column type must be `array<float>`, `array<double>`, `array<int>`, `array<byte>`).

- Databricks provides **a bge-large-en** model for embedding purposes. Embedding is a way to convert a text column to a vector. The BGE model from the University of Science and Technology of China specializes in natural language embedding. According to the BGE paper (`https://arxiv.org/pdf/2402.03216`), the model's features can be found in Figure 11-6.

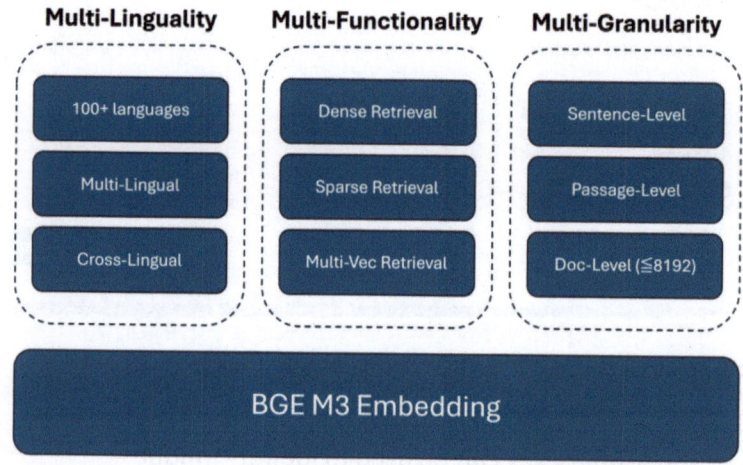

Figure 11-6.   bge-large-en model features

However, please note that this is optimized for text embedding, so if you need other embeddings, you might want to use a different model. The following link shows a list of Databricks curated models:

https://www.databricks.com/product/machine-
learning/large-language-models-oss-guidance

For a deep dive into the text-embedding, please refer to
the Massive Text Embedding Benchmark (MTEB) from
Huggingface:

https://huggingface.co/blog/mteb

- Finally, you can choose to sync the index, which is
the embedding column back by a Delta table, back to
a vector Databricks vector search, a vector database
optimized to store and retrieve embeddings. See
Figure 11-7.

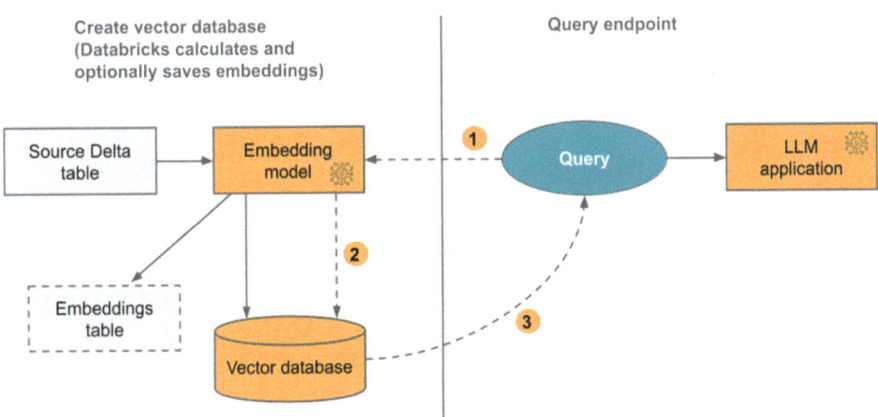

Figure 11-7. *Using Databricks to calculate embeddings*

You can choose to calculate the embedding automatically or provide
a precalculated column and optionally sync the embedding into the
vector database (Databricks vector search). As discussed, having a vector
database optimizes the way similarity search is calculated and hence
enhances the performance of RAG. See Figure 11-8.

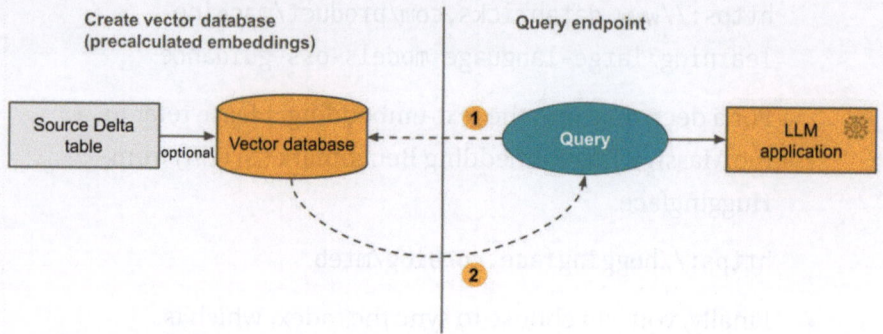

Figure 11-8. *Precalculating the embeddings*

Model Fine-Tuning

Usually, the steps in machine learning operations are required from end to end to ensure the best model is selected and the model is performing well. In LLMOps, however, most of the time it is sufficient to stop at RAG because high-quality prompt engineering and high-quality RAG are usually enough to get quality output on something that the model is trained for. But if the model is not trained on something specific, like the healthcare domain, and you want to ensure the model can adapt to the new domain, we will need to consider fine-tuning the model.

This is usually more costly and time-consuming than RAG, but from the deployment perspective, it becomes easier. Consider DBRX-base and DBRX-Instruct, where the latter is a fine-tuned model but we don't need to worry about maintaining the embedding and vector database.

Unlike RAG, fine-tuning is a model training process, which can consume a lot of resources. We must have a good understanding of the architecture of the neural network (large language model is a neural network) in order to train it properly.

An example using DeepSpeed for fine-tuning can be found here, which can take advantage of multiple GPUs for more resource-intensive tasks:

`https://github.com/databricks-academy/large-language-models`

Alternatively, a less resource intensive method is called *parameter-efficient fine-tuning* (PET) can also be used to fine-tune an LLM, the approaches are called LoRA, QLoRA, or IA3. Databricks has provided detailed discussions here:

`https://www.databricks.com/blog/efficient-fine-tuning-lora-guide-llms`

Model Pretraining

Pre-training is the most costly and would require the most effort to accomplish. Because everything will be created from scratch, one must create a model like traditional deep learning; only it will require perhaps billions of times more data and much more commodity hardware, which is not something a small-to-medium enterprise would want to do.

A Case Study of AI2's OLMo, a Truly Open-Source Large Language Model

The Open Language Model (OLMo), , is a collaboration between Databricks and Allen Institute for AI (`https://arxiv.org/pdf/2402.00838.pdf`). We will examine the requirements to re-create this model.

> **Dataset:** In traditional deep learning, the sample size required per category is about a few thousand. By comparison, the Dolma dataset is an open dataset of **3 trillion** tokens from a diverse mix of web content, academic publications, code, books, and encyclopedic materials (see Figure 11-9).

Source	Doc Type	UTF-8 bytes (GB)	Documents (millions)	GPT-NeoX tokens (billions)
Common Crawl	web pages	9,022	3,370	2,006
The Stack	code	1,043	210	342
C4	web pages	790	364	174
Reddit	social media	339	377	80
peS2o	STEM papers	268	38.8	57
Project Gutenberg	books	20.4	0.056	5.2
Wikipedia, Wikibooks	encyclopedic	16.2	6.2	3.7
Total		**11,519**	**4,367**	**2,668**

Figure 11-9. *Text content of the 3 trillion tokens (about 1 trillion words)*

Model training: Because the data volume is so huge, it can no longer fit in one GPU; it is required to distribute across multiple GPUs. In section 3.1 of the AI2 paper, it discusses the distributed framework in detail.

Model architecture: A proper model architecture must be implemented for the model. It is not prebuilt like foundation models. Section 2.1 of the AI2 paper discusses such an architecture for 1B, 7B, and 65B parameters.

Hardware: This might be the most expensive and most difficult part to achieve. Not to mention, there is currently very limited availability of high-end GPUs on the market; they are reserved for researchers who would deliver ultimate value to the company.

The OMLo model uses MosaicML with 27 nodes on the cluster, each consisting of 8x NVIDIA A100 GPUs with 40GB of memory and 800Gbps interconnect. In total, 216 GPUs will be required to pretrain this model.

Unless someone really understands the ins and outs of large language models and there is high ROI on these projects, organizations usually stop their GenAI journey at fine-tuning.

For further information, please refer to Chapter 13 where we discuss in great length how Databricks pretrained a model from scratch.

Model Governance

Let's discuss model governance.

MLflow Deployments Server

Databricks MLflow provides a deployment server (formerly AI gateway) for us to manage, govern, evaluate prompts, and switch models easily. Figure 11-10 illustrates how MLflow AI gateway is a bridge between LLMs and their use cases.

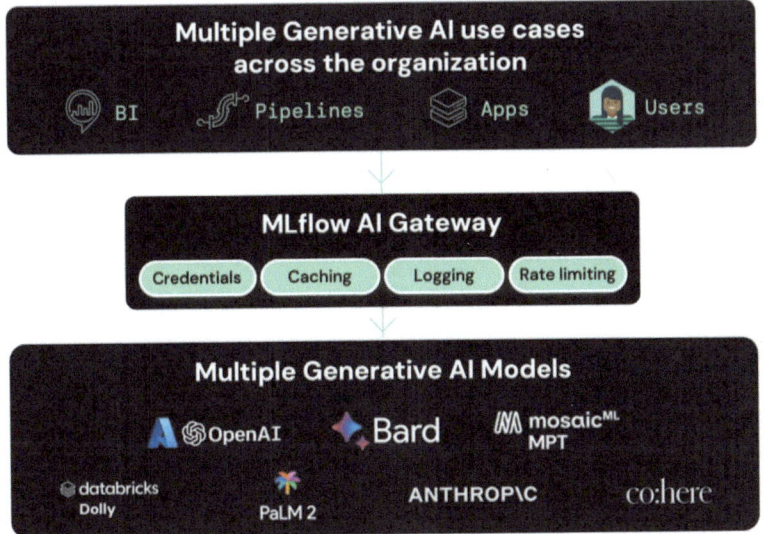

Figure 11-10. *Illustration of MLflow deployment gateway*

MLflow Deployments Server reduces the management overhead of managing multiple credentials for premium LLMs that would otherwise require different API keys. It also unifies different model inputs and outputs together to abstract the complexities behind the scenes to transform the input and parse the output.

For a list of supported models, please refer to the following:

`https://mlflow.org/docs/latest/llms/deployments/index.html#providers`

The credentials can also be managed in Databricks Model Serving. However, the advantage of MLflow is open source and not vendor-specific. On the Serving tab of Databricks sidebar, you can create a serving endpoint with stored credentials, as shown in Figure 11-11. In Figure 11-12, we can choose an entity that we want to serve.

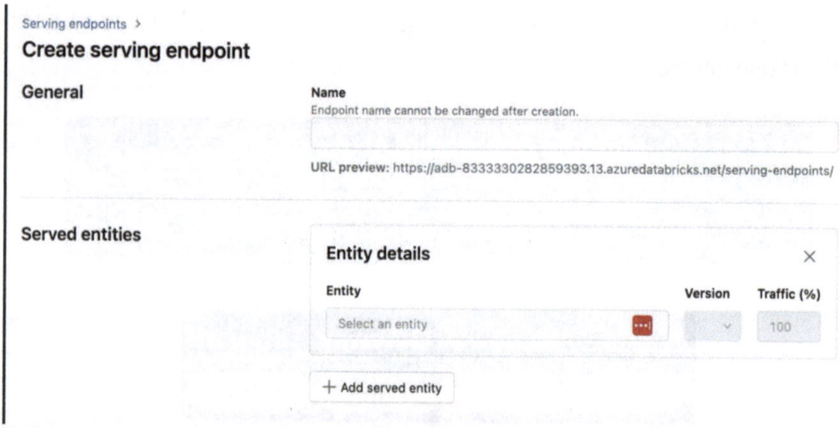

Figure 11-11. *Creating a serving endpoint*

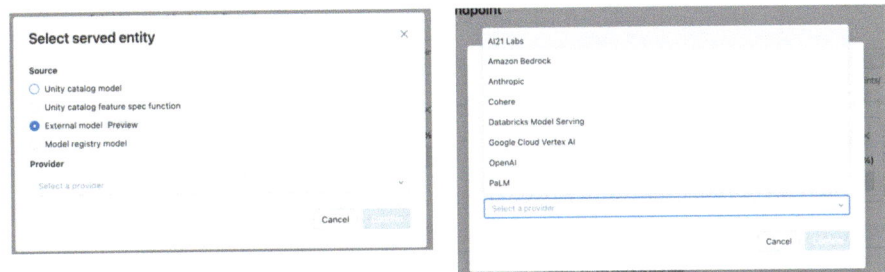

Figure 11-12. *Selecting a served entity, then saving the credentials*

Once a model is chosen, we can then enter the credentials; they can be retrieved from a secret store for the best of security. See Figure 11-13.

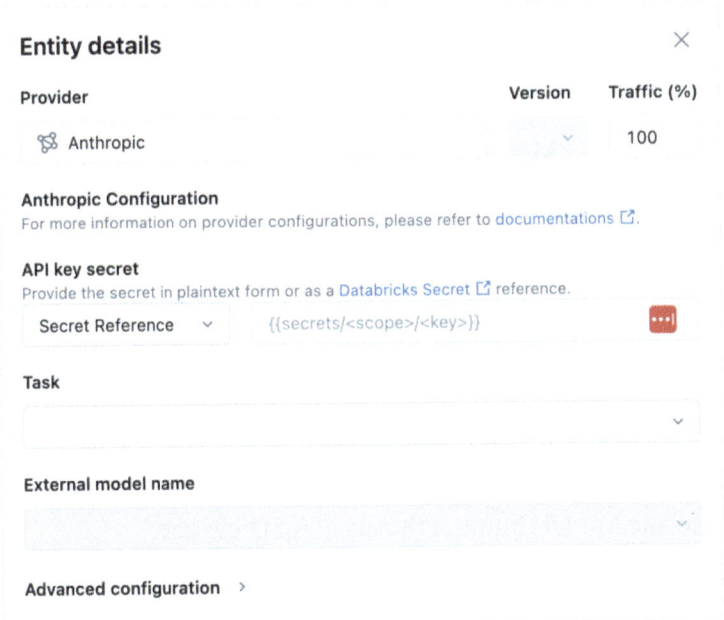

Figure 11-13. *An API key secret can be associated with a model entity*

Prompt evaluations can be done in two different ways; the most obvious way is of course evaluating prompts, which can be done using MLflow or Databricks' playground. But that does not allow you to do it at scale or understand the capabilities of the model. Standard benchmarks will provide greater insights into the model's strength. The most popular evaluation suites include but are not limited to the following:

- AI2 Wildbench (https://github.com/allenai/ WildBench)

  AI2 Wildbench is a carefully curated collection of 1,024 hard tasks from real users, which cover common use cases such as code debugging, creative writing, and data analysis. For more details of the dataset, please refer to the following page: https://huggingface.co/ datasets/allenai/WildBench

- EluetherAI LM Evaluation Harness (https://github. com/EleutherAI/lm-evaluation-harness)

  A holistic framework that assesses models on more than 200 tasks, merging evaluations like BIG-bench and MMLU, promoting reproducibility and comparability. It powers the popular Huggingface leaderboard (https://huggingface.co/spaces/HuggingFaceH4/ open_llm_leaderboard).

- Mosaic Model Gauntlet (https://github.com/ mosaicml/llm-foundry/blob/main/scripts/eval/ local_data/EVAL_GAUNTLET.md); see Figure 11-14

  Developed as part of the DBRX release by MosaicML, using an aggregated evaluation approach, categorizing model competency into six broad domains (shown below) rather than distilling to a single monolithic metric.

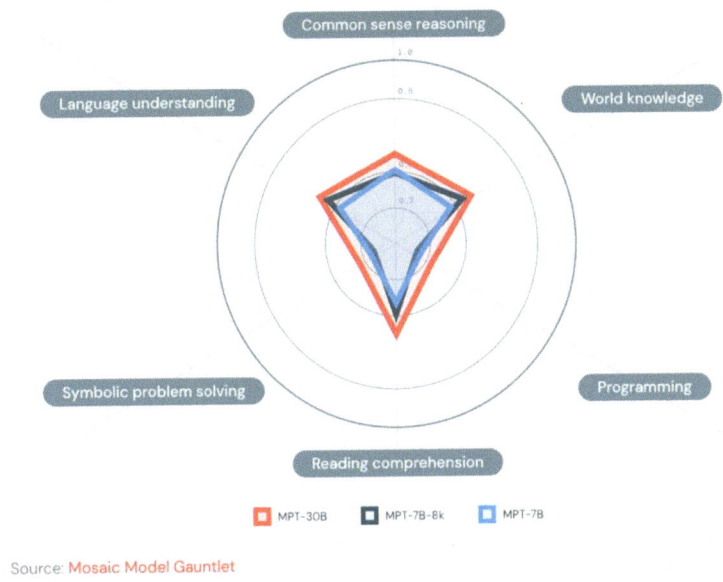

Source: Mosaic Model Gauntlet

Figure 11-14. *Mosaic AI Model Gauntlet*

LLM as a Judge

While human evaluation is powerful, LLM performance is being evaluated in domains where text is scarce or there is a reliance on subject-matter expert knowledge. In such scenarios, evaluating LLM output can be costly and time-consuming. For example, imagine gathering a group of medical specialists together to evaluate the correctness of an open-heart procedure. It will certainly not be easy.

Leveraging LLM as a judge is an idea to use a powerful model, say GPT4, to evaluate the performance of a fine-tuned, smaller domain-specific model. So often, organizations choose to deploy open-source alternatives to production to save costs. This is an opportunity to leverage a more powerful model in a limited capacity to ensure high quality. On the other hand, given enough examples, an LLM can also perform judging by itself like a machine learning model. The idea is to use judging in a systematic way to evaluate. See Listing 11-4.

Listing 11-4. Using LLM as a Judge

```
from mlflow.metrics.genai import EvaluationExample, answer_
similarity

# Create an example to describe what answer_similarity means
like for this problem.
example = EvaluationExample(
    input="What is MLflow?",
    output="MLflow is an open-source platform for managing
    machine "
    "learning workflows, including experiment tracking, model
    packaging, "
    "versioning, and deployment, simplifying the ML
    lifecycle.",
    score=4,
    justification="The definition effectively explains what
    MLflow is "
    "its purpose, and its developer. It could be more concise
    for a 5-score.",
    grading_context={
        "targets": "MLflow is an open-source platform for
        managing "
        "the end-to-end machine learning (ML) lifecycle. It was
        developed by Databricks, "
        "a company that specializes in big data and machine
        learning solutions. MLflow is "
```

```
        "designed to address the challenges that data
        scientists and machine learning "
        "engineers face when developing, training, and
        deploying machine learning models."
    },
)

# Construct the metric using OpenAI GPT-4 as the judge
answer_similarity_metric = answer_similarity(model="openai:/
gpt-4", examples=[example])
```

Figure 11-15 shows the result of the judging.

	inputs	ground_truth	outputs	token_count	toxicity/v1/score	flesch_kincaid_grade_level/v1/score	ari_grade_level/v1/score	answer_similarit
0	How does useEffect() work?	The useEffect() hook tells React that your com...	useEffect() is a React hook that allows you to...	53	0.000299	12.1	12.1	
1	What does the static keyword in a function mean?	Static members belongs to the class, rather th...	In C/C++, the static keyword in a function mea...	55	0.000141	12.5	14.4	
2	What does the 'finally' block in Python do?	'Finally' defines a block of code to run when ...	The 'finally' block in Python is used to defin...	64	0.000290	11.7	13.5	
3	What is the difference between multiprocessing...	Multithreading refers to the ability of a proc...	Multiprocessing involves the execution of mult...	49	0.000207	22.8	28.0	

Figure 11-15. *Judging results*

Model Packaging and Deployment

By now, you have learned how to develop an application using LLM and evaluate its performance interactively. Similar to MLOps, once we finish developing the machine learning model, we need to pack and deploy it via MLflow so the model can be reused.

MLflow offers several different standardized interfaces for LLM, including Huggingface, OpenAI, SBERT.net, and LangChain. With the standard interfaces, we can perform standard logging and monitoring like we do with ML models. You can also pack the pipeline into a PyFunc and make it easy for inference.

LangChain Flavor with MLflow

We talked about prompt engineering, we discussed using LangChain to create a prompt template. Let's take a look at how to pack this with MLflow and deploy it into production.

In the previous example, instead of separating the model for inference, we can use LangChain to chain both the prompt template and a model together. Of course, a chain can contain a lot more than these two components; we are just examining a quick start scenario here (see Listing 11-5).

Listing 11-5. Chaining an LLM and a Prompt Together

```
chain = LLMChain(llm=client, prompt=prompt)
```

In the "Prompt Engineering" section, the client is using an OpenAI interface. However, when chaining it with LangChain, we need to use the LangChain interface. Listing 11-6 illustrates how to use the LangChain interface in the code. Databricks documentation provides various ways to interact with the models; a quick reference can be found here: https://docs.databricks.com/en/machine-learning/model-serving/score-foundation-models.html#query-a-chat-completion-model.

Listing 11-6. Using LangChain to Process a Prompt

```
from langchain.llms import Databricks
from langchain_core.messages import HumanMessage, SystemMessage
```

```
def transform_input(**request):
  request["messages"] = [
    {
      "role": "user",
      "content": request["prompt"]
    }
  ]
  del request["prompt"]
  return request

llm = Databricks(endpoint_name="databricks-dbrx-instruct",
transform_input_fn=transform_input)
```

Next we can chain the LLM and prompt together. See Listing 11-7.

Listing 11-7. Chaining an LLM and a Prompt Together After Delcaration

```
prompt = PromptTemplate(
    input_variables=["input_model"],
    template="What is {input_model} model?"
)
chain = LLMChain(llm=llm, prompt=prompt)
```

With the previous chain, we can then log the chain like how we do it in ML models using MLflow. See Listing 11-8.

Listing 11-8. Logging a LangChain Model

```
mlflow.set_experiment("/Users/jason.yip@tredence.com/
DatabricksDIP")

with mlflow.start_run():
    model_info = mlflow.langchain.log_model(chain,
    "langchain_model")
```

Now once the model is logged, we can load it back using the PyFunc, as shown in Listing 11-9.

Listing 11-9. Loading a PyFunc model from the logged LangChain

```
loaded_model = mlflow.pyfunc.load_model(model_info.model_uri)
answer = loaded_model.predict({"input_model": "Mixture of
Expert"})
print(answer[0])
```

The model can now be accessed; view the Experiments tab on the sidebar. Inside the experiment, it also contains various tabs that are standard across all MLflow projects. See Figure 11-16 and See Figure 11-17.

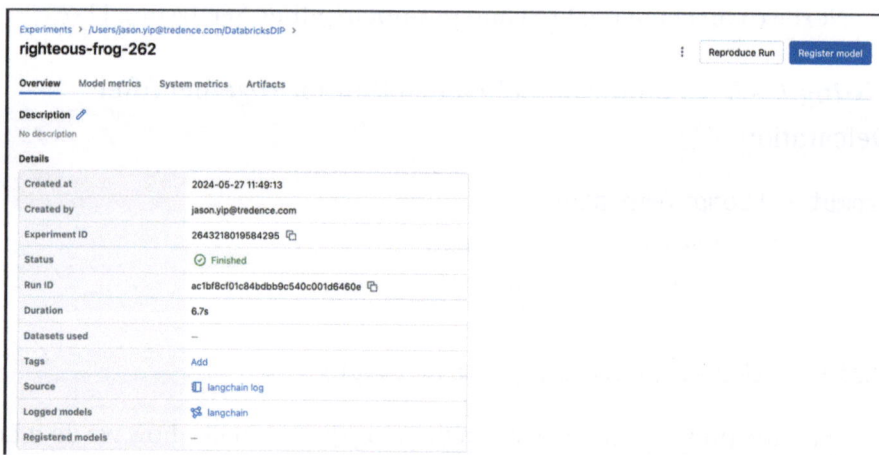

Figure 11-16. Model logged by MLflow, Overview tab

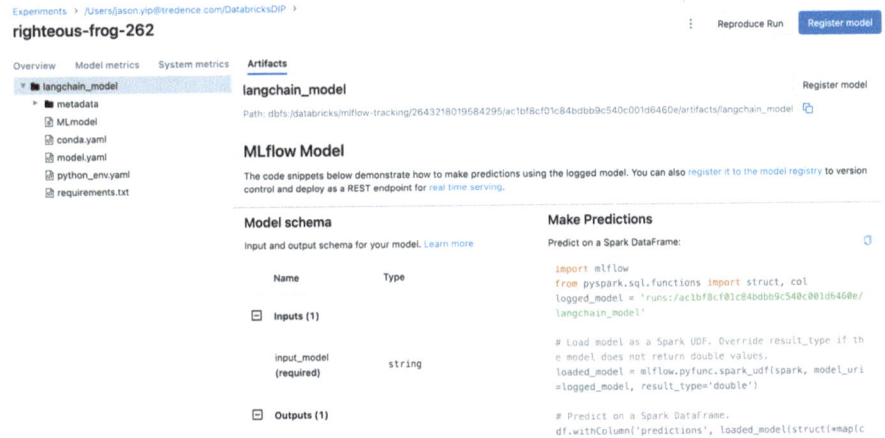

Figure 11-17. *Binaries of our logged model, Artifacts tab*

Conclusion

In this chapter, you learned about the differences between MLOps and LLMOps. While the life cycle is similar to each other, the focus on LLMOps is more about newly introduced components like vector indexes or a LangChain/Huggingface pipeline. Rarely do we need to train a new large language model like in machine learning. Instead, we take a pre-trained, aka base model, to enhance its knowledge by using retrieval augmented generation technique or fine-tuning a model by providing a domain-specific dataset. If there is a need to pre-train a model from scratch, Databricks' MosaicML platform is also capable of handling such a demanding task.

Finally, we can also use Databricks MLflow to continue packaging the LLMOps pipeline into artifacts and deploying it into production. However, we have to decide which flavor we want to use the model in for compatibility purposes. Databricks Model Serving and batch inference capabilities can be used to consume the model and generate outputs.

In the next chapter, we will put these components into practice and create a chatbot using the RAG technique.

CHAPTER 12

Mosaic AI Agent Framework: Creating Quality AI Agents

In this chapter, we will discuss the secret weapon for updating a large language model (LLM) with custom unstructured data, like PDF or PowerPoint. While most applications allow you to build a bot or GPT very easily, enterprises are looking for ways to evaluate the quality of the chatbot. This is where the AI Agent Framework comes in. We will not only discuss how to deploy a chatbot from end to end, but how to evaluate it with an LLM as a Judge or human feedback. These metrics will ensure data scientists who are already familiar with MLflow will be able to transition to LLM evaluation easily.

Without a doubt, there are a lot of components involved in setting up an application with a Retrieval Augmented Generation (RAG) workflow.

Databricks has simplified the deployment of this infrastructure by providing an accessible Python package via MLflow to get users up and running without a lot of manual intervention. More than that, the Mosaic AI Agent Framework also provides continuous logging and allows users to deploy a user interface to gather feedback, putting it all together so it can iterate quickly and get to business values in less time. Figure 12-1 demonstrates this simplified workflow.

N. Gupta and J. Yip, *Databricks Data Intelligence Platform*,
https://doi.org/10.1007/979-8-8688-0444-1_12

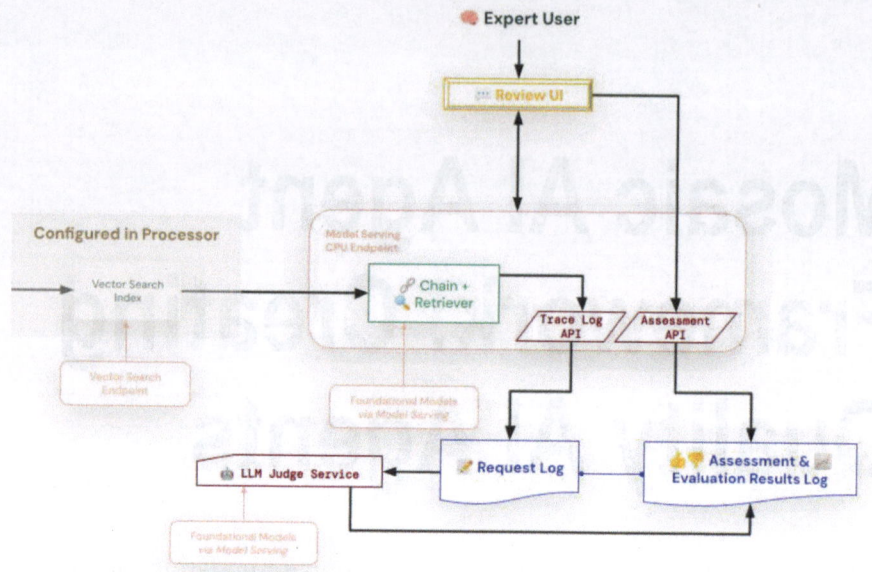

Figure 12-1. *AI Agent Framework workflow*

The AI Agent Framework supports unified logging, parameterizing, and tracing chains between the development and production and simplified UI deployment.

Let's do a walkthrough of a RAG application using the AI Agent Framework to see the differences.

Here are the main features that comes AI Agent Framework:

- Python dictionary or YAML file parametrization allows different configurations of a chain (prompt template, model and model config, etc.) for the selection of champion config

- MLflow logging on the model artifacts, and experiment tracking on evaluation metrics as well as deployment

Part 0: The Installations

The Mosaic AI Agent Framework is conveniently packaged as a Python library and can be installed, along with other libraries, using the command shown in Listing 12-1.

Listing 12-1. AI Agent Framework Installations

```
%pip install -U -qqqq databricks-agents mlflow langchain==0.2.1
langchain_core==0.2.5 langchain_community==0.2.4
```

The framework can then be referenced using the imports shown in Listing 12-2.

Listing 12-2. Imports for AI Agent Framework

```
import os
import mlflow
from databricks import agents

# Use the Unity Catalog model registry
mlflow.set_registry_uri('databricks-dip')
```

Part 1: LangChain Parametrization

Next, we need to provide our configuration so we can iterate different settings of our model. Adding mlflow.models.ModelConfig in MLflow allows settings to be configured easily using Python or YAML, as shown in Listing 12-3 and Listing 12-4.

Listing 12-3. Model Config with Python

```python
config_dict = {
    "prompt_template": "You are a hello world bot.  Respond
    with a reply to the user's question that is fun and
    interesting to the user.  User's question: {question}",
    "prompt_template_input_vars": ["question"],
    "model_serving_endpoint": "databricks-dbrx-instruct",
    "llm_parameters": {"temperature": 0.01, "max_tokens": 500},
}
model_config = mlflow.models.ModelConfig(development_
config=config_dict)
```

Listing 12-4. Model Config with YAML

```yaml
llm_parameters:
  max_tokens: 500
  temperature: 0.01
model_serving_endpoint: databricks-dbrx-instruct
prompt_template: 'You are a hello world bot.  Respond with a
reply to the user''s
  question that indicates your prompt template came from a YAML
  file.  Your response
  must use the word "YAML" somewhere.  User''s question:
  {question}'
prompt_template_input_vars:
- question
```

The config dictionary/file can then be used as shown in Listing 12-5.

Listing 12-5. Model Config Usage

```
config_file = "configs/rag_config.yaml"
model_config = mlflow.models.ModelConfig(development_
config=config_file)
model_config.get("prompt_template")
```

Model_config allows us to reuse the chain as it is without having to duplicate code. The YAML file can be used with any value that conforms the same chain setting, making it highly flexible.

Part 2: MLflow Evaluation

The Mosaic AI Agent Framework has extended the custom metric list and included a lot of new metrics. Similar to standard MLflow, these metrics will be computed automatically and logged on the Model Metrics tab in the experiment (see Figure 12-2).

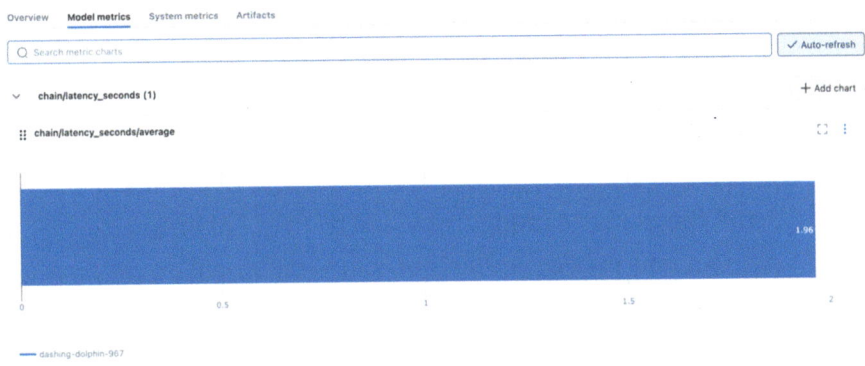

Figure 12-2. *Model metrics tab from MLflow Experiment*

In Chapter 10, we discussed the model_type parameter for evaluation, the Mosaic AI Agent Framework introduced a new model type called databricks-agent. See Listing 12-6.

Listing 12-6. databricks-agent model_type for Evaluation

```
eval_results = mlflow.evaluate(
    data=eval_set_df,
    model=logged_chain_info.model_uri,
    model_type="databricks-agent",
)
```

AI Agent framework also introduced new custom metrics as MLflow extension. The two types include LLM as a Judge and system statistics.

- **Aggregated metric values across the entire evaluation set:** Each row of the evaluation set is passed into an LLM, and a rating will be given on each output or given ground truth, or an expected retrieval context is provided (aka the document name). A full list of metrics is available in Databricks documentation:

  https://docs.databricks.com/en/generative-ai/ agent-evaluation/llm-judge-metrics.html

  With the numeric ratings provided by the LLM judge, we can now expect similar metrics in traditional machine learning like precision and recall.

- **Data about each question in the evaluation set:** Each row of the input will have an output rated, including but not limited to groundedness, correctness, relevancy to query and chunk, and chain statistics.

By default, all metrics will run during evaluation using `mlflow. evaluate()`, but they can be set in the YAML file to optionally run them. See Listing 12-7 for an example.

Listing 12-7. Metrics for Evaluation

```
builtin_assessments:
      - groundedness
      - correctness
      - relevance_to_query
      - chunk_relevance
```

Then run the evaluation harness; just run the input dataset with `mlflow.evaluate`. See Listing 12-8.

Listing 12-8. MLflow Evaluation

```
evaluation_results = mlflow.evaluate(
    data=eval_set_with_chain_outputs_df,
    model_type="databricks-agent",
)
```

The input dataset has the schema shown in Table 12-1.

Table 12-1. *Input Evaluation Dataset Schema*

Key	Type	Description
request_id	string	Unique identifier of this row in the evaluation set.
request	string	Input to the chain to evaluate, e.g., the user's question/query such as "What is RAG?"
expected_ retrieved_ context	array**	An array of objects containing the expected retrieved context for the request.
expected_ response	string	The ground truth (i.e., correct) answer to request.
response	string	The response generated by the chain being evaluated.
retrieved_ context	array**	The retrieval results generated by the retriever in the chain being evaluated. If multiple retrieval steps are in the chain, this should be the retrieval results that were put into the LLM's prompt.
trace	MLflow trace	MLflow Trace with the Chain's outputs.

*** The expected_retrieved_context and retrieved_context arrays expect
each array element to be a dictionary with the keys shown in Table 12-2.*

Table 12-2. *Array Structure from Table 1's Parameters*

Key	Type	Description
content	string	The contents of the retrieved context. Can be any string regardless of formatting e.g., HTML, Plain Text, Markdown, etc.
doc_uri	String	Unique identifier (URI) of the parent document where the chunk came from. e.g. dbfs:/Volumes/databricks_doc/spark.pdf

Based on this input schema, we can specify three different levels of parameters, and at each level, a subset of the metrics will be computed automatically. See Table 12-3.

Table 12-3. *Metrics for Different Levels of Datasets*

	Level A	Level B	Level C
Required data - Input Dataset			
Evaluation set: request	✓	✓	✓
Evaluation set: expected_response	X	✓	✓
Evaluation set: expected_retrieved_context	X	X	✓

(*continued*)

Table 12-3. (*continued*)

	Level A	Level B	Level C
Supported metrics - output metrics			
response/llm_judged/relevance_to_query_rating	✓	✓	✓
response/llm_judged/harmfulness_rating/average	✓	✓	✓
retrieval/llm_judged/chunk_relevance_precision/average	✓	✓	✓
response/llm_judged/groundedness_rating/average	✓	✓	✓
chain/request_token_count	✓	✓	✓
chain/response_token_count	✓	✓	✓
chain/total_token_count	✓	✓	✓
chain/input_token_count	✓	✓	✓
chain/output_token_count	✓	✓	✓
Customer-defined LLM judges	✓	✓	✓
response/llm_judged/correctness_rating/average	X	✓	✓
retrieval/ground_truth/document_recall/average	X	X	✓
retrieval/ground_truth/document_precision/average	X	X	✓

Part 3: Model Development

The AI Agent Framework provides an easy interface to deploy a chatbot as a review app for human feedback leveraging the chain that was just built. But before we dive into the app, as discussed in Chapter 11, we need to develop our model, in the case of a LangChain pipeline, and log the model as artifacts by using the mlflow.langchain.log_model() function. See Listing 12-9.

Listing 12-9. Model Logging with LangChain

```
with mlflow.start_run():
    # Log the chain code + config + parameters to the run
    logged_chain_info = mlflow.langchain.log_model(
        lc_model=chain_notebook_path,
        model_config=baseline_config,  # The configuration to
        test - this can also be a YAML file path rather than a
        Dict e.g., `chain_config_path`
        artifact_path="chain",
        input_example=input_example,
        example_no_conversion=True,
        extra_pip_requirements=[
            "databricks-agent"
        ],
    )
```

There is an important difference between the RAG artifacts compared to normal MLflow artifacts. By default, MLflow will "pickle" the LangChain objects, but as the complexity of the chain grows, often this process will fail. For more information, please refer to the FAQ section of the LangChain flavor of MLflow:

`https://mlflow.org/docs/latest/llms/langchain/index.html#faq`

So instead of "pickling" the chain, the AI Agent Framework opted to log the artifacts as code. That's why in the log_model() function, there is an lc_model parameter, which basically specifies the path of the chain notebook. While converting the notebook to a Python file is not needed, the notebook is required to support Python-only code. Otherwise the log will not be able to run successfully. An inspection of the Artifacts tab in the experiment reveals that a Python file is taking the place of the PKL file in the model, as shown in Figure 12-3.

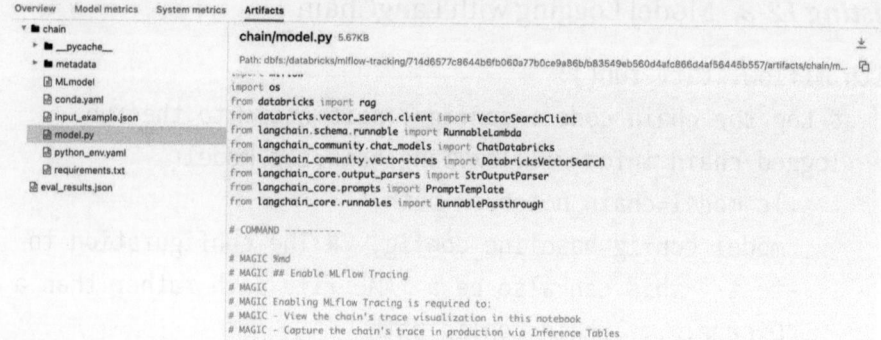

Figure 12-3. *LangChain pipeline logged as code*

Next, we will do a walkthrough of using this cutting-edge product to create a RAG application. Here's what a basic chain would look like:

1. Extract a user query from messages.

2. Retrieve relevant information using vector search.

3. Format the docs returned by the retriever into the prompt.

4. Generate a prompt for the language model.

5. Call the model endpoint with a prompt as input.

6. Parse the output into a string format.

Listing 12-10 shows the sample code of the chain from the previous logic.

Listing 12-10. Sample LangChain Pipeline

```
chain = (
    {
        "question": itemgetter("messages") |
        RunnableLambda(extract_user_query_string),
        "context": itemgetter("messages")
```

```
    | RunnableLambda(extract_user_query_string)
    | vector_search_as_retriever
    | RunnableLambda(format_context),
  }
  | prompt
  | model
  | StrOutputParser()
)
```

From the previous steps, the following components are configurable, and hence we can put them in a YAML file.

- Vector search endpoint/index

- Vector store-backed retriever (LangChain),

  `https://python.langchain.com/v0.1/docs/integrations/vectorstores/databricks_vector_search/`

- Chunk template

- Prompt template

- Model endpoint

Now consider one set of YAML configurations for our purpose. To increase readability, we can first create a spreadsheet of these settings and generate YAML files at a later stage; see Table 12-4.

Table 12-4. *Template for YAML File Configurations*

Vector search endpoint	Vector search index	LangChain vector store parameters	Chunk template	Prompt template	Model endpoint

Listing 12-11 shows an example of a YAML file, which should correspond to one entry of Table 12-4. We will call this file rag_chain_config.yaml.

Listing 12-11. rag_chain_config.yaml

```
chat_endpoint: databricks-dbrx-instruct
chat_endpoint_parameters:
  max_tokens: 500
  temperature: 0.01
chat_prompt_template: 'You are a trusted assistant that helps
answer questions based
  only on the provided information. If you do not know the
  answer to a question, you
  truthfully say you do not know.  Here is some context which
  might or might not help
  you answer: {context}.  Answer directly, do not repeat the
  question, do not start
  with something like: the answer to the question, do not add
  AI in front of your
  answer, do not say: here is the answer, do not mention the
  context or the question.
  Based on this context, answer this question: {question}'
chat_prompt_template_variables:
- context
- question
chunk_template: '`{chunk_text}`

  '

vector_search_endpoint_name: test
vector_search_index: unitygo.rag.gold_volume_databricks_
documentation_chunked_index
vector_search_parameters:
```

```
  k: 3
vector_search_schema:
  chunk_text: chunked_text
  document_source: doc_uri
  primary_key: chunk_id
```

By loading the file rag_chain_config.yaml, we can derive the code shown in Listing 12-12 for our chain.

Listing 12-12. Setting Up a Chat Endpoint Using a Configuration File

```
############
# Get the configuration YAML
############
model_config = mlflow.models.ModelConfig(development_
config="rag_chain_config.yaml")

############
# Connect to the Vector Search Index
############
vs_client = VectorSearchClient(disable_notice=True)
vs_index = vs_client.get_index(
    endpoint_name=model_config.get("vector_search_
    endpoint_name"),
    index_name=model_config.get("vector_search_index"),
)
vector_search_schema = model_config.get("vector_search_schema")

############
# Turn the Vector Search index into a LangChain retriever
############
vector_search_as_retriever = DatabricksVectorSearch(
    vs_index,
```

```python
        text_column=vector_search_schema.get("chunk_text"),
        columns=[
            vector_search_schema.get("primary_key"),
            vector_search_schema.get("chunk_text"),
            vector_search_schema.get("document_source"),
        ],
).as_retriever(search_kwargs=model_config.get("vector_search_
parameters"))

############
# Required to:
# 1. Enable the Review App to properly display retrieved chunks
# 2. Enable evaluation suite to measure the retriever
############
rag.set_vector_search_schema(
    primary_key=vector_search_schema.get("primary_key"),
    text_column=vector_search_schema.get("chunk_text"),
    doc_uri=vector_search_schema.get(
        "document_source"
    ),  # Review App uses `doc_uri` to display chunks from the
    same document in a single view
)

############
# Method to format the docs returned by the retriever into
the prompt
############
def format_context(docs):
    chunk_template = model_config.get("chunk_template")
    chunk_contents = [chunk_template.format(chunk_text=d.page_
    content) for d in docs]
    return "".join(chunk_contents)
```

```
############
# Prompt Template for generation
############
prompt = PromptTemplate(
    template=model_config.get("chat_prompt_template"),
    input_variables=model_config.get("chat_prompt_template_
    variables"),
)

############
# FM for generation
############
model = ChatDatabricks(
    endpoint=model_config.get("chat_endpoint"),
    extra_params=model_config.get("chat_endpoint_parameters"),
)
```

Before we log the chain into a model, we can optionally test it by invoking the model. Databricks also provides an interface of the LangChain pipeline, so any troubleshooting can be done within the platform. Listing 12-13 is for testing purposes and should not be in production.

Listing 12-13. Testing the Chat Endpoint

```
model_input_sample = {
    "messages": [
        {
            "role": "user",
            "content": "What is Spark?",
        }
    ]
}
chain.invoke(model_input_sample)
```

After we invoke the code, MLflow automatically generates a tracing interface for the LangChain flavor via `mlflow.langchain.autolog`. You can easily see the prompts, which models and retrievers were used, which documents were retrieved to augment the response, how long things took, and the final output. Figure 12-4 demonstrates this view that will be useful for troubleshooting.

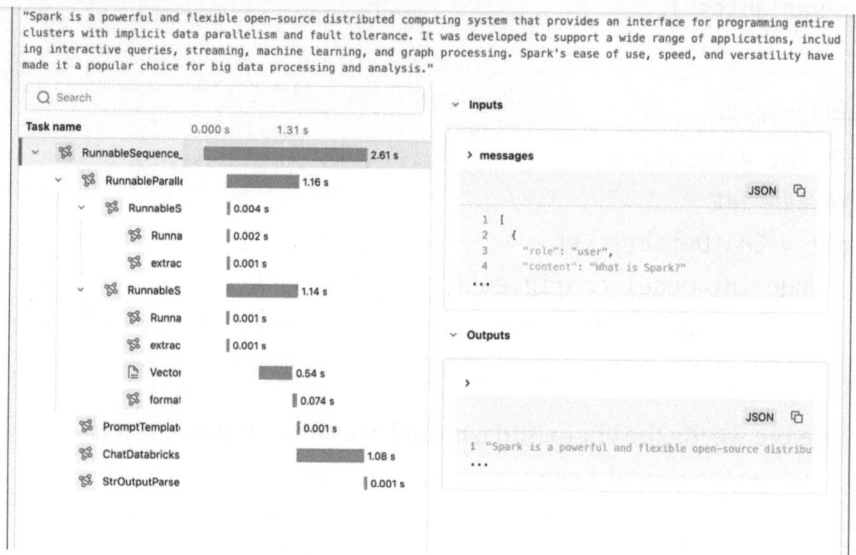

Figure 12-4. *LangChain trace interface*

When we combine everything together in a notebook, this will become the file `chain/model.py` in the logged artifacts.

Part 4: Deployment

The AI agent framework comes with an easy one-line chatbot deployment that can collect human feedback, greatly reducing the time required to develop an interface for humans to interact with.

First, we need to register our model. See Listing 12-14.

Listing 12-14. LangChain Model Registration

```
# Unity Catalog location
uc_model_fqn = f"{uc_catalog}.{uc_schema}.{model_name}"

# Register the model to the Unity Catalog
uc_registered_model_info = mlflow.register_model(model_
uri=logged_chain_info.model_uri, name=uc_model_fqn )
```

Next, the one-line deployment command looks like Listing 12-15.

Listing 12-15. One-Line Deployment Command

```
deployment_info = agents.deploy(model_name=UC_MODEL_NAME,
model_version=uc_registered_model_info.version)
```

Once we run the deployment command, it will take some time for the magic to work behind the scenes. We can check the status of the deployment using the command in Listing 12-16. At the end of the wait, the review app URL will be shown.

Listing 12-16. Getting Deployment Status

```
# Wait for the Review App to be ready
print("\nWaiting for endpoint to deploy.", end="")
while w.serving_endpoints.get(deployment_info.endpoint_name).
state.ready == EndpointStateReady.NOT_READY or w.serving_
endpoints.get(deployment_info.endpoint_name).state.config_
update == EndpointStateConfigUpdate.IN_PROGRESS:
    print(".", end="")
    time.sleep(30)

print(f"\n\nReview App: {deployment_info.review_app_url}")
```

To retrieve the deployed endpoints in general, we can use the command in Listing 12-17.

Listing 12-17. Retrieving Review App URLs

```
active_deployments = agents.list_deployments()

active_deployment = next((item for item in active_deployments
if item.model_name == UC_MODEL_NAME), None)

print(f"Review App URL: {active_deployment.review_app_url}")
```

Finally, following the URL, we can have a review app with a single line of code, but before that, we need to share the model with users so they can use the chatbot aka the Review App. See Listing 12-18 and Figure 12-5.

Listing 12-18. Sharing Permission to Users to Query the Model

```
user_list = ["user@databricks.com"]

# Set the permissions.  If successful, there will be no
return value.
agents.set_permissions(model_name=UC_MODEL_NAME, users=user_
list, permission_level=agents.PermissionLevel.CAN_QUERY)
```

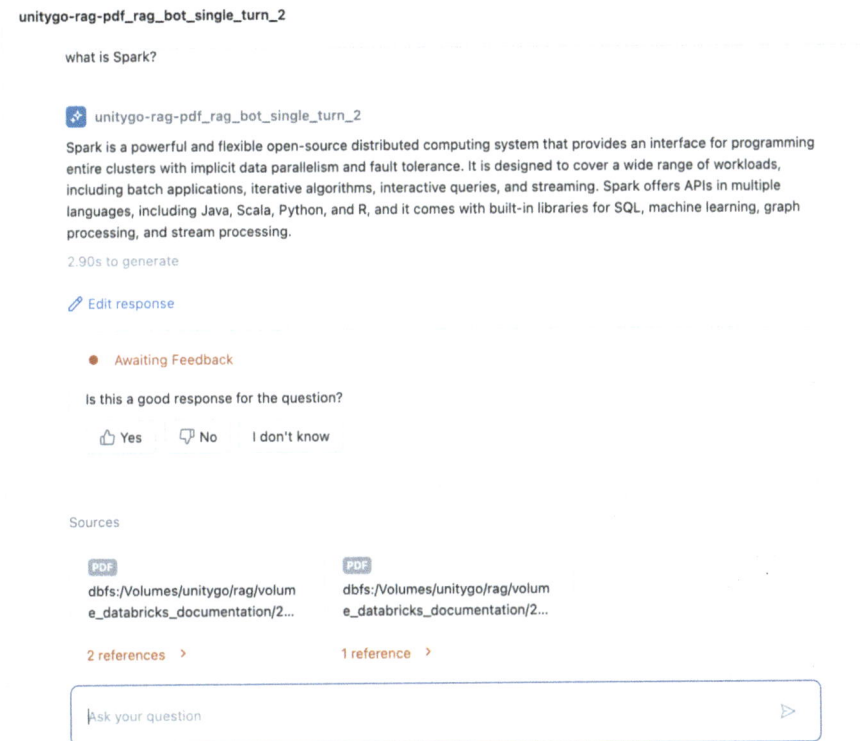

Figure 12-5. *Review app user interface*

Evaluation Example

To deliver high-quality RAG applications, Databricks recommends following an evaluation-driven approach to development. To start your development process, we suggest starting with 5 to 10 examples of questions that your users will expect your RAG application to answer correctly. Over the course of your development process, you will expand this evaluation set. The input schema can be found in the earlier "Part 2: MLflow Evaluation" section. Listing 12-19 is one example.

Listing 12-19. Setting the Evaluation Dataset

```
eval_set = [
    {
        "request_id": "97496aa16cefcde44bc4ad97f00b9f85",
        "request": "Did GPT-4's opinion response rate increase
        or decrease by June 2023?",
        "expected_response": "Decrease", # Optional
        "expected_retrieved_context": [ # Optional
            {
                "doc_uri": "dbfs:/Volumes/unitygo/rag/volume_
                databricks_documentation/2307.09009.pdf",
            }
        ],
    }
]
```

After we are able to evaluate the best chain configuration, then we can deploy the chain as an app for human review. Once again, `mlflow.evaluate()` will be used to test against the previous logged model with the eval set (see Listing 12-20).

Listing 12-20. Running Evaluation of a Model Based on the eval Dataset

```
    eval_results = mlflow.evaluate(
        data=eval_set_df,
        model=logged_chain_info.model_uri,
        model_type="databricks-agent",
    )
```

`eval_results` will contain the LLM judged metrics, and we can simply display them or save them into a table (see Listing 12-21).

Listing 12-21. Visualizing the Evaluation Results

```
display(eval_results.tables['eval_results'].
drop(columns=["trace"]))
```

Alternatively, they can also be found in the MLflow UI, as shown in Figure 12-6.

Figure 12-6. *Evaluation metrics logged into MLflow*

Deployment is only our first step; we need to collect critical feedback via evaluation and human feedback. Then there are quality knobs that we need to tune. The details of tuning the app is beyond the scope of this book and can be found on the Databricks' GenAI cookbook website:

```
https://ai-cookbook.io/nbs/3-deep-dive.html
```

Conclusion

Mosaic AI Agent Framework's product philosophy is underpinned by the following principles:

- **Quality through metrics:** Objective metrics are the cornerstone of quality assessment. Metrics provide indicators for evaluating the RAG application's quality and cost/latency performance and for identifying areas for improvement.

- **Comprehensive "always-on" logging:** Metrics work best if they can be computed for any invocation of the RAG app. Therefore, every invocation of the app, both in development and in production, must be logged. The log must capture all inputs and outputs, as well as the detailed steps that transform inputs into outputs.

- **Human feedback as the benchmark:** Collecting human feedback is costly, but its value as a quality measure is unmatched. RAG Studio is designed to make the collection of human feedback as efficient as possible.

- **LLM judges scale feedback:** Utilizing RAG LLM judges in tandem with human feedback accelerates the development loop, allowing for quicker development cycles without subsequently scaling the number of human evaluators. However, RAG LLM judges are not a substitute but, rather, an augment to human feedback.

- **Rapid iteration:** The cycle of creating and testing new versions of a RAG application must be quick.

- **Effortless version management:** Tracking and managing versions must be seamless, reducing cognitive load and letting developers concentrate on enhancing the application rather than on administrative tasks.

- **Development and production are unified:** The tools, schemas, and processes used in development should be consistent with those in production environments, ensuring a consistent workflow for quality improvement in development to deployment *with the same code base.*

Beyond LangChain

While this chapter did a walkthrough using LangChain and leveraged the interface mlflow.langchain.log_model(), the AI Agent Framework is not limited to LangChain. The pyFunc interface is available for any Python model. However, there are some customizations needed.

In other words, if you are using pyFunc, Databricks recommends using type hints to annotate the predict() function with input and output data classes that are subclasses of classes defined in mlflow.models. rag_signatures (see https://github.com/mlflow/mlflow/blob/master/mlflow/models/rag_signatures.py).

You can construct an output object from the data class inside predict() to ensure the format is followed. The returned object must be transformed into a dictionary representation to ensure it can be serialized.

The LangChain implementation in the MLflow source code provides an example of how to create such a customization:

https://github.com/mlflow/mlflow/blob/master/mlflow/langchain/output_parsers.py

You can find a custom pyFunc model in the MLflow documentation:

https://mlflow.org/docs/latest/traditional-ml/serving-multiple-models-with-pyfunc/notebooks/MME_Tutorial.html#2---Create-an-MME-Custom-PyFunc-Model

DBRX: Creating an LLM from Scratch Using Databricks

In this chapter, we will discusse a model that Databricks trained using Databricks, which is called DBRX. DBRX is a state-of-the-art large language model (LLM) trained from scratch on the Databricks and MosaicML platforms. At the time of model release, it outperformed established open-source models on language understanding (MMLU), programming (HumanEval), and math (GSM8K), as shown in Figure 13-1.

© The Editor(s) (if applicable) and The Author(s),
under exclusive license to APress Media, LLC, part of Springer Nature 2024
N. Gupta and J. Yip, *Databricks Data Intelligence Platform*,
https://doi.org/10.1007/979-8-8688-0444-1_13

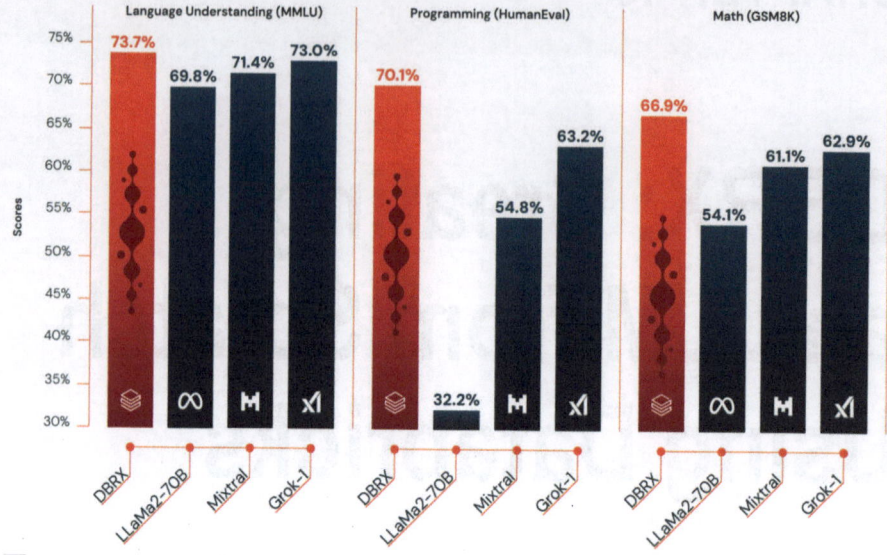

Figure 13-1. *DBRX performance versus established open-source models*

What Is DBRX?

While the world has moved on to better models by now, like Meta's LLaMa3-70B, there are many more objectives that DBRX is trying to accomplish.

- Allowing enterprises to own their model

  Databricks is a platform that you can use to do everything from end to end within your own network, or in the case of serverless there is a private link to the VPN. All the models are deployed and fine-tuned over internal data. The data and model stays within the customers' own Databricks environment. That makes the experience not only secure but also seamless. Now customers can also leverage DBRX for their tuning needs.

- Moving to production quickly

  With Databricks Model Serving and the serverless architecture, customers can easily serve their model on an API endpoint. To demonstrate how quickly one can do it end to end, Databricks designed and trained a model in three months and immediately made it available to all customers via the Foundation Model API. In other situations, developers can use Model Serving. Details can be found in the previous chapter.

- Bringing down the cost

  In the old days, Intel co-founder Gordon Moore published his famous Moore's law, which observed that the number of transistors on an integrated circuit would double every two years with minimal rise in cost. The founders of MosaicML are now predicting that the cost of pre-training an LLM will come down by a factor of four every year. For example, when looking at the cost of pre-training a Stable Diffusion model in late January 2023, it was $160,000. That cost was reduced by 75% by 2024. In the case of DBRX, it cost $10 million to train in 40 days and 3 months in total including R&D (see Figure 13-2).

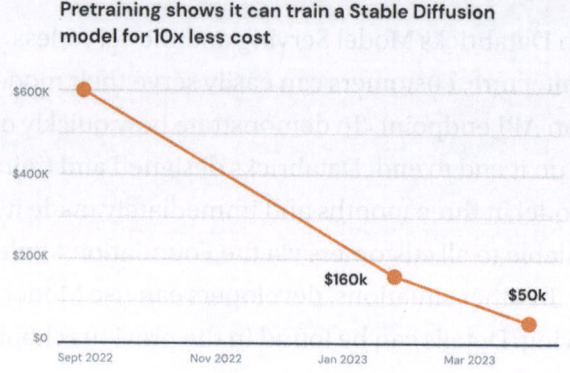

Figure 13-2. *The cost of model pre-training over time*

- Open-sourcing the tooling

   First, we must acknowledge the work that the open-source community has done to accelerate AI research by open-sourcing their model and paper. Take Meta's Llama as an example. They have started to open-source its code and weights since Llama 2. However, a few vendors went to great lengths to discuss their toolings and how they leverage open-source frameworks to train their models. Later in this chapter, we will discuss these toolings.

- Open-sourcing the model

   As discussed, many vendors open sourced the implementation source code of their LLM. Databricks didn't shy away from doing the same thing. The source code is also open-sourced on GitHub.

- Demonstrating end-to-end capabilities of Databricks for customers

  While not all customers need to pre-train an an LLM, but not all platforms are made available to customers to pre-train a LLM from scratch. Utilizing all the toolings available to customers, Databricks successfully trained DBRX from scratch.

- Allowing the community to fine-tune the model

  Like other open-source models, the weights of DBRX are available for download. For those who are not familiar with neural networks, the weights are the connectivity between the nodes in neural networks, and the weights are those learned in the training, so open-sourcing the weights means people can further fine-tune the network from the pre-trained weights, without having to do it from scratch.

The DBRX Benchmarks

First, we must understand that the world is working around the clock like never before to release the next best LLM and increasingly headed toward multimodal, in other words, support for text, audio. and photo. DBRX demonstrates that the Databricks infrastructure can train a best-in-class model at the time of release, which can compete against all open-source models. So naturally, there is another best-in-class open-source model by now. So take a look at a snapshot in time of how DBRX stands in the benchmarking race (see Table 13-1). Another reason we need to look at these benchmarks is that evaluation is a big part of building an LLM, as opposed to traditional machine learning, or deep learning, where there are standard metrics for evaluations.

Table 13-1. *Quality of DBRX Instruct and Leading Open Models*

Model	DBRX Instruct	Mixtral Instruct	Mixtral Base	LLaMA2- 70B Chat	LLaMA2- 70B Base	Grok-1
Open LLM Leaderboard (Avg of next 6 rows)	74.5%	72.7%	68.4%	62.4%	67.9%	—
ARC-challenge 25-shot	68.9%	70.1%	66.4%	64.6%	67.3%	—
HellaSwag 10-shot	89.0%	87.6%	86.5%	85.9%	87.3%	—
MMLU 5-shot	73.7%	71.4%	71.9%	63.9%	69.8%	73.0%
Truthful QA 0-shot	66.9%	65.0%	46.8%	52.8%	44.9%	—
WinoGrande 5-shot	81.8%	81.1%	81.7%	80.5%	83.7%	—
GSM8k CoT 5-shot maj@1	66.9%	61.1%	57.6%	26.7%	54.1%	62.9% (8-shot)
Gauntlet v0.3 (Avg of 30+ diverse tasks)	66.8%	60.7%	56.8%	52.8%	56.4%	—

As we can see, DBRX excels in many areas compared to other popular models. Considering that Grok 1 also uses the Mixture of Expert (MoE) architecture, Databricks excels in mastering the MoE architecture.

When DBRX is trained, Open AI provides GPT 3.5 for free. That's why we compared DBRX to some leading free models in its timing.

From Figure 13-3, we can see that DBRX is especially good at math and programming. It can also compete with other models in areas like truthfulness, scientific concepts, general knowledge, and common sense, demonstrating that DBRX is indeed a very powerful model that Databricks was able to train.

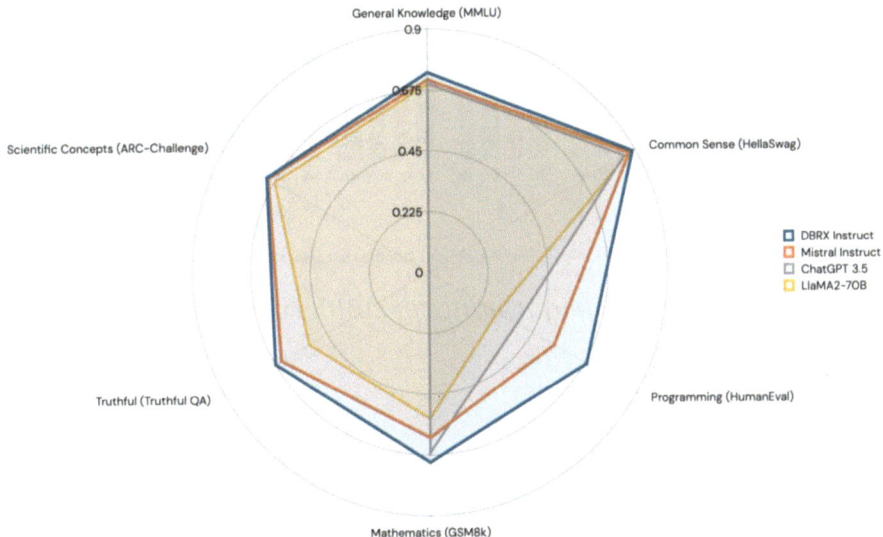

Figure 13-3. *DRBX benchmark against prominent models*

It goes without saying that the larger the model is, the slower it is able to operate. So there's always an argument about large models versus small models. DBRX, which has 132 billion total parameters, has achieved both at the same time. And thanks to the MoE architecture, only 36 billion parameters are active at the same time. Figure 13-4 illustrates the inference performance of DBRX compared to other MoE models in a similar parameter count dense model. Please note that Mixtral's eight experts only have 7 billion active parameters versus 36 billion in DBRX.

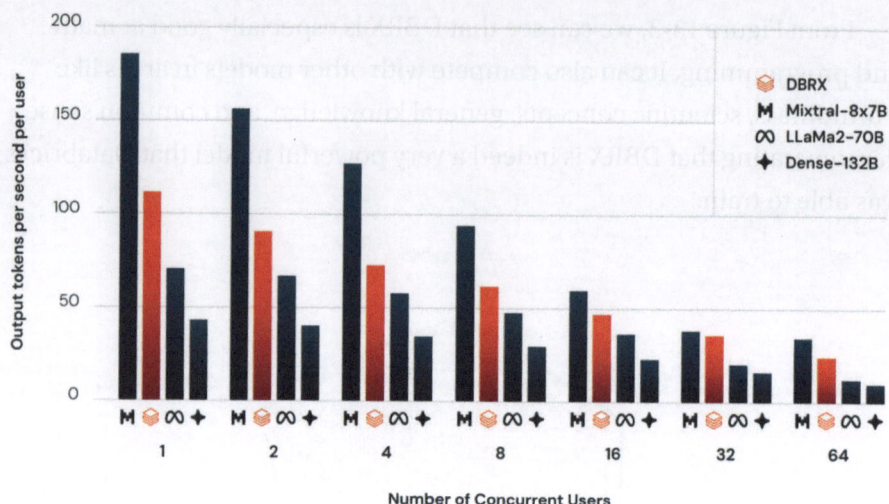

Figure 13-4. *Inference performance of DBRX compared to other models*

DBRX Architecture

According to Databricks, **DBRX is a transformer-based decoder-only LLM that was trained using next-token prediction**. If you are not a research scientist on natural language processing (NLP), this might sound confusing. Although this is not a textbook about NLP, we will introduce some concepts so you can follow along with future sections.

Jonathan Frankle, chief scientist of MosaicML, follows the motto "Attention is all you need." The following website simply yet powerfully explains the importance of the Attention mechanism: www. isattentionallyouneed.com.

Without diving too deep into the Attention mechanism, let's rewind back in time a little bit. If you have learned about deep learning, you might have heard about recurrent neural networks (RNNs). From Figure 13-5, we understand that we have an input sequence of words and the goal of the neural network is to learn how to process and predict patterns in data that comes in a series, such as text or speech.

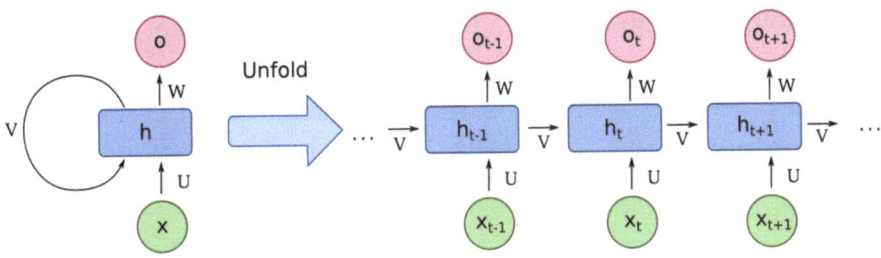

Figure 13-5. *RNNs(source:* `https://en.wikipedia.org/wiki/`
`Recurrent_neural_network)`

There have been many different improvements since the introduction
of RNN; one is the Transformer model, which was introduced to replace
RNNs using the Attention mechanism, from the famous paper "Attention
Is All You Need" by Vaswani et al. of Google. The paper can be found here:
`https://arxiv.org/abs/1706.03762`. The architecture of the Transformer
model can also be found in the paper. But for simplicity's sake, the paper
introduced using an encoder and decoder network with attention. An
attention function can be described as mapping a query and a set of key-
value pairs to an output, where the query, keys, values, and output are all
vectors. (See Figure 13-6.)

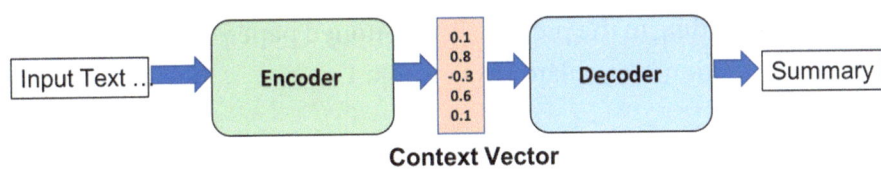

Figure 13-6. *Encoder-decoder network*

To improve efficiency, researchers explored using a decoder-only
network. The transformer-based decoder-only network generates the
next token based on the previous input autoregressively. Autoregressive
is a statistical term; for details, please refer to Wikipedia: `https://`
`en.wikipedia.org/wiki/Autoregressive_model`.

Finally, while Transformer and Attention are two different things (the former is a neural network architecture, the latter is a technique used to guide the processing of input data within that architecture), ever since the paper "Attention Is All You Need," they have been inseparable. They are what GPT used to storm the world.

Shortcomings of the Transformer Architecture

We often hear LLMs are expensive to train and run as well as has hallucinations. Welcome to the heart of the problem, which lies in the Transformer architecture. We will discuss a few of these issues so you can understand why the industry is trying to change the architecture.

Will there be a brand new architecture that replaces Attention on Transformers in the future? We certainly hope so. But for the time being, we know that the LLMs in the time of DBRX are largely relying on Transformers, and until we see a new industrial wave that makes it irrelevant, we still need to have some basic understanding of it. The following problems appear in the news most often:

1. Expensive computation

    According to the previously mentioned paper, Attention is calculated in Formula 1:

    $$Attention(Q, K, V) = softmax\left(\frac{QK^T}{\sqrt{d_k}}\right)V$$

    Formula 1: Attention equation, the foundation of all Large Language Models

    In Formula 1, the query (Q), key (K), and value (V) are generated from the input sequence to obtain the value A, which is the weight of the attention.

In plain English, the self-attention mechanism in Transformers has a computational complexity of $O(n^2)$ because it requires comparing every element in the input sequence to every other element, resulting in a quadratic increase in computations as the sequence length grows. Imagine doing a quadratic computation on the entire text of Wikipedia. This is only part of the inputs for LLMs.

2. Slow inference time

    The term *deep* in deep neural network (DNNs) refers to the number of layers in a complex neural network. Coupled with the activation function ReLu or Sigmod, as well as matrix multiplications in a long sequence of input text, we can imagine the work required to get meaningful outputs.

3. Limited context length, aka input length

    According to the "Attention" paper, "Since our model contains no recurrence and no convolution, in order for the model to make use of the order of the sequence, we must inject some information about the relative or absolute position of the tokens in the sequence." In other words, the input must be chunked in the training process. And the shorter the context window, the more overlapping will be required to avoid losing information while training. That also limits its ability to learn new data without re-training.

4.   Hallucinations

We know hallucinations have been a big problem
ever since ChatGPT went live. There is now
mathematical proof of the limitations of the
Transformer architecture. For details, please refer
to the paper "On the limitations of the Transformer
architecture" by Peng et al. (https://arxiv.org/
html/2402.08164v1).

Mixture of Experts

Traditional neural networks consist of many nodes fully connected to
each other. Deep learning or deep neural networks contain many layers.
For example, Microsoft's famous computer vision model ResNet-50,
which won the ImageNet competition in 2015, is a 50-layer convolutional
neural network. Because the nodes are densely connected (every node is
connected to every other node in every layer), they are also called *dense*
models, as illustrated in Figure 13-7.

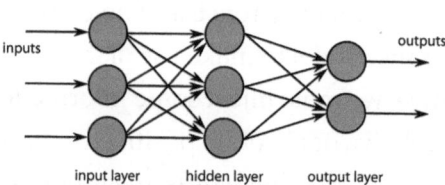

Figure 13-7. *Neural network architecture*

It has been discussed many times in this chapter that LLMs are very
large by nature. That's why researchers came up with a new architecture to
try to reduce the size of the model without losing performance by dividing
one big model into smaller models, which are called *experts*, but in fact
they are just smaller models with the same architecture. Traditional Moe
models (shown in Figure 13-8) divide a very large model into a subset of

large models; a routing strategy is employed to distribute the training for these smaller but still large experts. Because the models are still not able to fit into one machine, an inefficient routing strategy will lead to dropping tokens from the computation or wasting computation resulting in over/under trained experts.

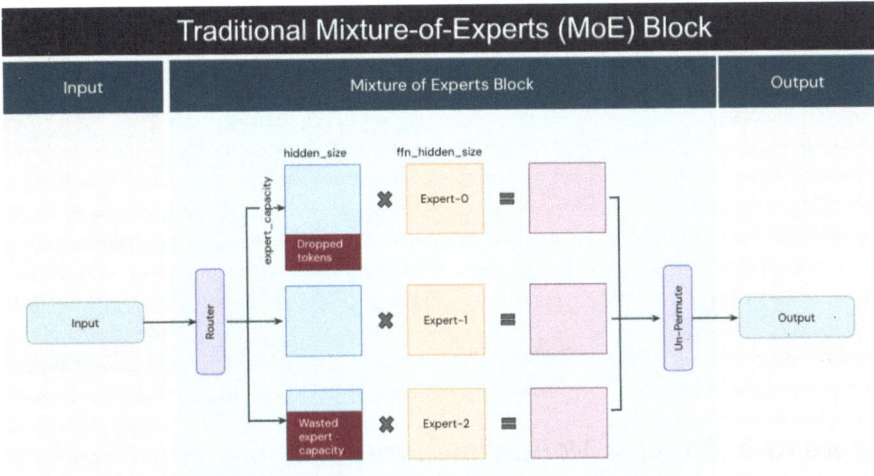

Figure 13-8. *Traditional MoE architecture*

MegaBlocks: Efficient Sparse Training with Mixture-of-Experts

Because of the inefficient routing strategy in traditional MoE training, Trevor Gale et al. proposed a new method called MegaBlocks. The idea is to group these experts and use a new efficient routing strategy to re-assign them at the hardware level instead of trying to train the experts separately. The original paper can be found at https://arxiv.org/abs/2211.15841. Experiments show that the architecture will never drop any tokens; hence, it's called Dropless blocks.

MegaBlocks, or Dropless MoE blocks (see Figure 13-9) is now an official open-source Databricks project. The GitHub repo can be found at `https://github.com/databricks/megablocks`.

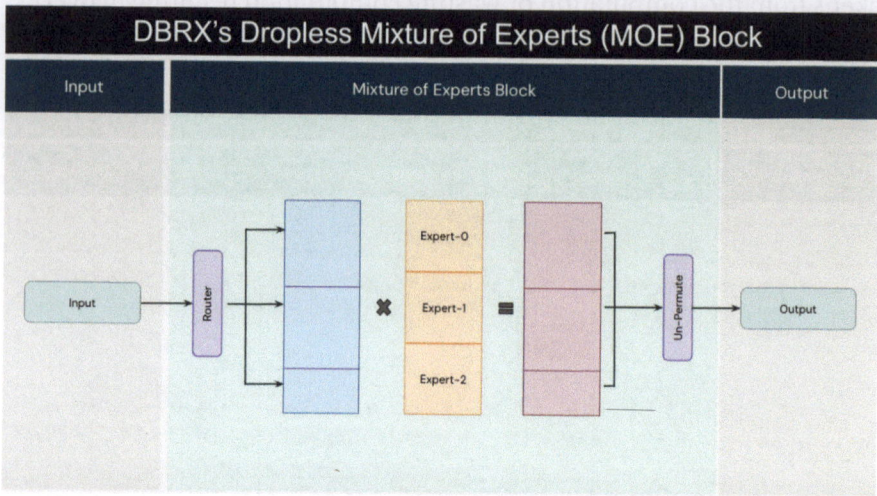

Figure 13-9. *Dropless MoE architecture*

Fine-Grained MoE

Using the dropless blocks, it enabled DBRX to divide the experts into even smaller models, known as fine-grained MoE. With smaller models, DBRX can use less active parameters at once and can still achieve good performance. In reality, DBRX has 16 experts and chooses 4, while Mixtral and Grok-1 have eight experts and choose two. This provides 65x more possible combinations of experts, and that's how DBRX can improve model quality.

The MosaicML Stack

The core model of DBRX packed a lot of innovations from fine-grained MoE to the continuous support of the MegaBlocks project. With every successful project, there is a backbone to support it. All the source code of this backbone can be found in the MosaicML GitHub repo: `https://github.com/mosaicml/`.

- **Composer:** Built on top of PyTorch, the Composer library makes it easier to implement distributed training workflows on large-scale clusters. Its tight integration with PyTorch means developers can easily abstract the complexity of distributed deep learning easily using this library. One can train models of any size including:

  - Large Language Models

  - Diffusion models

  - Embedding models (e.g., BERT)

  - Transformer-based models

  - Convolutional neural networks (CNNs)

- **StreamingDataSet:** If you have trained a model in PyTorch, you'd have used the IterableDataset. StreamingDataSet is the replacement of this library in a distributed form. Making the transition to distributed training seamless.

- **LLM Foundry:** Similar to Databricks' ML Ops stack (`https://github.com/databricks/mlops-stacks`), LLM Foundry has a focus on LLMs. The features of MosaicML's LLM Foundry including the following:

  - Focuses on scaling, optimizing, and managing the entire LLM life cycle, from training to deployment

  - Emphasizes automation, reproducibility, and collaboration for LLM development

  - Targets use cases like natural language processing, text generation, and multimodal processing

- **Evaluation Gauntlet:** Part of the LLM Foundry (`https://github.com/mosaicml/llm-foundry/blob/main/scripts/eval/local_data/EVAL_GAUNTLET.md`), the Evaluation Gauntlet is Databricks' new evaluation suite. The goal of this suite is to allow reporting benchmarks in different categories separately instead of being in one metric. The Eval Gauntlet encompasses 35 different benchmarks collected from a variety of sources, and organized into six broad categories of competency that good foundation models should have.

Distributed GPU Training

Composer wouldn't be successful without the help of the community. It has integration with various distributed training libraries including the following:

- Pytorch DistributedDataParallel (DDP)

- Pytorch Fully Sharded Data Parallel (FSDP)

- Microsoft DeepSpeed

Particularly, DBRX was trained using Pytorch FSDP. All of the previous libraries are already part of the Composer framework.

For details about how these libraries work, please refer to the documentation:

- `https://pytorch.org/tutorials/distributed/home.html`

- `https://deepspeed.ai/`

Model Serving

Nowadays there is no standard for developing LLM, and their interface is different from each other. That's why there are different projects to ensure the interoperability of the models so platforms like Databricks can integrate the Model Serving capabilities. Databricks works closely with these two libraries and provides support on DBRX. You may notice that not every model is supported in these projects, which really goes back to whether the creator of the LLM extended the support or if there is tremendous interest in the community to extend the support, in the case of open source projects.

- NVIDIA TensorRT-LLM (`https://github.com/NVIDIA/TensorRT-LLM`)

  Developed by NVIDIA, TensorRT-LLM allows you to use production-grade servers and build a Python API on top for model inference. This is powering the Databricks Model Serving API.

- vLLM (`https://github.com/vllm-project/vllm`)

  No GPU or simply not enough GPU powers? vLLM aims to allow everyone access to LLMs. With its quantization support, you can even run DBRX on a CPU.

Using DBRX on Databricks

Databricks has curated some models on its platform where users don't need to host the infrastructure by themselves. These are called the Foundation Model API. A full list of models can be found at https://www. databricks.com/product/machine-learning/large-language-models-oss-guidance.

Without a doubt, DBRX is one of the hosted models, and we can use it with the code shown in Listing 13-1.

Listing 13-1. Running DBRX on Local Using the Databricks Foundation Model API

```
import json
import os
from openai import OpenAI

# -------------------------------------------------------------
# Configurations
# -------------------------------------------------------------

# API Key
my_api_key = os.environ['DATABRICKS_TOKEN']

# Databricks Serving Endpoint
my_base_url = os.environ['DATABRICKS_SERVING_ENDPOINT']

# Configure your system prompt
my_system_prompt = "You are a chef of a 3-star Michelin
restaurant and have the credibility of some of the best chefs
such as Anthony Bourdain.  Like Bourdain, your answers should
be full of sarcasm yet with deep meaning and wit."
```

```
# Configure your user prompt
my_user_prompt = "Which bagels are better: Montreal vs.
New York?"

# Next we will configure the OpenAI SDK with Databricks Access
Token and our base URL
client = OpenAI(
    api_key = my_api_key,
    base_url = my_base_url
)

# Now let's invoke inference against the PAYGO (Pay Per Token)
endpoint
response = client.chat.completions.create(
    model="databricks-dbrx-instruct",
    messages=[
      {
        "role": "system",
        "content": my_system_prompt
      },
      {
        "role": "user",
        "content": my_user_prompt
      }
    ],
)

json_output = json.dumps(json.loads(response.json()), indent=4)
print(json_output)
```

With the Foundation Model API, we can try the model quickly. As mentioned, every model has its own input and output interface for various reasons. One most popular interface is the OpenAI API. Most likely one would already have an OpenAI API code in the test environment or even production environment. To minimize the code change required, Databricks supports the OpenAI Python SDK interface. All you need to do is to replace your OpenAI API key with a Databricks token and the base URL with a Foundation Model API endpoint!

Conclusion

In this chapter, we have introduced DBRX, Databricks' LLM, which was created in about three months with $10 million. According to standard evaluation suites as well as Databricks' new Evaluation Gauntlet, we learned that DBRX exceeds the performance of all the major open- source models of its time.

Databricks has demonstrated leadership in providing transparency on the process of building a fast and efficient LLM. Firstly, by open-sourcing the model code on GitHub as well as discussing the fine-grained Mixture-of-Expert (MoE) architecture publicly. Secondly, by taking ownership of the MegaBlocks project and keeping it open source. Along with the MosaicML tooling, the entire stack that's used for training is available to everyone. MosaicML also created a wrapper around some very popular frameworks in Pytorch and Microsoft DeepSpeed, ensuring compatibility of the code that others have developed when migrating to MosaicML.

To enhance accessibility from production workload to casual usage, Databricks has provided access via Foundation Model API and support for vLLM at launch time. The community has also initiated various quantization techniques to provide further access in different local environments.

Finally, DBRX's fast inference speed will allow enterprises to enhance the model using RAG and fine-tune it with internal proprietary data. Along with the entire Databricks stack, DRBX is enterprise-ready at launch.

CHAPTER 14

The Databricks Data Intelligence Platform

In the previous chapters, we learned about the Databricks lakehouse, which essentially means storing all your data in open storage in an open format with Unity Catalog providing a single governance layer and Databricks providing features to enable all use cases such as data engineering, data science, streaming, and warehousing. With the advent and popularity of GenAI and LLMs since 2023, Databricks has integrated them into its platform. The Databricks data intelligence platform (see Figure 14-1) combines the lakehouse platform and AI/LLMs to add the "data intelligence" engine that understands the uniqueness of your data and uses that understanding across everything in the platform.

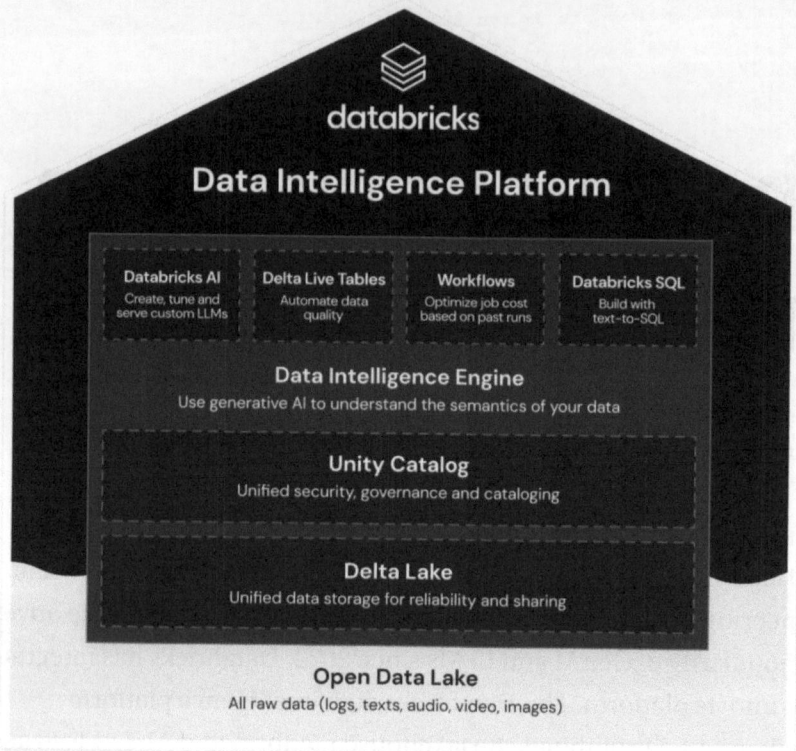

Figure 14-1. Databricks data intelligence platform

Thus, the Databricks data intelligence platform is a groundbreaking effort combining the power of AI and the lakehouse platform. Imagine a team of experts on the platform guiding you through every step of your data needs. There will be little room to go wrong, and you can get optimized speed and performance. This is the promise of the Data Intelligence Engine, which sits underneath the lakehouse platform.

In this chapter, we will examine key features of the Databricks data intelligence platform. We will begin by defining the data intelligence platform and how it evolved. Then, we will examine some of the key features, such as Databricks IQ, AI/BI Genie, etc.

Databricks IQ

Databricks IQ is at the heart of Databricks' data intelligence platform. Many people are using chatbots or co-pilots to assist with their work. However, most of these are trained on open data sources and have little context around your data.

To put this in perspective, the ideal co-pilot for organizations to be productive while working in Databricks or any other developer tools will need to meet the following requirements:

- Be within a secured environment so internal information is not being used to train the model that will ultimately be exposed to the general public

- Automatically learn about internal information and stay within the organization

- Understand human language and be able to translate to a programming language

- Lightning-fast performance so problems can be solved in seconds and not minutes

With that in mind, Databricks developed Databricks IQ, which is powered by Mosaic AI Model Serving. Let's look at what areas Databricks IQ can help us with.

- **Databricks Assistant:** This aims to help you understand how to write a query, troubleshoot, and find performance bottlenecks in the system, all powered by natural language understanding.

- **AI-powered governance:** This helps in a variety of tasks including generating comments for the metadata and providing lineage, automatic PII detection and masking, and AI security filtering with the eventual aim to learn how to give advice based on the Databricks AI Security Framework.

- **Search and discovery:** The Databricks platform can now return personalized results when you search for something on the platform. These results are enhanced by relying on recent and most-viewed content. Further, the search is more context-aware in the sense it provides results based on which part of the platform you initiated the search in.

- **AI/BI Giene:** With the ever-growing data in every organization, it will be impossible for an LLM to keep up with the knowledge. AI/BI Genie enables business users to interact with their data through natural language. It leverages GenAI to understand your data and underlying metadata and gives relevant and accurate answers based on that knowledge.

  Automated job tuning: Not all AI is related to a large language model. There are techniques called *deep learning* that can be used to tune the jobs automatically resulting in less time for human fine-tuning. This is called *predictive I/O*.

Deep Dive into Databricks IQ

In the following sections, we will look at each of these features in detail.

Databricks Assistant

Let's talk about the Databricks Assistant.

Generate Code in Any Language

Not everyone speaks code as their native language. Databricks supports several different languages, including SQL, Python, Scala, and R. There are times you will forget the syntax or simply need to extend a function. The old way was certain to scan through numerous blog posts or Stack Overflow to find your answer, and there would be lots of clicks and searches to get the final answer. What if someone is there just to tell you the answer?

Databricks Assistant can generate, explain, and fix SQL and Python code using natural language and is now available across all code editors in the Databricks platform including notebook and SQL editor (see Figures 14-2 and 14-3).

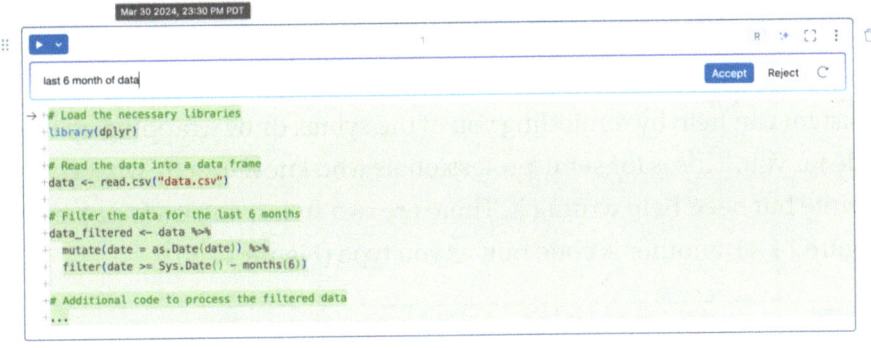

Figure 14-2. Code generation in cell

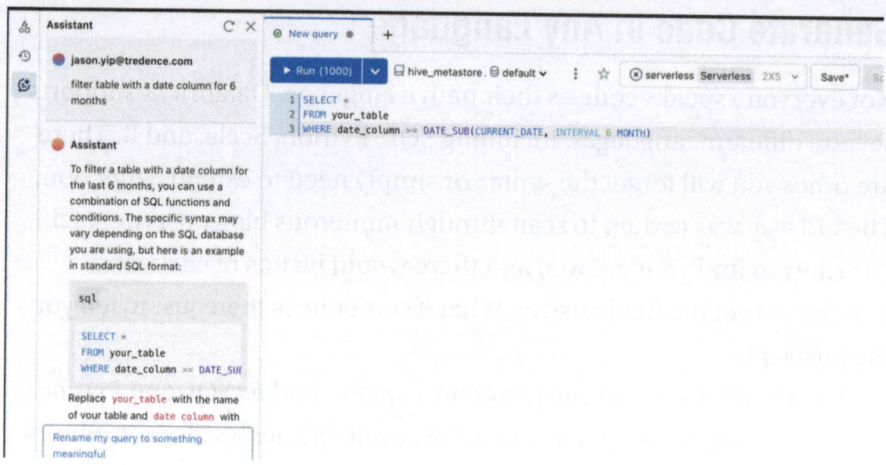

Figure 14-3. *Code generation in SQL editor*

Autocomplete Code or Queries

Whether you want IntelliSense or inline code completion, Databricks Assistant can help by reminding you of the syntax or by wrapping up the code for you. This is for semi-professionals who know exactly what code to write but need help writing it. There are two styles; one is via comment (Figure 14-4), another is code hint as you type (Figure 14-5).

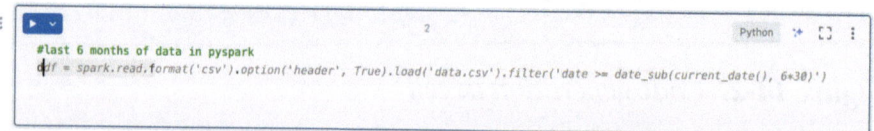

Figure 14-4. *Generating code based on comments*

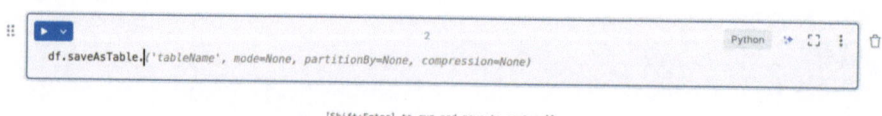

Figure 14-5. *Code completion*

Code Conversion

One of the most common use cases is to convert Python code into pySpark to take advantage of the distributed computing. If you were to use other tools, you need first to copy the code and paste it to other media, like a chatbot or a search bar. The assistant has direct access to the notebook and can understand the code and do the migration automatically. The answer can also be replaced with the existing cell with a click of a button. See Figure 14-6.

Figure 14-6. *Code conversion*

Code Explanation

Whether you don't understand the code or you want to explain your code to a business stakeholder who is interested in the business logic, you can ask the assistant to do it for you (see Figure 14-7). Having an English description of the code will help you understand it. And if needed, you can always resort to inline code generated to tweak the business logic.

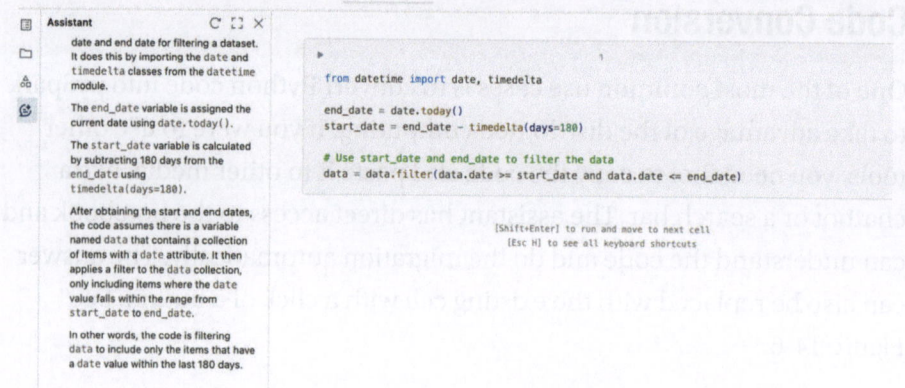

Figure 14-7. *Code explanation*

Code Fixing

While having a debugger is helpful, fixing the code will take a lot of time if you don't have a good handle on it. Databricks Assistant can explain where the error is coming from and also suggest a fix (see Figure 14-8). Best of all, you can collaborate with the LLM to find the best solution right inside the notebook without leaving the environment.

One thing to note is that the assistant will show up only when there is an actual error. Some application developers would use a try … catch block to catch the exception, which is a standard practice, but in these scenarios it will not trigger the assistant.

Figure 14-8. *Databricks Assistant suggesting code fix*

AI-Powered Governance

If you think Unity Catalog is the go-to tool for data governance, then you
are on the right track (see Figure 14-9). Delta Live Tables' data validation
capability, Unity Catalog's lineage information, Lakehouse Federation, and
auditing and access control are all perfect elements for data governance.
Coupled with its AI power, Unity Catalog will enable organizations to
govern more intelligently.

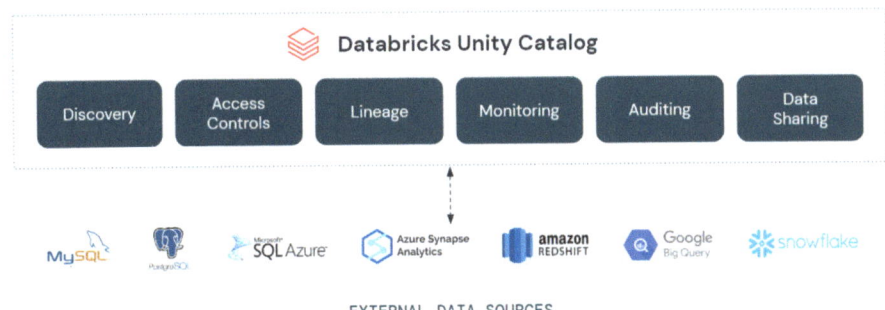

Figure 14-9. *Unity Catalog federated governance*

Let's dive into the AI powers that will help with the governance process.

- **AI-generated comments enhancements**

    Documentation has a love-and-hate relationship with developers. In certain cases there will be some initial effort for documentation, but as the number of data assets and tables grows, it will become hard to keep the documentation up-to-date. Although AI-generated comments are not bullet-proof, they can perform certain functions like a non-subject-matter expert (non-SME) would do toward the data, which is sampling the data and inferring the meaning based on the meaning of the table and the columns (see an example in Figure 14-10). Most importantly, the data dictionary can live with the data, instead of having to maintain a separate spreadsheet or stay in a system that requires due diligence to keep up-to-date.

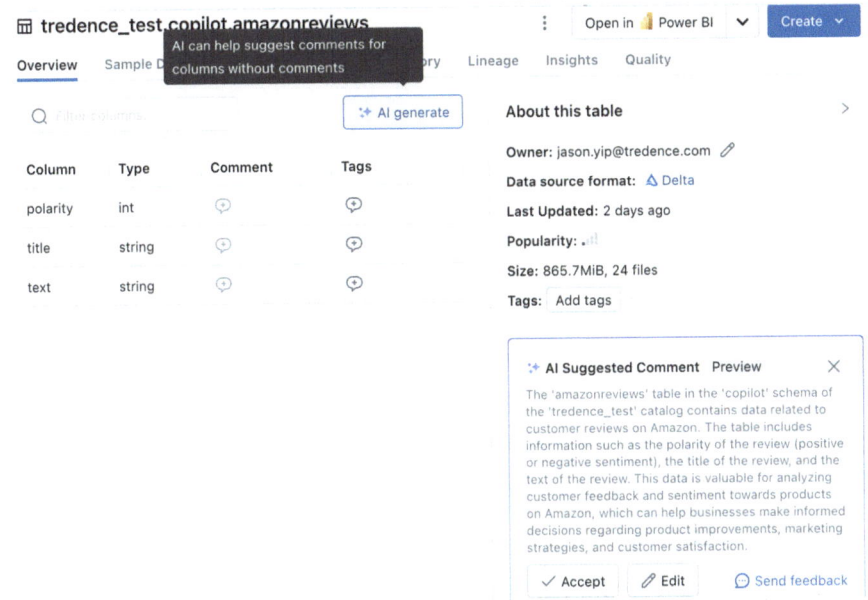

Figure 14-10. *AI-suggested comments for table description*

Transparency is at the heart of Databricks. The following article talks about the making of this AI feature and illustrates how it is not simply looking up from a dictionary:

```
https://www.databricks.com/blog/creating-
bespoke-llm-ai-generated-documentation
```

- **Lineage**

  As discussed in Chapter 5, Databricks provides lineage in two different ways: Delta Live Tables and Unity Catalog (Figure 14-11). While capturing the lineage is not a result of machine learning or a large language model, it plays a pivotal role as an input to the machine learning model so it can generate meaningful queries in AI/BI Genie and beyond.

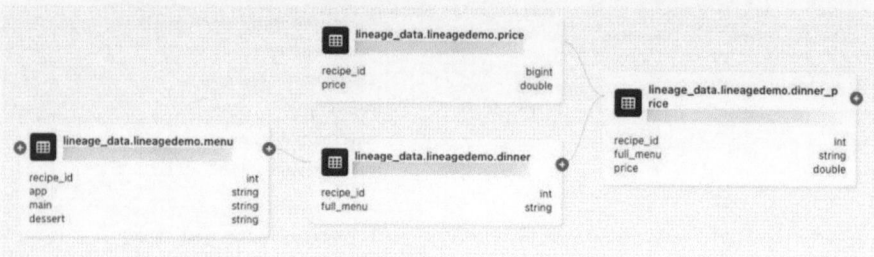

Figure 14-11. *Databricks lineage*

- **PII masking**

  Goodbye regular expression, hello artificial intelligence. With the requirements of compliance with the General Data Protection Regulation (GDPR) and California Consumer Privacy Act (CCPA) compliance, organizations are often required to identify columns containing PII and mask them accordingly.

  Previously, without the help of an LLM, regular expressions were often required to extract the patterns of email and street address; the process was error-prone. Machine learning models came along and tried to solve this problem, but it will require an extra layer of model processing and inferencing, either through a batch pipeline or through an API.

  Databricks serverless SQL comes with two very powerful functions designed for these scenarios: `ai_classify` and `ai_mask`.

ai_classify: What if the LLM is already a very good classifier? Is it possible to ask the LLM to classify if a column contains PII or not? When we think in this direction, we will have our answer. Consider the query in Listing 14-1.

Listing 14-1. AI Query in serverless SQL

```
SELECT ai_classify('my name is Jason, email address is jason@
email.com', ARRAY('contains PII', 'no PII')) as classification
union all
SELECT ai_classify('Today''s weather is awesome',
ARRAY('contains PII', 'no PII'))
```

The result, shown in Figure 14-12, is as you might expect.

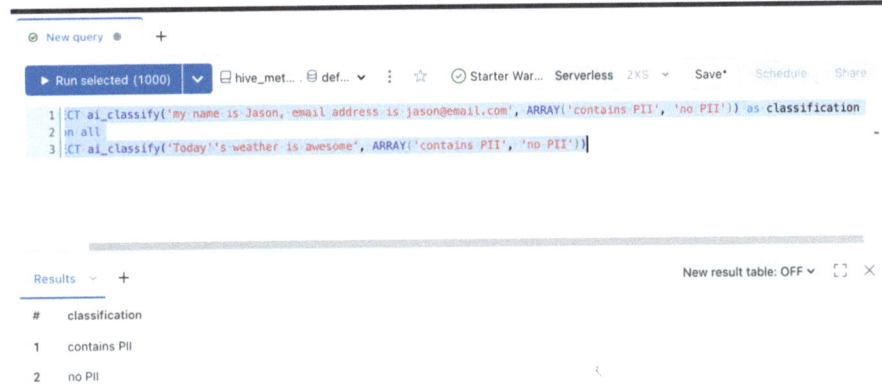

Figure 14-12. *DB SQL AI function: ai_classify*

ai_mask: Similarly, you can mask the sensitive columns by specifying what you wanted to mask. While it is not limited to PII, you can mask weather if you want, but from the PII perspective, it is a no-brainer. Listing 14-2 is an example with a name and an email address. Similar to regular expression searches, it will automatically match patterns for you. The result from Listing 14-2 can be seen in Figure 14-3.

Listing 14-2. ai_mask Function for Ease of PII Scanning

```
SELECT ai_mask('my name is Jason, email address is
jason@email.com', ARRAY('name', 'email')) as text
```

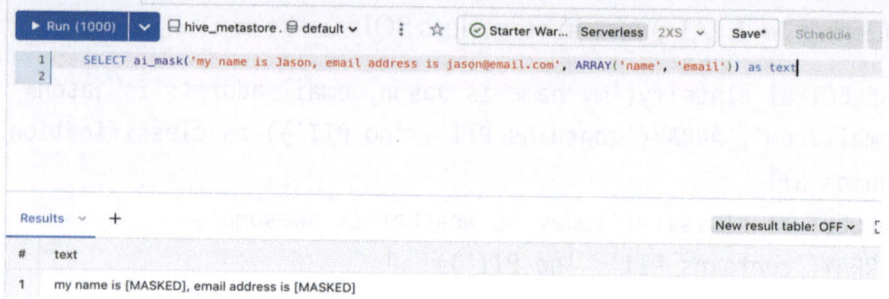

Figure 14-13. *DB SQL AI function: ai_mask*

- **AI security filtering**

    Content moderation is one of the hottest topics on the
    Internet. It started because social media companies
    needed to moderate their content in relation to
    hallucinations from LLMs and the accidental leak
    of profanity words. Databricks has included an API
    security filter (shown in Figure 14-14) either by setting
    a flag ("enable_safety_filter": True) in the API or a
    toggle in the Playground.

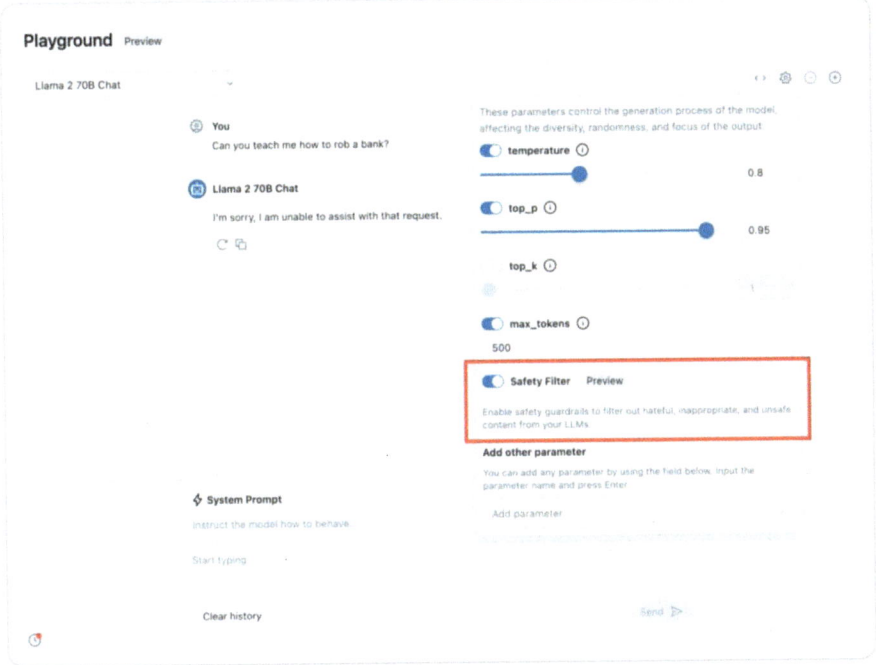

Figure 14-14. *AI security filter*

Behind the scenes, it is using Llama Guard's content moderation API. The Llama Guard paper can be found at. `https://ai.meta.com/research/publications/llama-guard-llm-based-input-output-safeguard-for-human-ai-conversations/`.

Llama Guard currently supports the following categories:

- Violence & Hate

- Sexual Content

- Guns & Illegal Weapons

- Regulated or Controlled Substances

- Suicide & Self Harm

- Criminal Planning

Additionally, Databricks also commits to providing a safe platform. You can find Databricks IQ's safety information at `https://docs.databricks.com/en/ databricksiq/databricksiq-trust.html`.

Beyond the guardrails provided, one can also set up custom guardrails, either through feature serving or another guardrail model. To get started with custom guardrails, check out this notebook demonstrating how to add personally identifiable information (PII) detection as a custom guardrail:

```
https://github.com/databricks/databricks-ml-
examples/blob/master/llm-models/safeguard/
llamaguard/Llama_Guard_Demo_with_Databricks_
marketplace_simplified_pii_detect.ipynb
```

- **AI security framework**

  Databricks AI Security Framework is a very comprehensive guide to CISOs and the guide to implementing Data and AI security in an organization. The whitepaper can be found here, and it contains a lot if valuable information:

```
https://www.databricks.com/resources/
whitepaper/databricks-ai-security-
framework-dasf
```

Search and Discovery

Databricks has been on a journey to enhance the user search experience on the platform. The search here refers to the growing number of data assets within the organization.

Intelligent Search

If you are familiar with GitHub's code search, you might think that Databricks is improving its offerings in terms of being able to search code. However, the search is not limited to code but other objects as well, including notebooks, workflows, etc. Figure 14-15 illustrates the different object types that Databricks search can search.

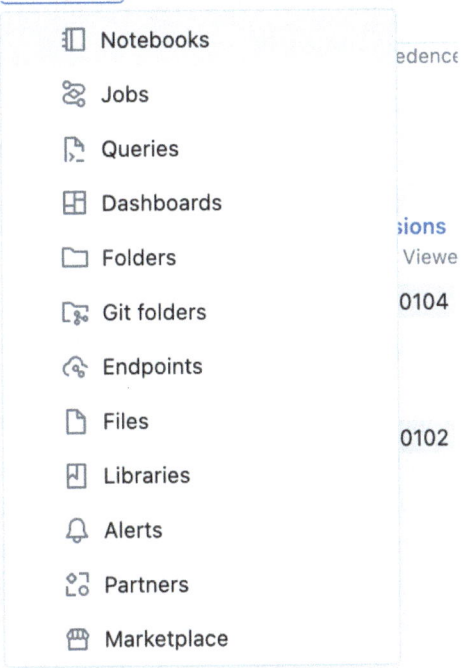

Figure 14-15. *Databricks object search*

So what are the capabilities in the new search experience?

- **Text search:** This primarily refers to code search. In addition to a single word, it can search words by including a double quote. It also supports escape quotes using a backslash.

- **Semantic search:** Search with meaning; you can ask questions like "how do I build a financial report?" Then it will return relevant financial tables

- **Search engine style filter:** You can filter by object types using search engine style like `type:table owner:me`.

- **Popularity:** Only the table is there; this doesn't mean it contains the right data. Popularity will ensure others use the objects returned.

- **Knowledge card:** For managed table only, search will present a knowledge card for top search results.

In the previous section of this chapter, we will look into AI/BI Giene, previously called Data Rooms.

AI/BI Genie (Previous Data Rooms)

AI/BI is a natural language Q&A experience that allows a nontechnical business users to ask questions in plain English and get their answers in either a table or a visualization. However, a key difference with AI/BI is that it uses agentic reasoning to continuously learn and improve to understand the nuances of your data and business semantics to deliver useful and contextual answers concerning your data.

To use Genie, the data should be in Unity Catalog, which provides fine-grained access control over the data so that no unintended leakage of sensitive data will happen in the Genie space environment; Serverless or Pro SQL Warehouses are required. Further, Genie is accessible to users with SQL entitlement. With these requirements in mind, let's move into some of the key aspects of using Genie. Figure 14-16 illustrates the architecture behind AI/BI Genie.

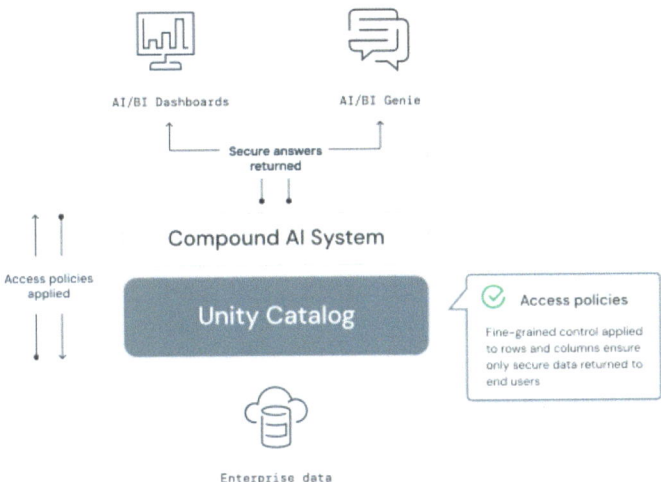

Figure 14-16. *Databricks AI/BI Genie architecture*

How to Set Up Genie

Let's look at an example of a large retail organization that wants its business users across different departments within the organization to use Genie. As a first step, the data owners and teams within the organizations that know most about the data will set up topic/context-specific Genie spaces (Figure 14-17). For example, POS Genie spaces contain tables that hold point-of-sales (POS) data, a finance space has all the financial data. Please note that a Genie space uses table and column names and descriptions to generate the equivalent SQL query based on the natural language query, which in turn runs on the data in the Unity Catalog.

New

Create a new space by giving it a name and define which tables make sense for the use case.

Title

> Genie-Test_Space

Description

> This room analyses Marketing and Finance Data

Default warehouse

> [Running] test_Wh   ∨

Tables

Choose tables to use for answering questions in the space. It is best to keep the scope for each space as small as possible. Data access is governed with the viewers Unity Catalog permissions.

Catalog	Schema	Table
dbdemos ∨	uc_lineage ∨	price ∨

dbdemos.abdi.sales	Remove
dbdemos.abdi.sales_country	Remove
dbdemos.uc_acl.customers	Remove
dbdemos.uc_acl.customers_drift_metrics	Remove
dbdemos.uc_lineage.price	Remove

Sample questions

Sample questions will be presented in new chat windows for users to ask the Space.

> E.g. What is our annual revenue?   +

Figure 14-17. *Creating an AI/BI Genie space*

After the Genie space is set, relevant tables and their associated metadata are brought in. It is important to note that your table metadata must be well documented with comments so Genie can understand the columns/tables that may be unclearly named and get more context. Further, one can create more focused views and remove unnecessary columns, resulting in cleaner data.

Next is to define business-specific terms using general instructions within your Genie spaces. Here, you can define unique jargon, logic, concepts, and KPIs in the given domain, and this knowledge will be used across all new questions. Further, you can iterate this over time as you see more questions come in or some new KPIs get developed, thus continuously teaching Genie .

Finally, if you already have SQL statements that were used to query tables in a specific Genie space, you can add them as well in "Save as Instruction" to teach the model how to answer specific questions. You can also keep examining the SQL statements generated by Genie. If you find them a bit off, you can save them, and Genie will learn from them for future questions.

Now your Genie space is all set to be used by your end user. Genie is designed to learn over time as it is used increasingly. One way it does this is by asking follow-up clarification for more context if the question is not clear, which enables it to capture more information from user prompts. Further, this new semantic knowledge can be saved as instructions to help Genie learn over time.

Figure 14-18 is an example of how we can immediately chat with our Genie space and get answers without knowing any coding. We can also visualize it from within the space (via Quick actions). The engine will get smarter over time, but the knowledge is there for everyone.

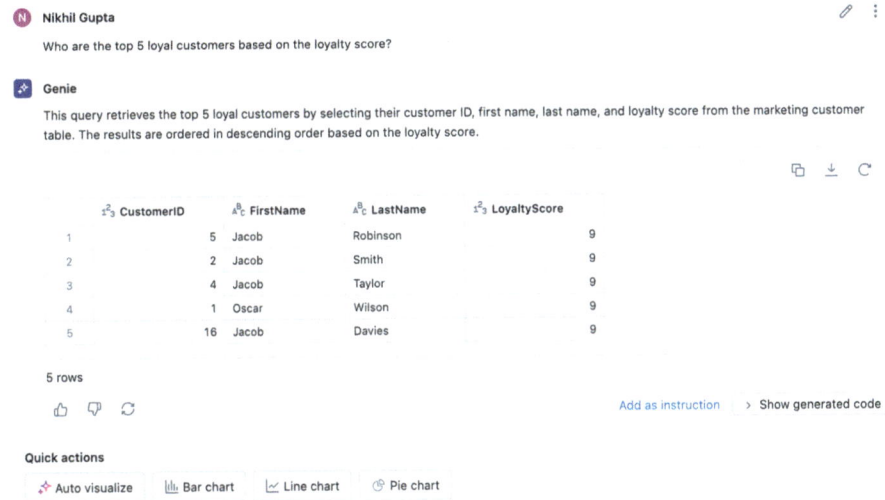

Figure 14-18. *Q/A with Genie in the space*

Conclusion

In this chapter, we defined the data intelligence platform as a combination of lakehouse and GenAI capabilities within the Databricks platform. Data intelligence is provided via a data intelligence engine called Databricks IQ. The platform has various features that enhance the user experience. Databricks Assistant can generate, fix, and explain Python and SQL code, helping developers increase productivity. Another feature is AI/BI Genie, which allows business users to ask questions in natural language about the data and get resulting tables and visualizations. We believe that Databricks will roll out many features like this over the next few years.

Databricks CI/CD

This chapter starts by understanding the concept of continuous integration/continuous deployment (CI/CD). Then we will move into Databricks repos and see how we can connect external Git repos to the Databricks workspace and illustrate the CI/CD process with regard to Databricks.

Finally, we will move into Databricks Asset Bundles, facilitating software engineering best practices such as source control, testing, and CI/CD.

What Is CI/CD?

A lot of development projects start with one person, and as soon as the work is released, either as open source or closed source, there will be issues and feature requests. The need to maintain stable and clean work therefore becomes increasingly important as development projects progress toward completion. Stable work ensures that each new release does not break the application or data pipeline. When users are asked for incremental features, they don't expect the existing features will break. As a result, a series of tests will need to run each time something changes (aka with each build). Clean work will allow developers to continue to build up the codebase without a problem. Even if we are talking about a single developer, it is important to keep a clean codebase because developers

© The Editor(s) (if applicable) and The Author(s),
under exclusive license to APress Media, LLC, part of Springer Nature 2024
N. Gupta and J. Yip, *Databricks Data Intelligence Platform*,
https://doi.org/10.1007/979-8-8688-0444-1_15

can often forget aspects of their own code very easily, especially when the codebase grows over time, or someone is multitasking. This process is called *continuous integration*.

Figure 15-1 illustrates this end-to-end flow. From the top, there is the development life cycle: Build, Test, and Merge. As the team continues to develop, we will need to integrate the work branches continuously and do regular releases. Regular updates will need to be deployed to production so the changes can be reflected as soon as the features are ready. In this chapter, we will dive deep into every step of the journey.

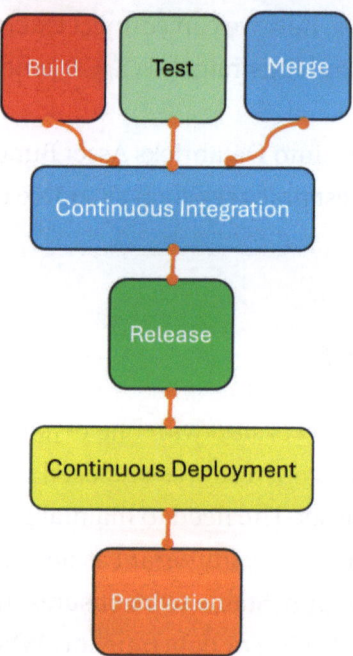

Figure 15-1. *End to end flow of CI/CD*

Once an application is ready to be tested (integration testing and regression testing) by developers and testers, they ensure it doesn't create any issues with the new release or simply deploy binaries to a server called a *release*. Many platforms offer these services, but the most common ones are Azure DevOps, GitHub, and GitLab (see Table 15-1). The process of releasing products to customers is called *continuous deployment*.

Table 15-1. *Comparison Between the Terms in Different Environments*

GitHub	GitLab	Azure DevOps
GitHub actions	GitLab CI/CD	Azure DevOps pipeline

An excellent example is Databricks' MLOps stack, which contains pre-written actions or workflow (`https://github.com/databricks/mlops-stacks`). Table 15-2 captures the whole structure of the repository. As we can see, the code and pipelines are abstracted out as individual components so they can easily be integrated into different parts of the workflow, which CI/CD will manage. Code modularization and abstraction are key to a successful CI/CD strategy.

Table 15-2. *Components Within Databricks' MLOps stack, Which Includes CI/CD*

Component	Description	Why It's Useful
ML Code	Example ML project structure (training and batch inference, etc.), with unit-tested Python modules and notebooks	Quickly iterate on ML problems, without worrying about refactoring your code into tested modules for productionization later
ML Resources as Code	ML pipeline resources (training and batch inference jobs, etc.) defined through Databricks CLI bundles	Govern, audit, and deploy changes to your ML resources (e.g., "use a larger instance type for automated model retraining") through pull requests, rather than ad hoc changes made via UI
CI/CD (GitHub Actions or Azure DevOps)	GitHub Actions or Azure DevOps workflows to test and deploy ML code and resources	Ship ML code faster and with confidence: ensure all production changes are performed through automation and that only tested code is deployed to prod

Stages of CI/CD

Before we examine the specific features within Databricks, it is helpful to understand the stages or flow of the CI/CD process to map the components with the flow (see Figure 15-2).

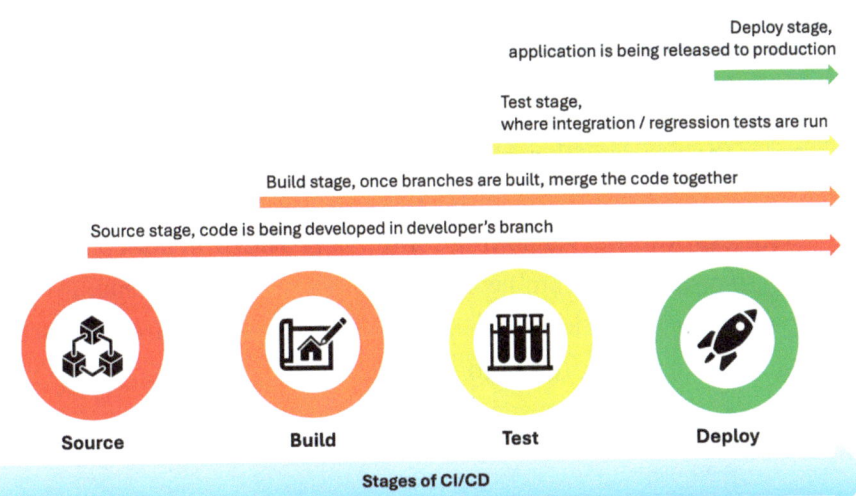

Figure 15-2. *Stages of CICD*

1. **Source stage:** This is where we develop our code. A best practice is to make changes in a feature branch and work on them until mature. If the feature work lasts a long time, it is recommended that the code be periodically synced with an integration branch.

2. **Build stage:** This stage is where the feature is complete and ready to combine, aka merge, with a stable, aka integration, branch.

3. **Test stage:** This stage aims to run all automated testing before going into manual testing, aka user acceptance testing.

4. **Deploy stage:** If the feature or change meets the expectations of quality and human evaluation, we can deploy it to production.

Introduction to Databricks Repos

The Databricks repo (shown in Figure 15-3) enables developers to synchronize their code with external Git providers. Developers do not need to leave the Databricks environment to commit their code. Along with Databricks' internal versioning control, it provides an added layer of security to protect the team's work. Moreover, it allows the team to promote the code from a lower environment to a higher environment such as from dev to staging to production.

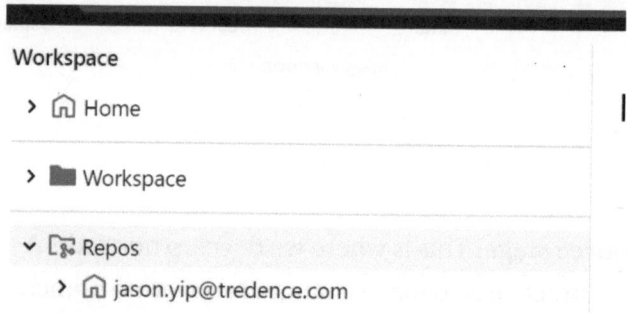

Figure 15-3. *Databricks repo*

Figure 15-4 shows an example of how to move code from development to production. Please note that this is not limited to ML Ops, but in this chapter, we will use it as an example to illustrate the CI/CD process. Later in this chapter, we will explain the workflow in detail, but as we can see in the orange squares (Figure 15-4), there are three different environments: the development workspace, the Staging workspace, and the Production workspace.

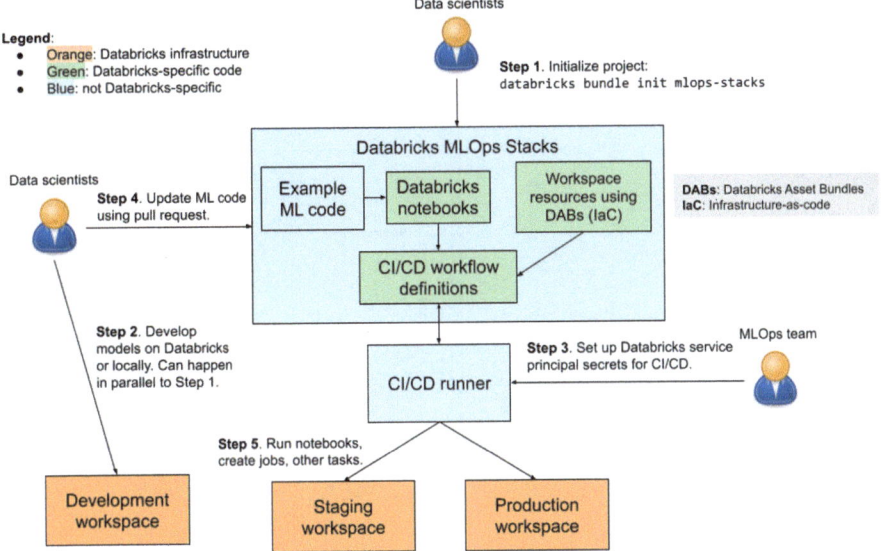

Figure 15-4. Databricks CI/CD process

Databricks supports both cloud and on-prem Git providers. From the list in Figure 15-5, we can see that the support is very comprehensive, including, but not limited to, the most popular providers: GitHub, GitLab, and Azure DevOps.

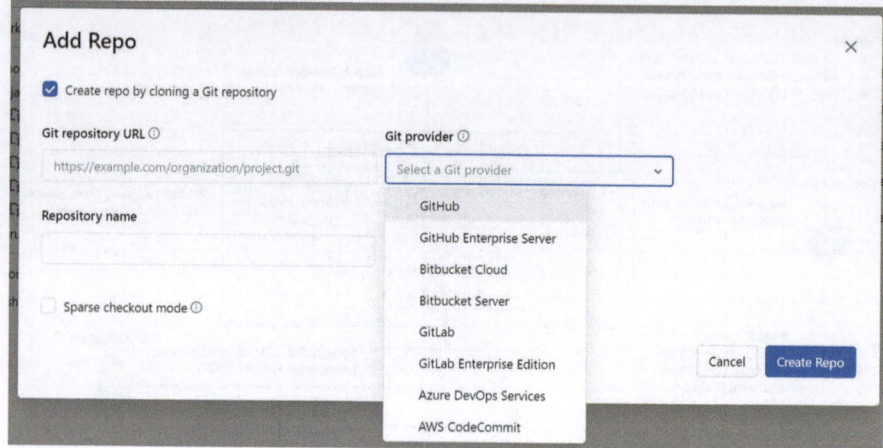

Figure 15-5. *Adding the repo to the Databricks workspace*

The following is a list of Git providers supported by Databricks:

Cloud Git providers supported by Databricks

- GitHub, GitHub AE, and GitHub Enterprise Cloud

- Atlassian BitBucket Cloud

- GitLab and GitLab EE

- Microsoft Azure DevOps (Azure Repos)

- AWS CodeCommit

On-premises Git providers supported by Databricks

- GitHub Enterprise Server

- Atlassian BitBucket Server and Data Center

- GitLab Self-Managed

- Microsoft Azure DevOps Serve

Databricks UI vs. Git Terminologies

Databricks repos allow users to manage their Git repositories from the Databricks Workspace UI. Users can pull code from repos, make changes, and then push it back to the Git repo. If you want to move code to higher environments, you would use the continuous deployment functionality of your respective Git repos.

If someone is new to CI/CD, it can be intimidating to understand so many terminologies. We will use a live example to explain via the Databricks UI, shown in Figure 15-6.

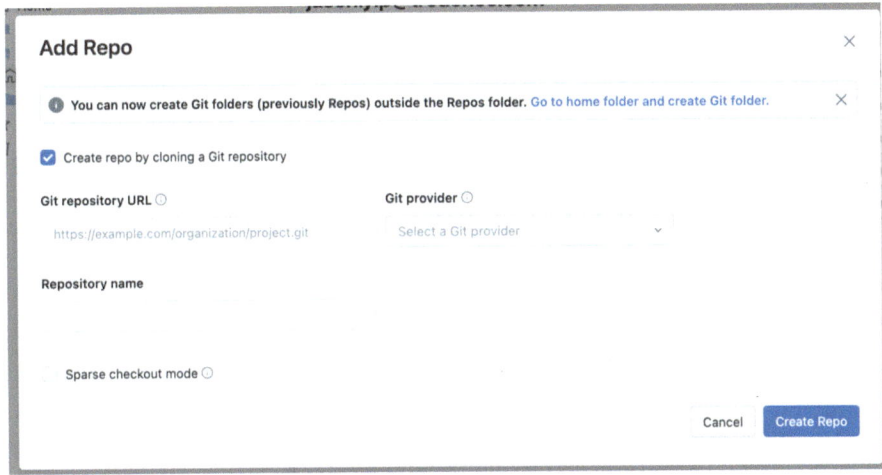

Figure 15-6. Adding a repo to Databricks through the UI

Clone: The very first action after identifying a Git repo is to clone it. As the name suggests, the whole purpose is to clone the repository from a remote location to a local destination. In Databricks, it is called Add Repo or using the Create Repo button. See Figure 15-7.

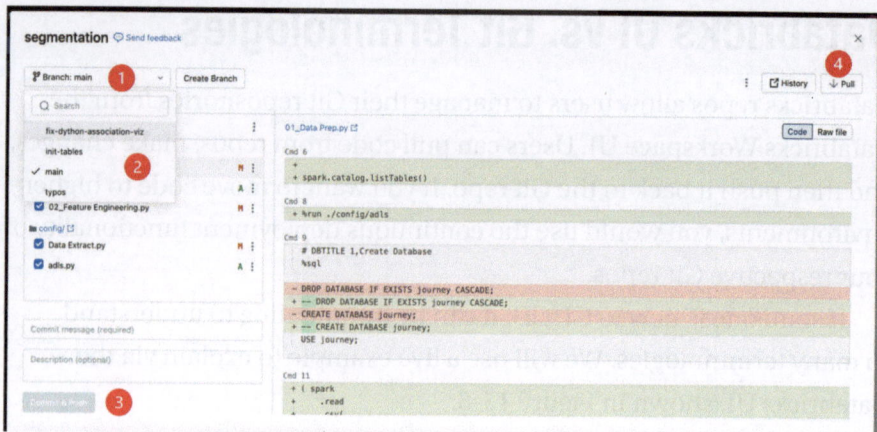

Figure 15-7. *The code check-in process*

1. **Branch:** A branch is used to hold a snapshot of the whole codebase. It is similar to versioning of a file, but instead it is versioning of the whole project. The usual branch names include the "main" branch, which is the most stable and up-to-date version of the repo, "feature" branches refer to a versioning of a specific feature you are developing, and "release" branches refer to the version of the release and are used to archive historical releases.

2. **Checkout:** Once a repo has been cloned, a *checkout* switches between different branches. However, in the Databricks UI you can simple click the branch name and there will be a drop-down to switch to a different branch.

   Commit: Once the changes are ready, the action *commit* will be used to publish them locally. It is important to note that the commit does not publish the changes to the remote location; in this case, it refers to the repo from which we clone the source code.

Push: *Push* is the action of publishing the changes available in the local repo to a remote repo. This action often involves a *merge* conflict.

3. **Commit & Push:** In Databricks, there is a button called Commit and Push, combining these two actions, because most of the time a commit is followed by push.

4. **Pull:** *Pull* is the action of retrieving the latest changes from a remote repo to a local repo. This action often involves a *merge* conflict.

   Merge: As its name suggests, *merge* is an action to merge new code into existing code. Whether it is from local to remote or from remote to local, a merge can happen, but so often if our local branch is too old, merge conflicts will occur. That means two of your commits modified the same line in the same file, and Git doesn't know which change to apply. This is called a *merge conflict*. That's why it is a recommended practice to do the work in a feature branch and then merge the branch back into a development branch so that the changes can be saved in a safe place in the case of a conflict. This process usually involves a pull request, which can be done in the Git provider interface.

   Rebase: When the commit history from two branches diverges, merging two different branches becomes difficult as there are many merge conflicts. Rebase is used to apply all commits one at a time, resulting in a cleaner history. However, the process can be challenging.

   Reset: Sometimes, when confusion occurs, it is best to reset the branch to an earlier history and rework the changes all over again. In the case of emergency breaking changes, a reset can always save the day. See Figure 15-8.

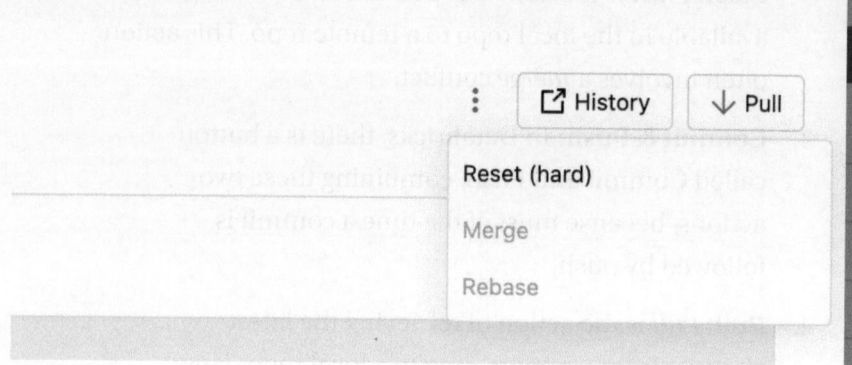

Figure 15-8. *Git reset, merge, and rebase*

Databricks Asset Bundles

According to Databricks, "Databricks Asset Bundles are a tool to facilitate the adoption of software engineering best practices, including source control, code review, testing, and continuous integration and delivery (CI/CD), for your data and AI projects."

In very simple terms, Databricks has provided some best practices of code and folder structure as well as deployment instructions (YAML files) for a team to work together seamlessly. These YAML files specify the artifacts, resources, and configuration of a Databricks project and are called Databricks Asset Bundles. These are useful during development and CI/CD processes. You can use the Databricks CLI to validate, deploy, and run Databricks Asset Bundles.

Teams can also customize their own template according to internal best practices. This will streamline the development standard so it is consistent across teams. Currently, Databricks provides four common templates for teams to use. To use these templates, we only need to leverage the Databricks CLI, for example using `databricks bundle init mlops-stacks`. Table 15-3 outlines the templates and their respective purpose.

Table 15-3. *Databricks Bundle Templates*

Template	Description
default-python	A template for using Python with Databricks. This template creates a bundle with a job and Delta Live Tables pipeline. See default-python.
default-sql	A template for using SQL with Databricks. This template contains a configuration file that defines a job that runs SQL queries on a SQL warehouse. See default-sql.
dbt-sql	A template that leverages dbt-core for local development and bundles for deployment. This template contains the configuration that defines a job with a dbt task, as well as a configuration file that defines dbt profiles for deployed dbt jobs. See dbt-sql.
mlops-stacks	An advanced full stack template for starting new MLOps Stacks projects.

Case Study: Databricks MLOps Stack

The Databricks MLOps stack provides some best practices in machine learning on the Databricks platform. Teams can use this template to deploy data science projects to production easily. This chapter aims to show the practical usage of CI/CD. For more information about MLOps, please refer to Chapter 9. See Figure 15-9.

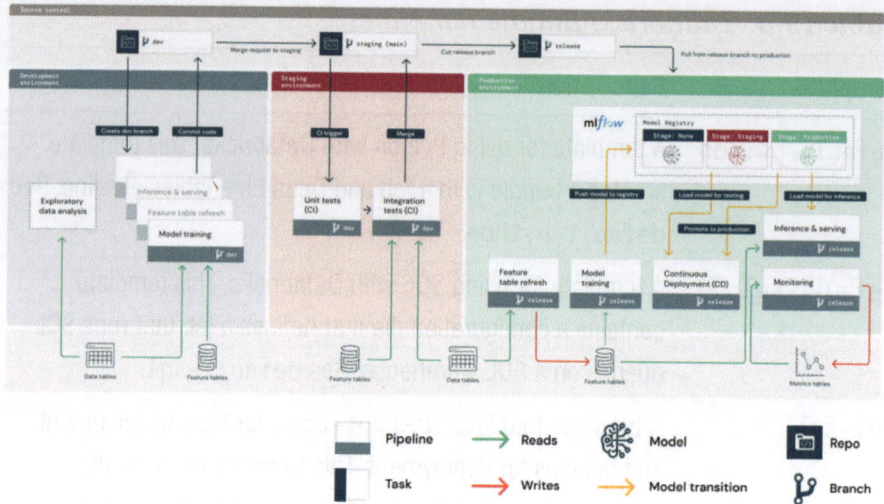

Figure 15-9. *Databricks MLOps stack using CI/CD in the workflow*

The MLOps stack is also an open-source project, so developers can explore it before initializing it in their local repository: `https://github. com/databricks/mlops-stacks`.

Here is the flow step-by-step:

Step 1: Initialize the project. The prerequisite is to install the Databricks CLI and configure it. Details can be found here: `https://docs. databricks.com/en/dev-tools/cli/index.html`.

Figure 15-10 features some sample output of what the command looks like.

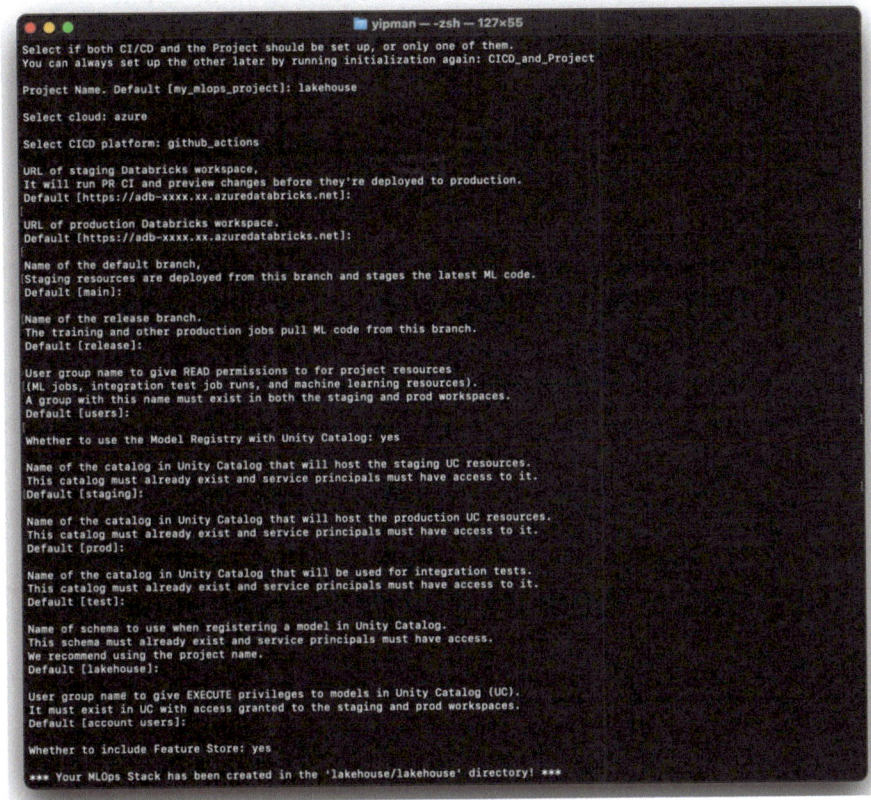

Figure 15-10. *Initializing the MLOps stack*

The command will generate the files shown in Figure 15-11 (these files are also available on GitHub).

```
yipman@jasons-mbp lakehouse % ls
README.md            deployment          project_params.json   resources      validation
__init__.py          feature_engineering pytest.ini            tests
databricks.yml       monitoring          requirements.txt      training
yipman@jasons-mbp lakehouse %
```

Figure 15-11. *Files generated or cloned by the Databricks CLI*

Step 2 is to develop the model. Databricks comes with MLlib, which can be used as a starter for model building. Data scientists can also bring in external libraries or build their own neural network architecture, which are all possible within Databricks. Keep in mind that the `feature_engineering` folder and the `training` folder will be responsible for building and training the model. Examples of how to use MLlib can be found here:

`https://docs.databricks.com/en/machine-learning/train-model/mllib.html`

Step 3: Setting up CI/CD is not just checking in the code. That's one of the steps. The bundle comes with `cicd.tar.gz`, and extracting the content contains the `.github` and `.azure` folders. They are workflows that will trigger the CI/CD, as explained in the intro section. More about GitHub actions can be found here: `https://github.com/features/actions`.

Step 4: As explained in the Git terminology section, once the code is checked in, the best way to collaborate is to update the code using pull requests. A pull request is an action after you finish committing your code to a feature branch, and creating a pull request is an ask to merge into the development branch, which usually triggers a code review (CR).

Step 5: Once the code is merged, GitHub actions will deploy the code to a staging environment as specified in the YAML file of the bundle. In terms of software development, this is called a *build*. However, since the output of the ML project isn't a binary itself, we will run the entire pipeline to ensure the data is refreshed, the model is trained, inferences are generated, and potentially the dashboard is refreshed with the latest predictions.

Step 6: When the team and stakeholders can verify the results, ML operators can trigger another GitHub action to deploy the pipeline to the production environment. It is critical that we don't automatically deploy the pipeline to production once the job runs successfully. Even if all the tests are passed, it is still a good idea to have humans review the results. For example, we need to ensure that the model does not generate biases for the sake of responsible AI.

`https://www.databricks.com/blog/helping-enterprises-responsibly-deploy-ai`

All these steps, except step 6, can be summarized in Figure 15-12.

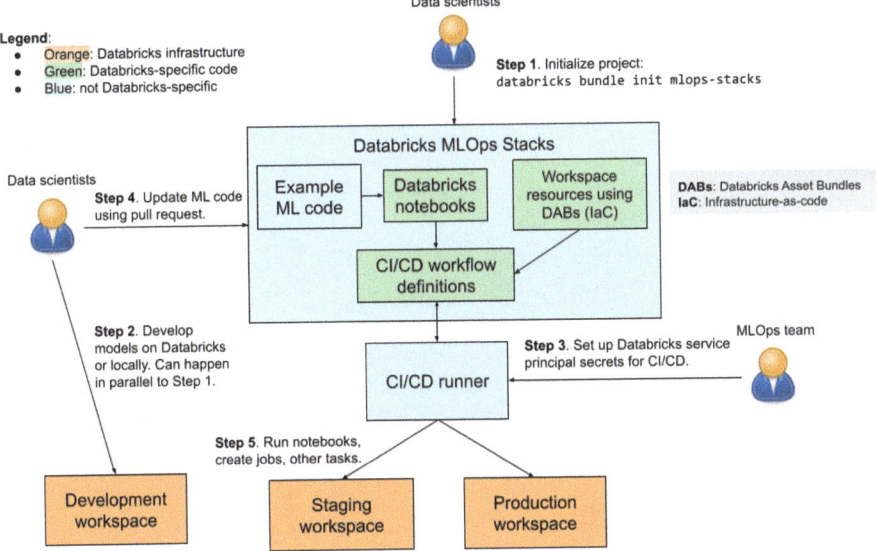

Figure 15-12. *Sample CI/CD workflow in the MLOps stack*

Conclusion

In this chapter, we learned the flow of CI/CD and the different stages. These concepts are generic no matter what tool you use. We also discussed how we can leverage these concepts in Databricks using different tools and how Databricks represents them in the user interface. It's important to understand the core concepts. Then, we are in the driver's seat and can look for the specific functions instead of trying to follow where the user interface design might lead.

Finally, we have look at a real-life case study using Databricks' open-source project MLOps stack, which is a generic name but is an actual project from Databricks. It might look more complicated than someone would learn for CI/CD, but it is always beneficial to go through a real-life scenario. Once you grasp this scenario, you are then ready to work with a team in real life.

CI/CD is a core strategy to keep the team productive and collaborative. It is imperative to master these skills to push any projects into production and beyond.

CHAPTER 16

Databricks Pricing and Observability Using System Tables

In this chapter, we will look into how pricing for running workloads on Databricks works. It is important to be able to calculate the costs involved in running solutions on Databricks. We will see what factors determine the pricing model and recommend which compute SKU should be used for running your specific workloads.

Then we will look at the concept of observability and how you can do observability on the Databricks platform using system tables. One of the most common ways to implement observability prior to Unity Catalog was an internally developed utility tool called Overwatch. However, for this chapter, Overwatch is out of scope.

Costs Associated with the Databricks Platform

Almost all the costs associated with Databricks are related to the compute resources being used. Since Databricks decouples storage and compute, the storage (which is provisioned in your cloud account) costs are

© The Editor(s) (if applicable) and The Author(s),
under exclusive license to APress Media, LLC, part of Springer Nature 2024
N. Gupta and J. Yip, *Databricks Data Intelligence Platform*,
https://doi.org/10.1007/979-8-8688-0444-1_16

directly paid to the cloud provider. Further, for compute, which is again provisioned in your cloud account, the costs can be divided into two parts—Databricks costs and cloud compute costs.

Cloud compute costs refer to the underlying hardware, such as virtual machines, disks, etc., within the customer's cloud account. The cloud compute is billed separately from the Databricks costs and is again paid directly to the cloud provider. It is important to note that the pricing model changes slightly with serverless compute, which we will discuss that later in this chapter.

For this chapter we will mostly look at the Databricks cost for running the compute resources in the lakehouse platform. But before we do that, let's look at some of the cloud costs components involved in running workloads on Databricks.

Cloud Infrastructure Costs

First we will break down the costs associated with the lakehouse platform:

- **Storage costs:** Within the lakehouse platform, data is stored in cloud storage (e.g., S3 on AWS, ADLS on Azure). The storage costs are paid directly to the respective cloud provider. The charges for storage are normally usage-based; i.e., they depend upon the amount of data being stored.

- **Compute costs:** The cloud compute costs are the cost of using the cloud compute infrastructure (VMs or EC2). The cloud infrastructure costs include costs for the virtual machines (VMs), disks, etc., that are paid directly to the cloud provider. Since Databricks clusters are ephemeral, the cloud provider charges for the duration for which the VMs have been deployed in the Databricks cluster.

- **Networking costs:** There are some networking
  costs involved while deploying/running workloads
  within workspaces. Some of these are the costs of IP
  addresses, NAT gateways, load balancers, and private
  links (if enabled). Further, if the data and workspace are
  in different regions, there are egress costs associated
  as well. All of these costs are again paid directly to the
  cloud provider.

Next, we will learn how to calculate Databricks' DBU costs and
Infrastructure costs.

Databricks Pricing

Let's look at the pricing in more detail now.

What Are Databricks Units?

A *Databricks unit* (DBU) is a normalized unit of processing power.
Databricks consumption is through clusters (job or all-purpose compute),
and SQL warehouses or serverless is priced in terms of DBUs. DBUs are the
underlying unit of consumption within the platform. However, the billing
is based on per-second usage.

Next, we will look into what factors determine the DBU consumption
of Databricks compute. These are the three key factors that influence the
cluster price:

- **Compute size and type:** This is the size of VMs one
  chooses both as the worker and the master node in
  the cluster. Depending upon the VM size, the number
  of DBUs change as well. Further, the number of DBUs

consumed depends upon whether Photon is enabled on the cluster. The compute size and type determines the number of DBUs that are consumed.

- **Product SKU:** The Product SKU determines the amount that would be charged per DBU-hr. For example, the per second price is DBU-hr / 3600. There are several different SKUs for the compute resources. This includes the all-purpose cluster, jobs cluster, DLT cluster, SQL warehouse (Classic and Pro), serverless, etc. Generative AI has a slightly different way to calculate the cost, but it is also based on DBUs. Depending upon the SKU being used, the dollars charged for the DBU-hr varies.

- **Account tier:** This is the Databricks account pricing tier in which the workspace runs, and one can select Standard, Premium, or Enterprise for AWS or Standard or Premium for Azure. Depending upon on the tier, the number of DBUs charged varies.

After looking at some of the levers that determine the DBU consumption, let's move on and look at an example to calculate the pricing of a cluster in dollar value. In the following example, we assume that we have a nine-node cluster, and together with the master node we have a total of 10 VMs that power this cluster, as shown in Figure 16-1.

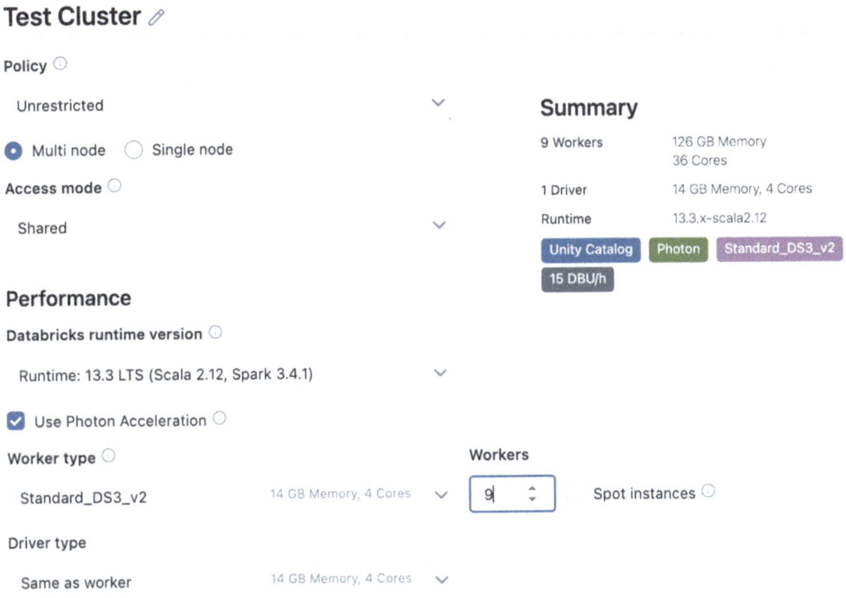

Test Cluster ✎

Policy ⓘ

Unrestricted ⌄

◉ Multi node   ○ Single node

Access mode ⓘ

Shared ⌄

Performance

Databricks runtime version ⓘ

Runtime: 13.3 LTS (Scala 2.12, Spark 3.4.1) ⌄

☑ Use Photon Acceleration ⓘ

Worker type ⓘ

Standard_DS3_v2   14 GB Memory, 4 Cores ⌄

Driver type

Same as worker   14 GB Memory, 4 Cores ⌄

Summary

9 Workers	126 GB Memory 36 Cores
1 Driver	14 GB Memory, 4 Cores
Runtime	13.3.x-scala2.12

Unity Catalog | Photon | Standard_DS3_v2

15 DBU/h

Workers

9 ⌄   Spot instances ⓘ

Figure 16-1. *Databricks cluster configuration*

In the **Summary** box in Figure 16-1, you can see that the cluster of this configuration would consume 15 DBUs/hr. Next, we can see how much that will cost in terms of dollar amount. Figure 16-2 shows the Azure Databricks pricing page. (Please note that this pricing is as of writing the book. For actual and most current pricing, visit `https://www.databricks.com/product/pricing`).

Workload	DBU prices—standard tier	DBU prices—premium tier
All-Purpose Compute"	$0.40/DBU-hour	$0.55/DBU-hour
Jobs Compute"	$0.15/DBU-hour	$0.30/DBU-hour
Jobs Light Compute	$0.07/DBU-hour	$0.22/DBU-hour
SQL Compute	.	$0.22/DBU-hour
SQL Pro Compute	.	$0.55/DBU-hour
Serverless SQL	.	$0.70/DBU-hour
Serverless Real-Time Inference	.	$0.082/DBU-hour

Figure 16-2. *Azure Databricks pricing information*

The cluster shown in Figure 16-1 is an all-purpose compute and is created in a Premium Databricks account. Therefore, referring to Figure 16-3 this cluster is priced at $0.55 DBU-hr. Since the cluster is consuming 15 DBU/hr, we can easily calculate that the price for running this cluster would be as follows:

Total DBU cost: 15* $0.55 = $8.25/hr.

Further, let's calculate the price of the VMs that are being used for the cluster. Since this is Azure Databricks and the VMs used are Standard_DS3_V2, let's go into the VM pricing page and find the costs for running the 10 VMs for 1 hr.

DS3 v2	4	14 GiB	28 GiB	$0.2930/hour	$0.2538/hour	$0.2003/hour	$0.0293/hour	+
					~13% savings	~31% savings	~90% savings	

Figure 16-3. *Pricing of DS3 v2*

Total VM cost: 10 * $0.2930 = $2.93/hr

For Azure:

`https://azure.microsoft.com/en-us/pricing/details/virtual-machines/linux/`

For AWS:

https://aws.amazon.com/ec2/instance-types/

For GCP:

https://cloud.google.com/compute/docs/general-
purpose-machines

Therefore, the user has to pay a total price of $8.25 + $2.93 = $11.18/
hr. Further, as noted earlier, the Databricks DBU cost would be paid to
Databricks and the VM cost of $2.93 would be paid to the cloud provider.

In the previous example, we have enabled Photon acceleration. Please
note that the number of DBUs required for Photon engine are 2x higher.
Therefore, if we disable Photon, the cost would be $4.125.

A logical question we would ask is, should we enable Photon to pay a
premium price? It is important to note that although Photon appears to
be two times as expensive, the performance will be roughly 3x higher than
without it. Therefore, for most workloads, Photon do give a better price/
performance than workloads running without Photon enabled. Figure 16-4
illustrates that when running a sample NYC Taxi query, the performance
is three times faster, and in our experience, the performance guarantee is
quite consistent.

Databricks SQL comes with Photon free of charge. We discussed
Photon in details in Chapter 8.

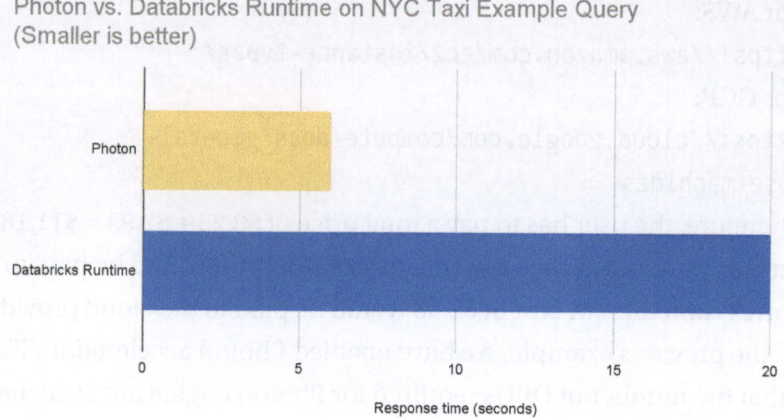

Figure 16-4. *Photon vs. Databricks Runtime on NYC taxi example query (source: https://www.databricks.com/blog/2021/06/17/ announcing-photon-public-preview-the-next-generation- query-engine-on-the-databricks-lakehouse-platform.html)*

Before we move on, let's discuss another important aspect around the pricing of jobs compute. Referring to the previous example, let's assume we had spun up a jobs cluster instead of the all-purpose cluster. If we refer to Figure 16-3, the jobs compute is $0.3 DBU/hr, which is almost 50% less than the all-purpose compute. The cost for running the same cluster would be 15* $0.3=$4.5, and the cloud compute costs of $2.93 remain the same. Therefore, it is strongly recommended that all automated jobs always utilize the job clusters.

☆ **BEST PRACTICE** ☆

During Databricks Data + AI Summit 2024, Nvidia CEO Jensen Huang announced the completion of a five-year project with Databricks to accelerate Photon with GPU. We will discuss the serverless SQL warehouse in the next section. It provides the best price-performance for data engineering workload and will drive down total cost of ownership.

SQL Warehouse Pricing

In this section, we will learn about pricing of Databricks' SQL warehouse compute and how pricing differs from other compute in Databricks.

There are two types of SQL warehouse computes: Classic and Serverless. Classic is similar to interactive clusters, where the DBU cost and underlying infra cost are paid separately, but Photon is included in the price. Second, Serverless is one price all inclusive.

As discussed earlier, Databricks fully manages the underlying cloud compute instances. Therefore, rather than having two separate charges (i.e., the DBU compute cost and the underlying cloud compute cost), the user pays only a single charge to Databricks for both. The concept of calculating the pricing of Classic and Pro SQL warehouses remains the same as discussed earlier. We will look into how we calculate the pricing of serverless SQL warehouse compute.

Next, we will look at how to calculate the cost of the serverless SQL warehouse. In Figure 16-5 we have an X-Large cluster size that will consume 80 DBU/hr.

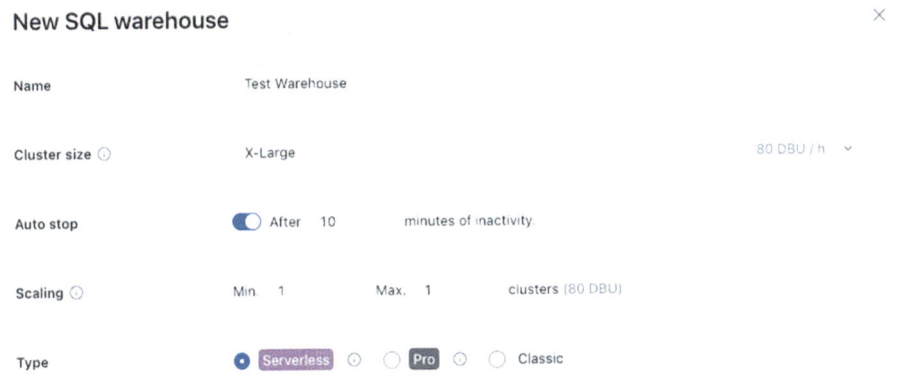

Figure 16-5. *Creating a new SQL warehouse*

According to the Databricks pricing page, Classic SQL is $0.22/hr, Pro is $0.55/hr, and Serverless is $0.7/hr. Therefore, the total cost for the example X-Large serverless SQL warehouse for one hour would be $0.70 * 80 = $56. This is the total cost, including the underlying infrastructure. This applies only to Serverless as the other tiers will require users to pay for underlying infrastructure.

Databricks SQL does not allow you to choose the infrastructure, unlike interactive and job clusters. However, Databricks has carefully worked on the best-suited VMs for each cloud and carefully tuned the performance to give users the best price/performance for their analytical SQL workloads. To understand what Databricks chose for the underlying infra, please refer to the following:

> Azure: https://learn.microsoft.com/en-us/
> azure/databricks/compute/sql-warehouse/
> warehouse-behavior

> AWS: https://docs.databricks.com/en/compute/
> sql-warehouse/warehouse-behavior.html

> GCP: https://docs.gcp.databricks.com/en/
> compute/sql-warehouse/warehouse-behavior

Databricks Cost Management Best Practices

In this section, we will look into some of the best practices for cost management on the Databricks Platform.

1. **Cluster Policies**

    Cluster policies allow users and groups to follow pre-defined rules when configuring or spinning up clusters. With cluster policies, admins can limit the size of clusters and define the type of VMs that can be used while creating the clusters. Further, admins can even set the max DBUs per hour on the clusters. Therefore, by enforcing cluster policies, admins can manage compute costs on the platform.

2. **Cluster Access Controls**

   Cluster access controls allow admins to control which users can create clusters. The cluster-level permissions give control over whether a user can attach to, manage, or restart a cluster. As a best practice, cluster creation abilities should be given admins who can manage and govern access to the clusters.

3. **Cluster Autoscaling and Cluster Termination**

   Databricks cluster autoscaling automatically adds and removes worker nodes in response to changing workloads to optimize resource usage. Autoscaling makes it easier to achieve high cluster utilization as one does not need to provision the exact number of nodes to match the workloads. This not only can enable the workloads to run faster than an under-provisioned cluster but also helps reduce the overall costs as compared to a statically sized cluster due to better resource utilization.

   Cluster auto-termination terminates a cluster after a specified inactivity period. As a best practice, always enable auto-termination for all-purpose clusters to prevent these clusters running overnight or over weekends.

4. **Spot Instances**

When you are requesting certain types of clusters, you are requesting new virtual machines from the cloud provider. Cloud providers also have the need to maximize their resource utilization. So spot instances allow you to utilize idle compute available in the region of the cloud provider, up to the price that you can optionally specify in advance. One thing to note about spot instances is that the cloud provider can recall spot instances at any time even when your job is running. So spot instances can be deployed in some long-running jobs where if the cluster restarts midway the job continues from where it left. Further, you can also configure it to fall back into on-demand instances

5. **Instance pools**

Similar to spot instances. Instance pools apply to the workspace level where administrators can pre-allocate some popular instances of virtual machines, either via on-demand or spot. Then when clusters are starting, there is no need to acquire them from the cloud provider, speeding up the time to start the cluster. However, administrators must be careful not to pre-load so many virtual machines because once they are loaded, the cloud provider will charge the account.

6. **Cluster Tags**

   Cluster tags can be associated with Databricks
   Clusters to attribute cost for chargeback purposes.
   For example, if there are multiple BUs in an
   organization, each BU can tag their clusters.

For in in-depth analysis of different cluster strategies, please refer to
the Databricks website: `https://docs.databricks.com/en/compute/`
`cluster-config-best-practices.html`.

Databricks Observability: System Tables

Observability is a key aspect of modern cloud data platforms. In simplistic
terms, observability is how well one can understand the IT system
from its generated outputs, such as logs, metrics, and traces. Therefore,
observability gives admins an approach to optimizing and controlling their
platforms based on the data they generate.

Some of the typical use cases platform admins might be interested in
doing the following:

- Monitoring costs

- Monitoring security and audit

- Monitoring platform usage/pipeline states

- Data observability and optimization

- Performance/resource utilization

Databricks' system tables integrated within Unity Catalog provides
curated datasets that enable users to query and answer these use cases.

Introduction to System Tables

Let's understand how one can utilize system tables for observability of the Databricks platform.

System tables are a Databricks-hosted analytical store for operational and usage data. They are fully integrated with Unity Catalog.

System tables can be used for monitoring and analyzing the performance, usage, and behavior of Databricks platform components. By querying these tables, users can gain insights into how their jobs, notebooks, users, clusters, ML endpoints, and SQL warehouses are functioning and changing over time. This historical data can be used to optimize performance, troubleshoot issues, track usage patterns, and make data-driven decisions.

System tables provide a means to enhance observability and gain valuable insights into the operational aspects of Databricks usage, enabling users to better understand and manage their workflows and resources. Based on the schemas/tables available as of writing the book, one can work toward solving/answering the following use cases:

- Cost and usage analytics

- Efficiency analytics

- Audit analytics

- GDPR regulation

- Service-level objective analytics

- Data quality analytics

System tables are available to customers who have **Unity Catalog** activated in at least one workspace in their account. This is needed to enable system tables for the account. *The data one sees in these tables is collected across all the workspaces in the account irrespective of whether*

Unity Catalog is enabled on the workspaces. However, system tables would be visible and queried only in the workspace that has Unity Catalog enabled.

Since system tables are governed by Unity Catalog, you need at least one Unity Catalog–governed workspace in your account to enable system tables. That way you can map your system tables to the Unity Catalog metastore. System tables must be enabled by an account admin. You can enable system tables in your account either by using the Databricks CLI or by calling the Unity Catalog API in a notebook.

The system tables are organized within a catalog named system, which is a fundamental component of every Unity Catalog metastore. Inside this catalog, you'll find schemas such as access and billing that house the system tables. These tables offer a comprehensive view of your Databricks environment, enabling you to make informed decisions and optimizing resource allocation. See Figure 16-6. It is important to note that the billing schema is enabled by default, but others have to be enabled manually.

For details of the system table schema, please refer to the Databricks documentation:

```
https://docs.databricks.com/en/administration-guide/system-
tables/index.html
```

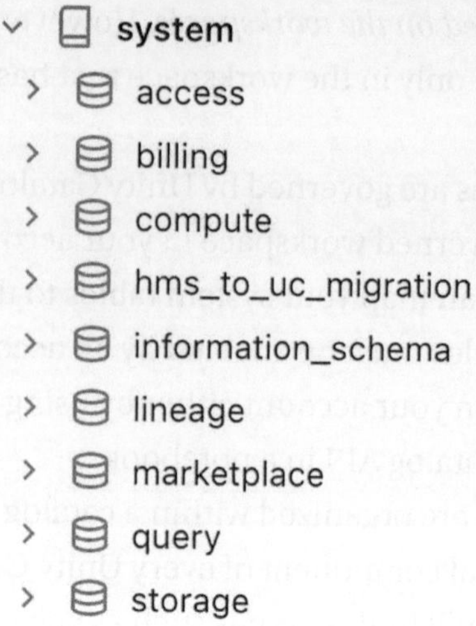

∨ 🖥 **system**
> 🗄 access
> 🗄 billing
> 🗄 compute
> 🗄 hms_to_uc_migration
> 🗄 information_schema
> 🗄 lineage
> 🗄 marketplace
> 🗄 query
> 🗄 storage

Figure 16-6. *Databricks system table catalogs*

System table access is governed by Unity Catalog. By default, no users have access to system tables. To grant access, a metastore admin or other privileged user must grant USE and SELECT permissions on the system schemas.

Common Schemas/Tables Available with System Tables

These schemas/tables are available with the system tables:

- **Audit logs:** Includes records for all audit events across your Azure Databricks account.

- **Billing usage:** Includes records for all billable usage across your account. Each usage record is an hourly aggregate of a resource's billable usage.

- **Table lineage:** Includes a record for each read or write event on a Unity Catalog table or path.

- **Workflow:** Allows you to view records related to job activity in your account. Further, you can join jobs system tables with billing tables to monitor the cost of jobs across your account.

These system tables provide valuable insights into the activities, resource utilization, and data lineage within your Databricks account and can be used for historical KPI tracking, monitoring and alerting, and forecasting expected usage for an intelligent lakehouse. There are many more schemas such as Pricing, Cluster, SQL Warehouse, etc., that users can analyze to ascertain the operational health of the Databricks platform.

System Table: Billing Usage Example

In the data and AI era, when there is data, there is AI. The granularity of the billing table is detailed enough to use as an input for a time-series forecast model. Databricks has built a demo, and the notebooks are available here:

```
https://notebooks.databricks.com/demos/uc-04-system-tables/
index.html#
```

Figure 16-7 illustrates that we can use cluster SKU and workspace ID along with the historical cost trend as training data to predict the future cost and feed into a dashboard for monitoring purposes.

Figure 16-7. *Predictive analysis of utilization and pricing*

You can build your custom solutions by leveraging the monitoring tables for predictive analysis and achieve greater savings in terms of cluster pooling, termination time, and beyond.

Conclusion

In this chapter, we looked at how to calculate the costs associated with Databricks. There are two types of costs associated with Databricks compute: cloud compute costs associated with VMs that are paid directly to the cloud provider and DBU costs that are paid to Databricks. We looked into how to calculate costs for various compute SKUs like interactive clusters, jobs clusters, and serverless SQL warehouses.

Then we moved into observability on the Databricks platform using system tables.

System tables in the Unity Catalog provide great insights to administrators who want to dig deeper into the platform, such as audit logs, pricing, and lineage. We have also demonstrated that, beyond a maintenance report, teams can create predictive analytics with the data, making it great for the finance team to do budgeting.

CHAPTER 17

Databricks Platform Security and Compliance

In this chapter, we will start by looking into the Databricks platform architecture. We will then move into Databricks workspace deployment and deep dive into topics like VNET injection and No Public IP (NPIP). Further, we will look into encryption and access control features. Finally, we will review an important tool called Security Analysis Tool developed by Databricks, which, when executed on a Databricks workspace, helps identify gaps in workspace security with recommended best practices and gives pointers to admins on how to resolve those deficiencies.

Please note that for this chapter we have used Azure Databricks as our reference to explain the concepts using single-cloud terminology, but the same concepts exist in both AWS and GCP.

© The Editor(s) (if applicable) and The Author(s),
under exclusive license to APress Media, LLC, part of Springer Nature 2024
N. Gupta and J. Yip, *Databricks Data Intelligence Platform*,
https://doi.org/10.1007/979-8-8688-0444-1_17

Databricks Architecture

Databricks is a hybrid PaaS general-purpose, cloud-agnostic compute platform. Let's decode this a bit. The term *hybrid PaaS* means deploying a data plane (virtual network and VMs) in your cloud service provider account while Databricks manages a multitenant control plane, as shown in Figure 17-1.

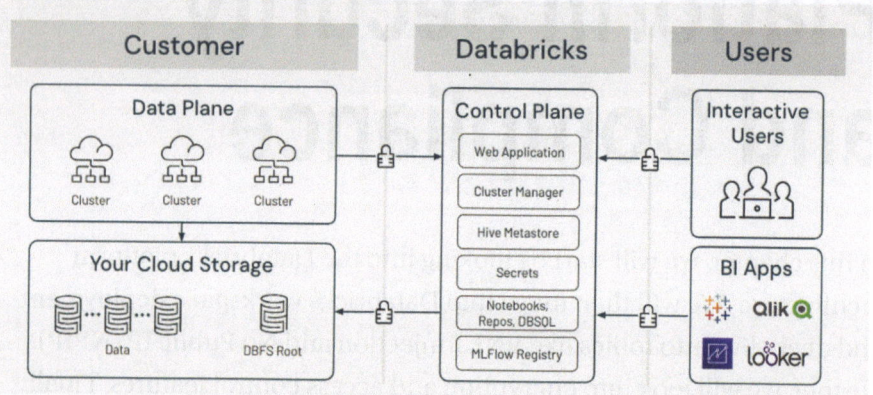

Figure 17-1. *Databricks data plane control plane architecture*

Next, we will further drill down into what the control plane and data plane are. The control plane contains all the back-end services such as WebApp, Cluster Manager, notebooks, workflow jobs, etc., and is managed by Databricks. On the other hand, a data plane is where you process and manage your data. The clusters/VMs get spun up in the data plane and connect to your storage account, where your data resides. Therefore, there is no need to send a copy of data to Databricks for processing, and as a result, there is no duplication of data required. Another advantage is that since data resides in your cloud storage, you can access it with or without Databricks.

Further, the data plane provides a natural isolation as it runs in your cloud account. It is important to note that the control and data planes always communicate over Azure Backbone.

Azure Databricks Deployment

In this section, we will examine the planning that needs to be done before Databricks deployment and some related best practices. Again, to reiterate, we have used Azure Databricks to explain the terminology, but the concepts are similar for AWS and GCP.

Capacity Planning

Within the Databricks workspace, you can spin up multiple clusters at a time for data processing. However, there is a limit to the number of clusters/nodes that can spin concurrently inside the workspace, and this is dependent on the size of the VNET and corresponding subnets selected during workspace deployment. Figure 17-2 showcases how the number of nodes that can be spun up in the workspace depend upon the size of virtual networks/subnets created.

Enclosing Vnet CIDR's Mask where ADB Workspace is deployed	Allowed Masks on Private and Public Subnets (should be equal)	Max number of nodes across all clusters in the Workspace, assuming higher subnet mask is chosen
/16	/17 through /26	32766
/17	/18 through /26	16382
/18	/19 through /26	8190
/19	/20 through /26	4094
/20	/21 through /26	2046
/21	/22 through /26	1022
/22	/23 through /26	510
/23	/24 through /26	254
/24	/25 or /26	126

Figure 17-2. *Nodes per virtual network calculations*

It is also important to note that once the size of the subnets is selected and the workspace has been deployed, we cannot resize the workspace or the subnets. In that case, a new workspace has to be deployed, and all the artifacts might need migration. As a best practice, if this is your first Databricks deployment, start with the /24 or /23 workspace. Once you size your workloads and jobs, you can always spin up a larger workspaces thereafter.

VNET Injection or Bring Your Own VNET

The default deployment of Azure Databricks workspace is a fully managed service on Azure. However, if you want customization and control over your environment, you can deploy the Databricks data plane in your own virtual network. For several reasons, you want to use your own VNET/subnets (known as VNET injection) to deploy your Databricks workspace.

First, this lets you connect Azure Databricks to other Azure services (such as Azure Storage) more securely using service endpoints or private link implementation. Next, you can connect to on-premises data sources with Azure Databricks via Express Route, taking advantage of user-defined routes. VNET injection allows you to connect Azure Databricks to a network virtual appliance to inspect all outbound traffic and take actions according to allow and deny rules. Finally, you can configure Databricks to use custom DNS and set up network security group (NSG) rules to specify egress traffic restrictions.

Hence, with the ability to fully manage your deployment, it is strongly recommended that Azure Databricks be deployed using VNET injection or in your own VNET/subnet.

Now let's move further to see how VNET injection works. The first step is to create a VNET if you don't have an existing one. Within the VNET there needs to be two dedicated nonoverlapping subnets per workspace that need to be created. The IP ranges for these VNET and subnets in Figure 17-2 determine the number of concurrent clusters you can spin.

By default, the subnets are named "public" and "private". Please note that these subnets cannot be shared with other applications. As a recommended practice, you should have a single workspace per VNET. Figure 17-3 shows the parameters required for the VNET-injected workspace in Azure Portal.

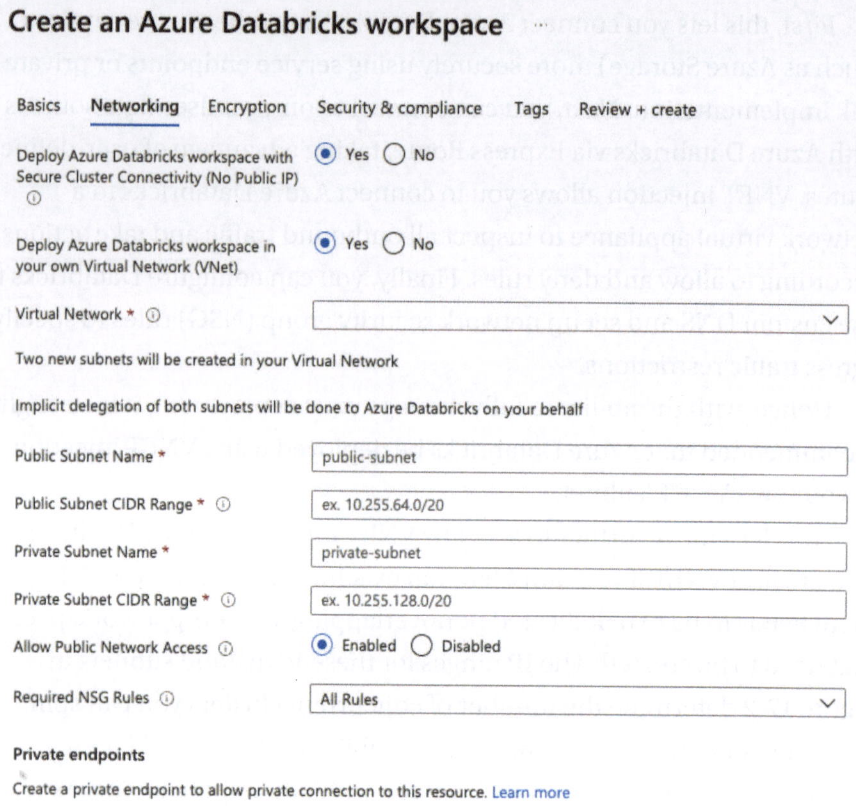

Create an Azure Databricks workspace

| Basics | **Networking** | Encryption | Security & compliance | Tags | Review + create |

Deploy Azure Databricks workspace with
Secure Cluster Connectivity (No Public IP)
ⓘ

◉ Yes ○ No

Deploy Azure Databricks workspace in
your own Virtual Network (VNet)

◉ Yes ○ No

Virtual Network * ⓘ

[∨]

Two new subnets will be created in your Virtual Network

Implicit delegation of both subnets will be done to Azure Databricks on your behalf

Public Subnet Name *

[public-subnet   ]

Public Subnet CIDR Range * ⓘ

[ex. 10.255.64.0/20   ]

Private Subnet Name *

[private-subnet   ]

Private Subnet CIDR Range * ⓘ

[ex. 10.255.128.0/20   ]

Allow Public Network Access ⓘ

◉ Enabled ○ Disabled

Required NSG Rules ⓘ

[All Rules   ∨]

Private endpoints

Create a private endpoint to allow private connection to this resource. Learn more

Figure 17-3. *Azure Databricks parameters*

Secure Cluster Connectivity (No Public IP/NPIP)

In Figure 17-3, the first checkbox below Networking is called Secure
Cluster Connectivity (No Public IP). Let's discuss what No Public IP (NPIP)
means and why it should be selected while deploying your workspace.

When Databricks is deployed without Secure Cluster Connectivity, the
Databricks control plane initiates an inbound connection to cluster(s).
As discussed earlier, each VM in a cluster requires one public and one
private IP. Thus, in this case, the traffic between the control and data plane

uses public IPs. Not only this, but ports need to be open on the firewall to enable this connection, which might be an issue for the enterprise infosec teams.

The Secure Cluster Connectivity, or NPIP, feature aims to solve the public IP issue. With NPIP, each cluster initiates a connection to the control plane's hosted secure cluster connectivity relay during cluster creation (Figure 17-4). This results in a data plane (the VNET) with no open ports, and classic compute plane resources have no public IP addresses for their nodes. The two subnets required for the workspace are now both private.

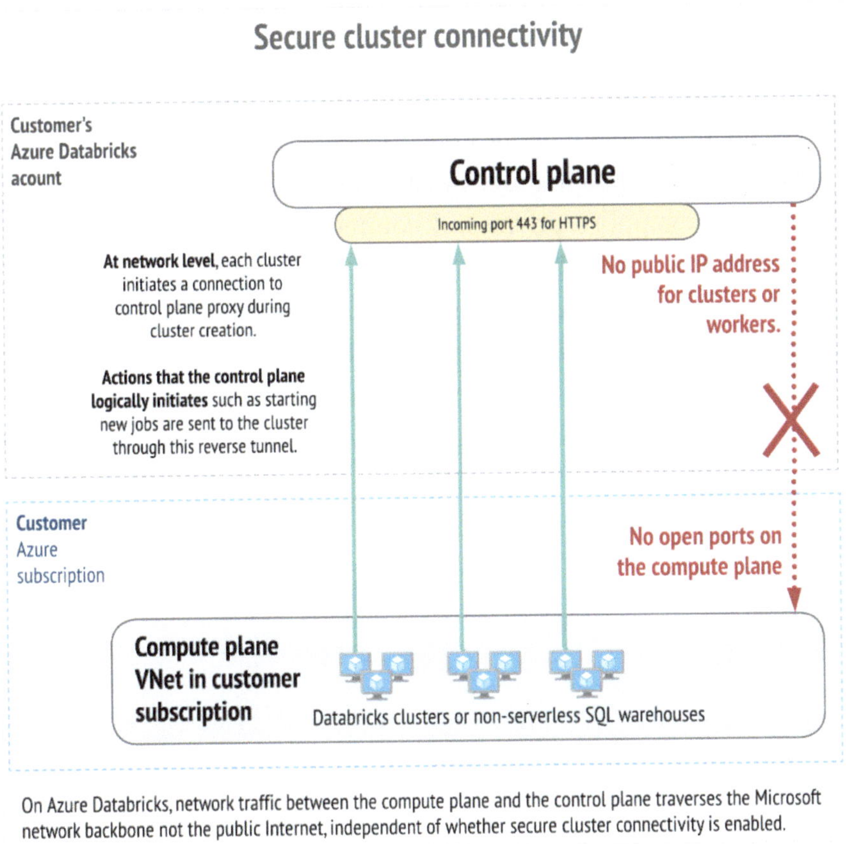

Secure cluster connectivity

On Azure Databricks, network traffic between the compute plane and the control plane traverses the Microsoft network backbone not the public Internet, independent of whether secure cluster connectivity is enabled.

Figure 17-4. *Secure cluster connectivity between control plane and data plane*

Therefore, with secure cluster connectivity enabled, customer virtual networks have no inbound open ports from external networks, and Databricks cluster nodes have no public IP addresses. This configuration is strongly recommended for all Azure Databricks workspaces because it significantly reduces the attack surface and hardens the security posture.

Azure Private Link for Back-End and Front-End Connections

After VNET injection and No Public IP (NPIP), Databricks introduced the Private Link feature. Azure Private Link provides private connectivity from Azure VNETs and on-premises networks to Azure services without exposing the traffic to the public network.

With the private link feature, illustrated in Figure 17-5, Azure Databricks now supports private link connectivity for two main in-transit connections in the data plane and control plane architecture. The first connection is from the user or front-end (including notebooks, REST API, JDBC/ODBC, and Databricks Connect) to the workspace control plane. The second connection is between the data plane to control plane. For both these connections, you can set up private endpoints while deploying the Databricks workspace.

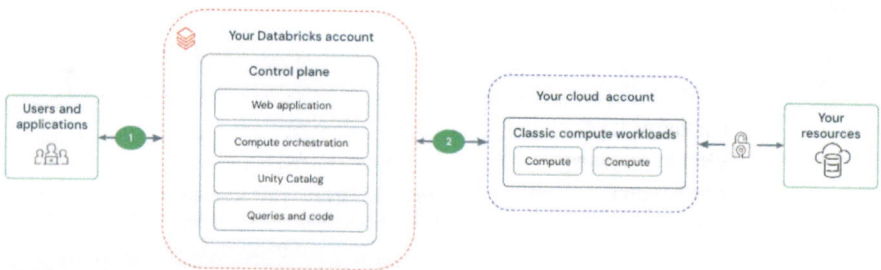

Figure 17-5. *Private link security*

You can give instructions on how you can set private link connectivity for Databricks deployment here: `https://learn.microsoft.com/en-us/azure/databricks/security/network/classic/private-link`.

After reviewing some of the main security features to consider while deploying a Databricks workspace, let's move on to the next section, which deals with encryption and auditing.

Encryption and Auditing

One important aspect of platform security is encryption and auditing. Let's first examine the default encryptions available within the Databricks platform.

First, all in-transit traffic is encrypted by default. Therefore, the control-plane data plane and user-control plane traffic is encrypted by default. Also, if you are communicating with other Azure services, that traffic is encrypted as well.

You can enable further encryptions in your Databricks deployment. The first is intra-cluster spark traffic, i.e. , data movement within your VMs in a cluster. Normally, this is not necessary to enforce (except for specific data processing use cases) because there is a performance degradation when this feature is enabled. The second encryption you can enable is the encryption of shuffle disks on compute workers.

Next, we will move on to learn about another very important feature: customer-managed keys, which can be used to encrypt artifacts in both the control plane and the data plane.

Customer Managed Keys

All managed services in the Databricks control plane are encrypted by default at rest. Optionally, you can add customer-managed keys (illustrated in Figure 17-6) for these managed services to control access to some services in the control and data planes. Some of the services where encryption via customer-managed keys can be done in the control plane are notebooks, notebook results, secrets stored by the secret manager, DBSQL queries, repo credentials, and PAT tokens.

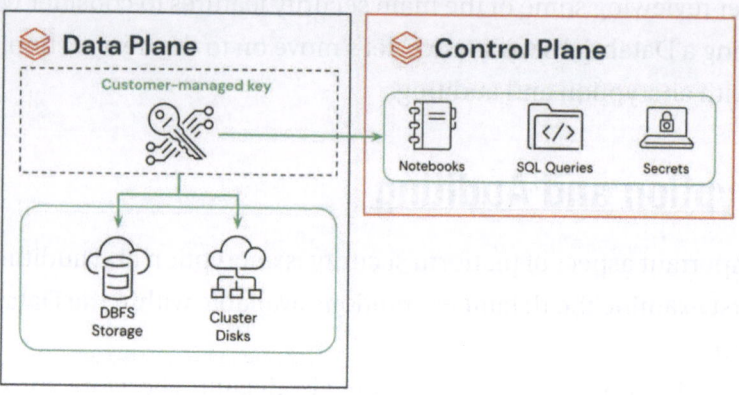

Figure 17-6. *CMKs*

Moving on to data plane artifacts, you can also use customer-managed keys to encrypt the DBFS root store and Azure-managed disks. The following page gives the steps to enable this feature in your Databricks workspace:

`https://learn.microsoft.com/en-us/azure/databricks/security/` `keys/customer-managed-keys`

To use this feature, you must first store your encryption key in the Azure key vault in your cloud. Similarly, Azure Databricks creates data-encrypting keys rooted in the customer key in the control plane. Applications now use customer-managed keys to encrypt and decrypt all data/artifacts. As a best practice, customers should develop policies to enable their key rotations.

To conclude, customer-managed keys give you full control over the keys used to encrypt data in the control and data planes.

Identity and Access

In this section, we will examine some of the features related to identity and access control on the Databricks platform.

SSO and Multifactor Authentication

Databricks provides security features such as single sign-on for user authentication. SSO enables you to authenticate your users using your identity provider (OKTA, AAD, etc.). It is highly recommended that SSO should be configured for enhanced security.

Further, once the SSO is enabled, you can enable multifactor authentication, again via your identity provider. In Azure Databricks, SSO in the form of Microsoft Entra ID-backed login is available in the account and workspaces by default.

Azure Databricks also supports Microsoft Entra ID conditional access, which allows administrators to control where and when users can sign in. Conditional access policies can restrict sign-in to your corporate network or require multifactor authentication (MFA).

IP Access Lists

IP access lists (see Figure 17-7) allow you to restrict access to Databricks accounts and workspaces based on the user's IP address. By default, users can connect to Databricks from any IP address. This might not be a best practice especially when the user accesses Databricks via the public/shared Internet like in a cafe.

When IP access lists are configured, it restricts the IP addresses that can authenticate to Databricks by checking if the user or API client is coming from a known good IP address range such as a VPN or office network. Further, if a user is moved from an established session to a bad IP address, the Databricks connection will not work, and workspace access will be denied. Thus, this gives you comprehensive control over which networks their workspaces can be accessed from.

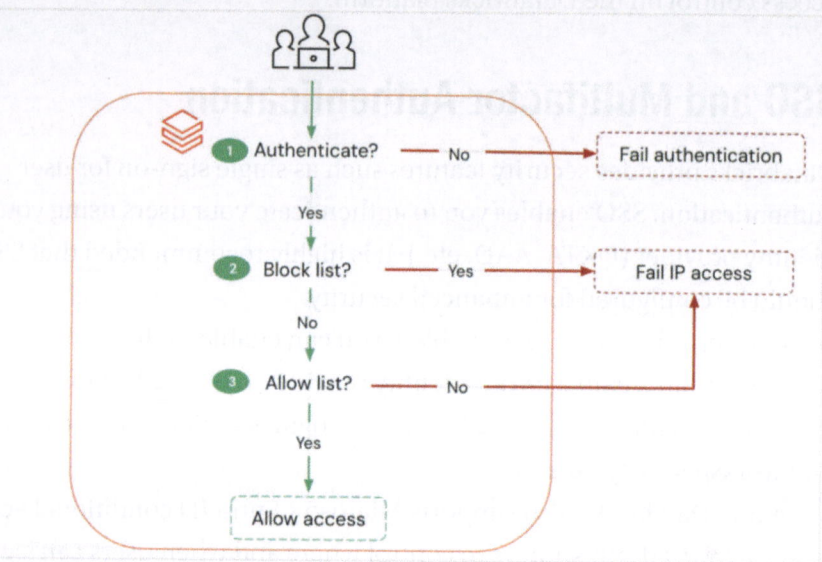

Figure 17-7. *IP access list*

An IP access list can be configured via the Databricks CLI or using the IP Access List API. Let's see an example in Listing 17-1.

Listing 17-1. IP Access List API Payload

```
{
  "label": "Office VPN",
  "list_type": "ALLOW",
  "ip_addresses": [
    "192.168.100.0/22"
  ]
}
```

This will allow users within the IP range to access Databricks. The rest of the IP addresses will be blocked.

Role-Based Access Control

In Databricks, you can use access control lists (ACLs) to configure permission to access workspace-level objects such as clusters, notebooks, etc., as shown in Figure 17-8. These ACLs are administered by workspace admins to users via the UI or the Permissions API. Workspace admins have the `CAN MANAGE` permission on all objects in their workspace, allowing them to manage permissions on all objects in their workspaces. Further, they can give/revoke access to Databricks workspace-level objects as needed.

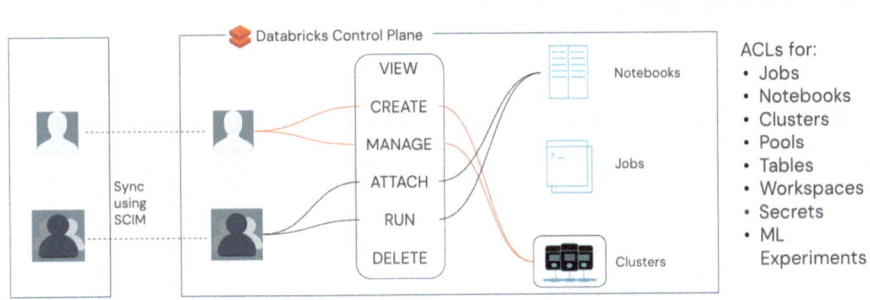

Figure 17-8. *ACLs*

Figure 17-9 provides a snapshot of notebook ACLs as an example. Different ACLs can be administered depending on the user's role. The following lists all ACLs to different Databricks objects:

https://learn.microsoft.com/en-us/azure/databricks/security/auth-authz/access-control/

Ability	NO PERMISSIONS	CAN READ	CAN RUN	CAN EDIT	CAN MANAGE
View cells		x	x	x	x
Comment		x	x	x	x
Run via %run or notebook workflows		x	x	x	x
Attach and detach notebooks			x	x	x
Run commands			x	x	x
Edit cells				x	x
Modify permissions					x

Figure 17-9. Notebook ACLs

Thus, workspace ACLs allow admins to provide appropriate access to Databricks objects to users.

Token Management API

Databricks Personal Access Tokens (PAT) are user-created tokens within Databricks. Users can create tokens through the UI or using the token API. While creating access tokens, users can mention the expiration date when the token will expire. If the field is left blank, PAT tokens never expire.

As illustrated in Figure 17-10, the Token Management API is built on top of the PATs by providing a stronger API for administrators to enable secure usage. It also gives admins the ability to turn off or disable PAT tokens. Using the permission API, admins can control which user is allowed to create tokens. The API also enables administrators to view and delete tokens from users in a workspace. Finally, administrators can set policies such as maximum token lifetime and more.

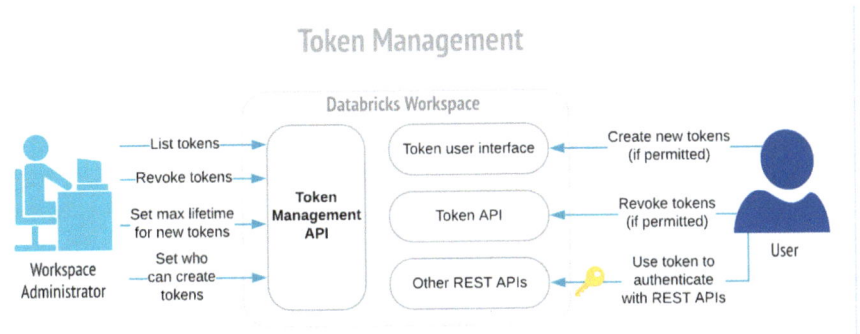

Figure 17-10. *Token Management APIs*

Let's look into a quick example (Listing 17-2) of using the Token Management API to set token permission. The API request lets you set a Can_Use permission to a user, group, or service principal.

Listing 17-2. Payload for Token Management API

```
{
  "access_control_list": [
    {
      "user_name": "string",
      "group_name": "string",
      "service_principal_name": "string",
      "permission_level": "CAN_USE"
    }
  ]
}
```

To conclude, admins must manage the PAT tokens created for user authentication. The Token Management API enables admins to do this seamlessly.

In the next section of this chapter, we will examine the Security Analysis Tool, a utility developed by Databricks that gives users a mechanism to see if the security features for their Databricks deployment follow security best practices.

Security Analysis Tool

Customers need to assess and reassess the security of the deployed architecture. Even if the initial deployment was well architected and all security features were taken into consideration, over time as newer features get released and configuration drift might happen, which could lead to data breach. To assess and monitor the security health of the deployed workspaces, Databricks launched the Security Analysis Tool (SAT), illustrated in Figure 17-11. SAT programmatically measures your workspace configuration against Databricks' security best practices. Thereafter, the reported deviations are ranked by severity, and links are provided to explain how you can extend your security to meet the Databricks requirements.

Figure 17-11. *Benefits of SAT*

SAT consists of a set of notebooks and libraries that collect details of the workspace using REST APIs. These notebooks run in Databricks workflows and can be scheduled or run manually. The notebooks' results are saved in a Delta table for historical reference. Finally, SAT comes with a prebuilt dashboard (Figure 17-12) that displays the latest results from the Delta table. Administrators, security analysts, and auditors can now assess their Databricks security posture from the comfort of a single screen.

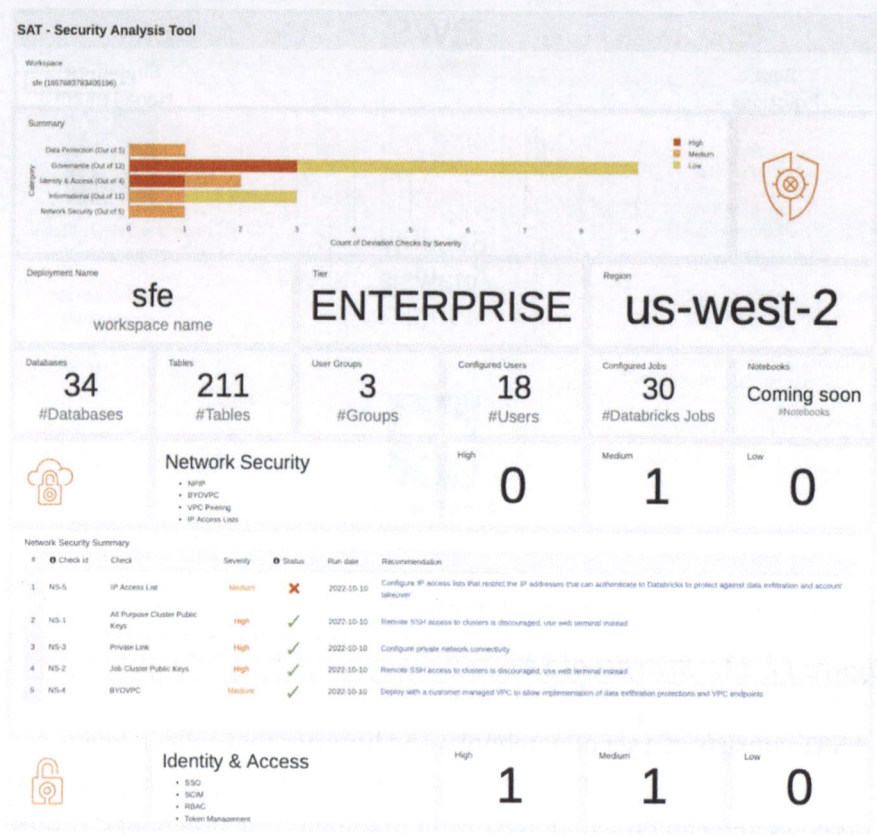

Figure 17-12. *SAT Dashboard report*

The SAT dashboard gives you information on certain dimensions. The first is Workspace Security Summary, which is a high-level summary of the findings by severity. The second dimension is workspace stats, such as users, databases, tables, etc. Then, it moves into individual Security Category details, which contain not only a summary of the deviation counts but also a table of security violations and links to documentation to fix the violations.

Databricks Security Best Practices

In this section, we will examine some of the security best practices in addition to the features we discussed earlier in the chapter.

- Do *not* use Databricks FileSystem (DBFS) as a storage layer as by default it is accessible to all workspace users. Use cloud storage for all data and with Unity Catalog enabled to manage access to tables and volumes.

- Back up and automate code deployment via CI/CD so you can integrate code scanning, better provide for permissions, perform linting, and more. Databricks repos enable you to move notebooks to your Git repos.

- Always monitor audit logs for user activities within the workspace. Audit logs are provided via system tables.

- Manage secrets and credentials via Databricks secret management or external systems like Azure Key Vault. Avoid entering credentials directly in notebooks, but reference them from the secret manager.

- Use service principals to run production workloads. You can configure service principals and generate PATs for service principals.

- Databricks Security and Trust Center (`https://www.databricks.com/trust`) provides extensive direction around the latest security features and best practices. Please refer to it as and when needed.

Conclusion

In this chapter, we examined key features related to Databricks security and compliance. We started by learning about the control plane/ data plane architecture. Then, we moved on to key security features recommended for Databricks deployment: VNET injection, Secure Cluster Connectivity (NPIP), and private link.

Next, we looked at key features that users can implement for encryption such as customer-managed keys (CMKs), which allow users to encrypt certain assets in both the control and data planes.

Then we moved into identity and access and discussed SSO and multifactor authentication, IP access lists that allows users from certain IP addresses to access Databricks workspaces, and token management for managing PAT tokens.

Finally, we discussed an excellent utility by Databricks: the Security Analysis tool. This tool allows users to assess their security with respect to Databricks' best practices and take appropriate measures based on the recommendations generated by the tool.

CHAPTER 18

Spark Structured Streaming: A Comprehensive Guide

Many people think of streaming as some very low-latency continuous real-time events like Twitter feeds or IoT devices; while that was the original use case, streaming has evolved over the years to allow integration with other non-real-time tables. In this chapter, we will first go back in time to visit Spark Streaming; then we will look at the latest Databricks Structured Streaming engine and how to use Delta Live Tables to process streaming. Apache Spark offers two popular streaming processing engines: Spark Streaming and Structured Streaming. While both engines are designed for real-time data processing, they have distinct architectures, advantages, and use cases.

Spark Streaming

Spark Streaming is the traditional streaming engine that uses the Resilient Distributed Dataset (RDD) API. It processes data in micro-batches, where each batch is processed as a whole. This approach allows for low-latency processing and high throughput. In micro-batching, illustrated in Figure 18-1, data is processed in small batches, e.g., 1,000 rows at a time. Spark Streaming uses a write-ahead log, which only means it will keep track of the count or offset before it writes to ensure disaster recovery. However, with this process, the batch writing will become sequential and result in hundreds of milliseconds of latency between batches.

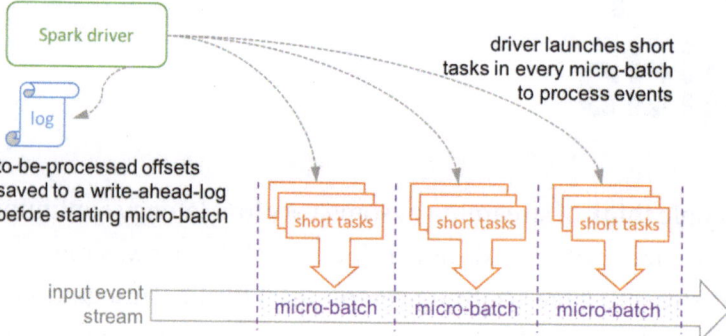

Micro-batch Processing uses periodic tasks to process events

Figure 18-1. *Micro-batch processing*

The high-level architecture of Spark Streaming consists of the following components:

- **Data source:** The source of the data stream, such as Kafka, Kinesis, or Flume

- **Receiver:** The component that receives the data from the data source and hands it over to Spark Streaming

- **Spark Streaming (engine):** The core engine processes the data stream

- **Processing:** The component that performs various operations on the data stream, such as filtering, mapping, reducing, joining, and so on

- **State management:** The component that manages the state of the streaming application, including checkpoint management

- **Output sink:** The component that writes the processed data to a target system, such as a file, database, or messaging system

Figure 18-2 shows the workflow.

Figure 18-2. *Spark Streaming workflow*

With the process in mind, let's explore, using Listing 18-1, how to build a Spark streaming application in Scala, Spark's native language.

Listing 18-1. Spark Streaming Example Using Scala

```scala
import org.apache.spark.SparkConf
import org.apache.spark.streaming._

object SparkStreamingExample {
  def main(args: Array[String]) {
    val conf = new SparkConf().setAppName("SparkStreaming
    Example")
    val ssc = new StreamingContext(conf, Seconds(10))
    // 10-second batch interval
```

```
    // Create a DStream from socket stream
    val lines = ssc.socketTextStream("localhost", 9999)

    // Split lines into words
    val words = lines.flatMap(_.split(" "))

    // Count words
    val wordCounts = words.map(x => (x, 1)).reduceByKey(_ + _)

    // Print word counts
    wordCounts.print()

    ssc.start()
    ssc.awaitTermination()
  }
}
```

While we appreciate the low latency brought by Spark Streaming, it is not easy to fit into today's rapid data engineering requirements due to the following reasons:

- Complex programming model

  As illustrated, creating an application requires a few steps, especially handling RDD and ultimately performing map and reduce operations.

- Requires manual state management

  Because streaming applications run 24/7, keeping track of the progress is important. There are multiple ways to handle states:

  a. **Checkpointing:** Spark Streaming can checkpoint the application's state at regular intervals, allowing it to recover from failures and resume processing from the last checkpointed state.

b. **Windowing:** Spark Streaming provides
    windowing operations (e.g., `window()`,
    `reduceByKeyAndWindow()`) that allow you to
    manage state over a sliding window of data.

c. **UpdateStateByKey:** Spark Streaming
    provides the `updateStateByKey()` method,
    which allows you to update the state of a key-
    value pair based on new data.

d. **Stateful transformations:** Spark Streaming
    provides stateful transformations (aka
    `mapWithState()`) that allow you to maintain
    state across batches of data.

While these operations are largely deprecated,
they are the foundation for Structured Streaming.
Understanding these operations will make
the transition into the enhancements easier
later. To read more about the operations of
stateful operations, please visit the following
Databricks blog:

`https://www.databricks.com/blog/2016/02/01/`
`faster-stateful-stream-processing-in-apache-`
`spark-streaming.html`

- Limited support for event-time processing

  Micro-batch processing in Spark Streaming focuses on
  the data that arrives within a specific time window, but
  it will do so only when the watermark reaches the event
  time of the late data, not immediately, resulting in a late
  arrival situation that is not ideal.

Structured Streaming

Structured Streaming is a newer streaming engine that uses the Dataframe/Dataset API, which is the strong foundation of Spark SQL. It processes data continuously, allowing for exactly-once guarantees and robust state management.

Beyond putting a structure (dataframe) into the streaming data, Structured Streaming is designed to address the following challenges:

- **Providing end-to-end reliability and correctness guarantees:** When failure occurs, batch processing is required to restart from its last successful batch, which is not hard to imagine why. With the increasing demand of streaming, pipelines must be continuously monitored and automatically mitigated to ensure highly available insights are delivered in real time.

- **Performing complex transformations:** In addition to streaming systems, data can often come in as flat file formats (CSV, JSON, Avro, etc.) that often must be restructured, transformed before being ingested into a bronze table. Structured streaming is designed to process and transform these data with minimal latencies.

- **Handling late or out-of-order data:** As discussed, there is a challenge in processing late arrival data because one must wait until the next batch is finished before processing the late arrival. We will discuss how the new architecture, called continuous processing, will be able to address this issue.

What Is Continuous Processing?

Instead of performing micro batches one after another, continuous processing (see Figure 18-3) is now tightly integrated with Spark. It launches a set of long-running tasks at the same time that keep reading, processing, and writing data continuously. This means as soon as new data is available, it gets processed and written right away, so the delay between when the data comes in and when it's ready is very short—just a few milliseconds.

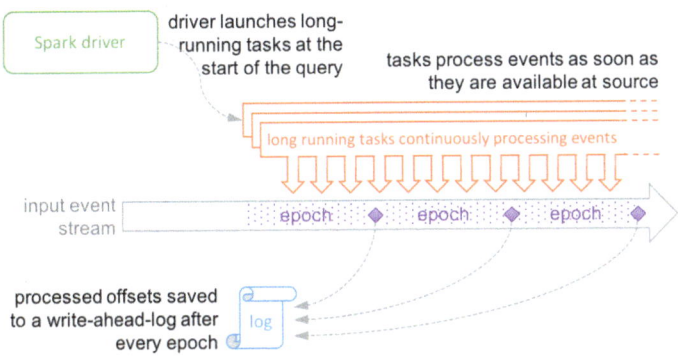

Continuous Processing uses long-running tasks to continuously process events

Figure 18-3. Continuous processing

Spark uses a special technique called the Chandy-Lamport algorithm to track how the processing is going. It adds special markers to the data stream, called *epoch markers*. When a task sees one of these markers, it tells the main computer (called the *driver*) where it stopped (offset) processing. The tasks report back asynchronously, in other words, without waiting for the task to finish; then the driver writes down all the offset in parallel, so the progress can be kept track of without waiting for the batch to finish. This all happens in the background, so the tasks can keep going without stopping, and everything stays fast and efficient.

Triggers

While Structured Streaming now supports continuous processing, it doesn't mean micro-batch processing becomes obsolete immediately. It is not hard to imagine that by spinning up an always-on Spark instance, it will incur a lot of resources as well as cost, so unless the real-time requirement is mandatory, like in a credit card transaction scenario where you cannot afford keeping the customer waiting for a batch to finish, there is always an option to fall back into micro-batch processing. That's where Trigger mode comes into the picture.

The following are the different trigger modes:

- Default mode (no trigger is specified)

  If the trigger option is not specified, then by default, the query will be executed in micro-batch mode.

- Fixed interval micro-batches
  (`trigger(processingTime = "1 second")`)

  As the name states, the micro-batch will be triggered in the interval specified. Since micro-batching is a sequential operation, if the previous batch cannot finish in the specified interval, the next batch will wait for the batch to finish before processing.

- `Available-now micro-batch`
  (`trigger(availableNow=True)`)

  If you were resuming from a streaming process, you can use this option to process all the batches in the queue. This trigger will stop on its own.

- Continuous with fixed checkpoint interval
  (`trigger(continuous = "1 second")`)

  The query will be executed in the new low-latency,
  continuous processing mode. The regular checkpoint
  will be written into the checkpoint location.

Output Modes

Similar to write modes in the Spark dataframe, the most initiative ones
are "overwrite" mode and "append" mode. Spark Structured Streaming
also has these two modes, which can be called using `writeStream.`
`OutputMode()`. There is an additional mode called "update" in Structured
Streaming that is more applicable to grouped aggregations on a sliding
window, which we will discuss later in this chapter. These are brief
descriptions of the modes:

> **Complete:** Similar to the overwrite mode, on every
> trigger, everything will be rewritten again, but it does
> not delete old data, so there will be duplications.
> However, this can be useful for aggregations, so we
> don't lose any count as a result of late arrival.
>
> **Append:** As its name suggests, all the data will be
> appended on every trigger. But late arrivals need
> to be handled properly for aggregations. This is the
> default mode.
>
> **Update:** Mainly applies to aggregations. This mode
> will put intermediate results in memory and update
> the aggregations once the threshold is reached for
> late arrival.

Windowed Grouped Aggregation

Structured Streaming offers a way to group aggregations together by windowing (sliding window or tumbling window), say for five minutes, similar to the groupBy operator. Imagine we are converting a timestamp column to a time range, say 12:00 to 12:10, and for every trigger, it will aggregate the events and save it into a table. In the table shown in Figure 18-4, we can notice "cat" changed to 2 from 12:05 to 12:10 because another "cat" arrived at 12:07.

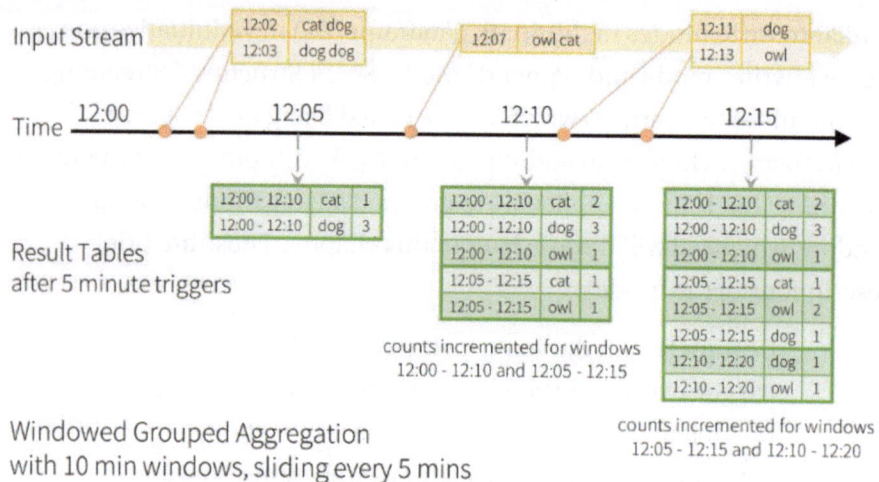

Figure 18-4.   Windowed grouped aggregation

State Management

When you need to handle state management all by yourself, Structured Streaming comes with a checkpointing option. Databricks recommends always specifying this option to ensure the job can be recovered in case of failure, as shown in Listing 18-2.

Listing 18-2. Spark Streaming Checkpoint

```
streamingDataFrame.writeStream
  .format("parquet")
  .option("path", "/path/to/table")
  .option("checkpointLocation", "/path/to/table/_checkpoint")
  .start()
```

However, there are times that require more advanced stateful processing. That's where the new operators `mapGroupsWithState()` and `flatmapGroupsWithState()` come into the picture. These operators allow you to maintain state for a group of target audience, and the key to group them together might not be in sequence; hence, they are arbitrary. For example, for a class of users in a geographic location or spending threshold, instead of applying on an individual basis, the grouping key can be a state name, but some data can come in the form of a city name. These techniques are helpful to ensure late data can be tagged to a specific group for analysis. See Figure 18-5.

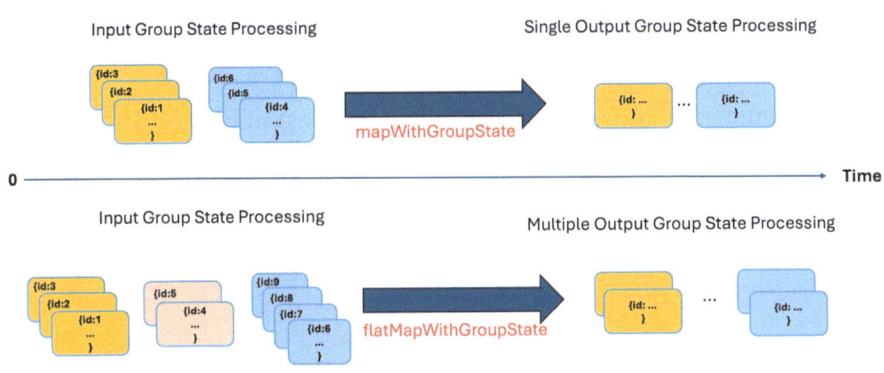

Arbitrary Stateful Processing
in Structured Streaming

Figure 18-5. *Arbitrary stateful processing in Structured Streaming*

Late-Arrival Handling: Watermark

Structured Streaming uses a **watermark** to control the threshold for how long to continue processing updates for a given state entity. A watermark (see the example in Listing 18-3) is a threshold timestamp used to track the latest event time of the data processed so far. Any data arriving with an event time older than the watermark is considered late and can be either ignored or processed separately.

Listing 18-3. Spark Streaming Watermark

```
from pyspark.sql.functions import window

(df
  .withWatermark("event_time", "10 minutes")
  .groupBy(
    window("event_time", "5 minutes"),
    "id")
  .count()
)
```

Earlier we discussed a new output mode called Update, which is useful for aggregations along with a watermark.

Update mode will continue to update the count on every trigger until after the watermark threshold is reached, as shown in Figure 18-6.

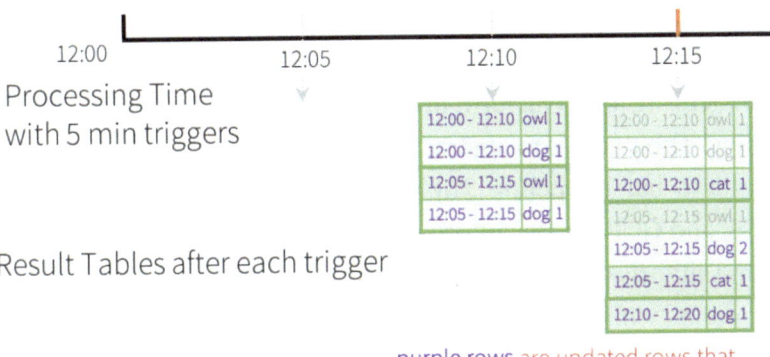

Figure 18-6. *Update mode in aggregation*

Append mode will write the data into a table only after a threshold is reached (see Figure 18-7).

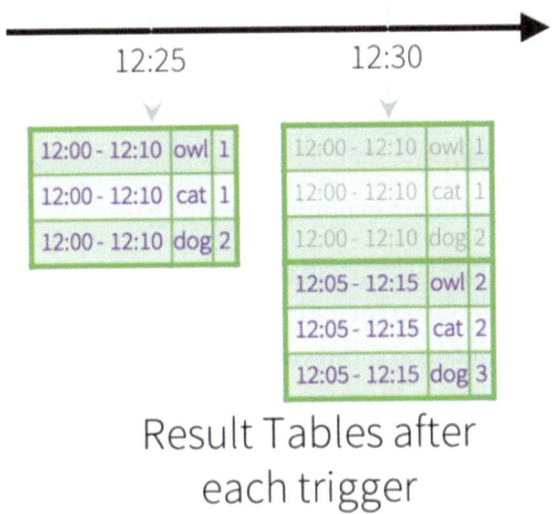

Figure 18-7. *Append mode in aggregation*

Auto Loader

Auto Loaders are a special form of streaming using micro-batch processing. Their goal is to abstract the complexity of loading file arrivals using a micro-batch pipeline. The CloudFiles protocol can monitor for ADLS, S3, and GCP for you automatically without the need to set up a file trigger in Databricks or an external system.

Combined with Delta Live Tables, Auto Loaders provide the following advantages:

- Autoscaling compute infrastructure for cost savings

  Choose from serverless or enhanced autoscaling. These two options can optimize streaming workloads by de-allocating resources that are not used quickly.

- Data quality checks with expectations

  Similar to the Great Expectation library, you can specify validation conditions in DLT and write outliners to an exception table.

- Automatic schema evolution handling

  By default, the stream will fail, and new columns will be added to the target table so logic can be applied to them if necessary. But you can also choose from different options, like rescuing the columns (by not failing), failing without adding new columns, or ignoring the new changes.

- Monitoring via metrics in the event log

  With Delta Live Tables, you can quite literally monitor everything from streaming progress to record counts, resource allocations, autoscaling activities, user audit logs, and many more possibilities.

Project Lightspeed

Announced in 2022, Project Lightspeed aims to enhance the capabilities of Structured Streaming. The following are the goals of this project:

- Improving the latency and ensuring it is predictable

  Advanced offsetting and state management capabilities are part of this project just to ensure that more state management scenarios and more responsive offsets are covered. Query performance optimization is also in scope in this project.

- Enhancing functionality for processing data with new operators and APIs

  Multiple state operators are introduced, along with Python support for the state API.

- Improving ecosystem support for connectors

  This goal is to improve support for connectors such as Amazon Kinesis and Google Pub/Sub.

- Simplifying deployment, operations, monitoring, and troubleshooting

  With the increased popularity of Python, Databricks made sure that a new Python query listener is introduced and supported in an observability API

Advanced State Management

State management has been discussed throughout this chapter. However, Databricks has continued to improve it because state management is so important. With Project Lightspeed, joining multiple state operations together is now possible, which was not previously possible. Let's consider a real-life scenario.

Use Case: E-commerce Operation

Let's consider the first scenario where an e-commerce system wants to serve ads to customers browsing their website. After getting the logs from the data center, we need to filter on specific products that contain promotions. With `mapGroupsWithState`, we can do some targeted grouping for the ads. At the same time, these filtered products can also feed into a knowledge graph in the second route for cross-selling product recommendations. Without diving into details, we can imagine the importance of being able to chain through these operations instead of separating them into different pipelines. This scenario can be found in the paper at `https://par.nsf.gov/servlets/purl/10277558`. Figure 18-8 provides an illustration.

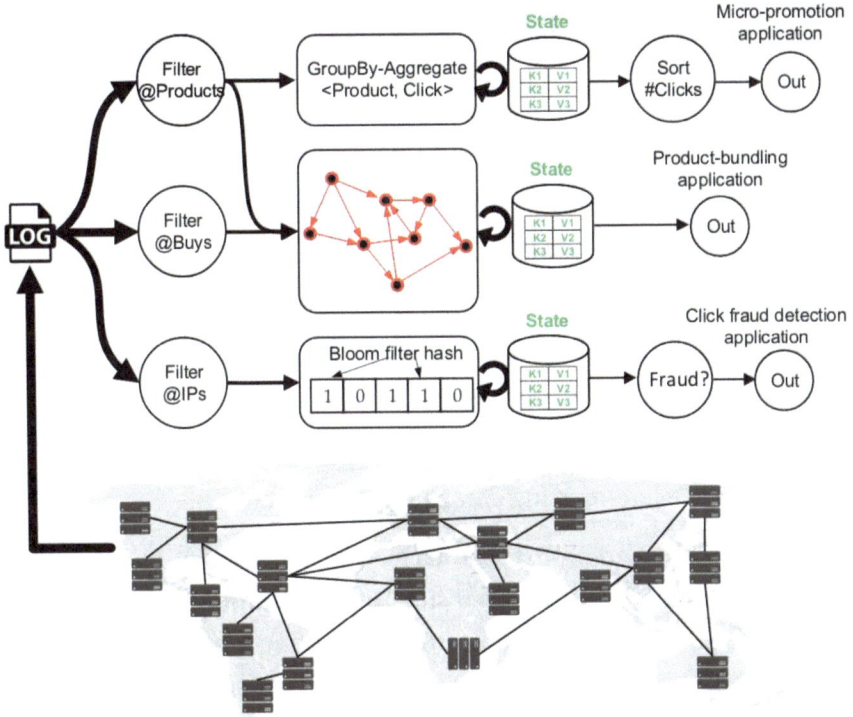

Figure 18-8. *Sample e-commerce streaming workflow*

It is also worth noting that the paper discussed the shortcoming of Spark Streaming. The DStream approach is what we initially covered. Fast-forward to Project LightSpeed; writing streaming applications is never easier with Structured Streaming.

For full details and updates on Project Lightspeed, please refer to the following blog:

```
https://www.databricks.com/blog/project-lightspeed-update-
advancing-apache-spark-structured-streaming
```

Structured Streaming Best Practices

These are some best practices:

- **Use Dataframe instead of Dataset:** Dataframe is optimized for streaming workloads.

- **Specify trigger intervals:** Control the frequency of streaming data processing.

- **Use Update mode for aggregation:** Efficiently update aggregates instead of recalculating.

- **Leverage watermark for event-time processing:** Handle late-arriving data and manage state.

- **Monitor and adjust resources:** For micro batching, use Spark UI's structured streaming monitor for detailed monitoring and troubleshooting. The streaming UI provides real-time statistics, so if anything is out of the ordinary, say when the processing rate spikes, we can take action immediately to determine if it was a cyberattack or due to some trending news. Figure 18-9 illustrates this interface. For an in-depth case study, please refer to this Databricks announcement: https://www.databricks.com/ blog/2020/07/29/a-look-at-the-new-structured- streaming-ui-in-apache-spark-3-0.html.

Streaming Query Statistics

Running batches for **26 minutes 56 seconds** since **2020/05/21 22:44:28** (**505** completed batches)

Name: myQuery
Id: 9d0dd219-995d-423e-b97d-36931110bb27
RunId: 04e101e6-8bf8-4165-9099-9104c951ba3b

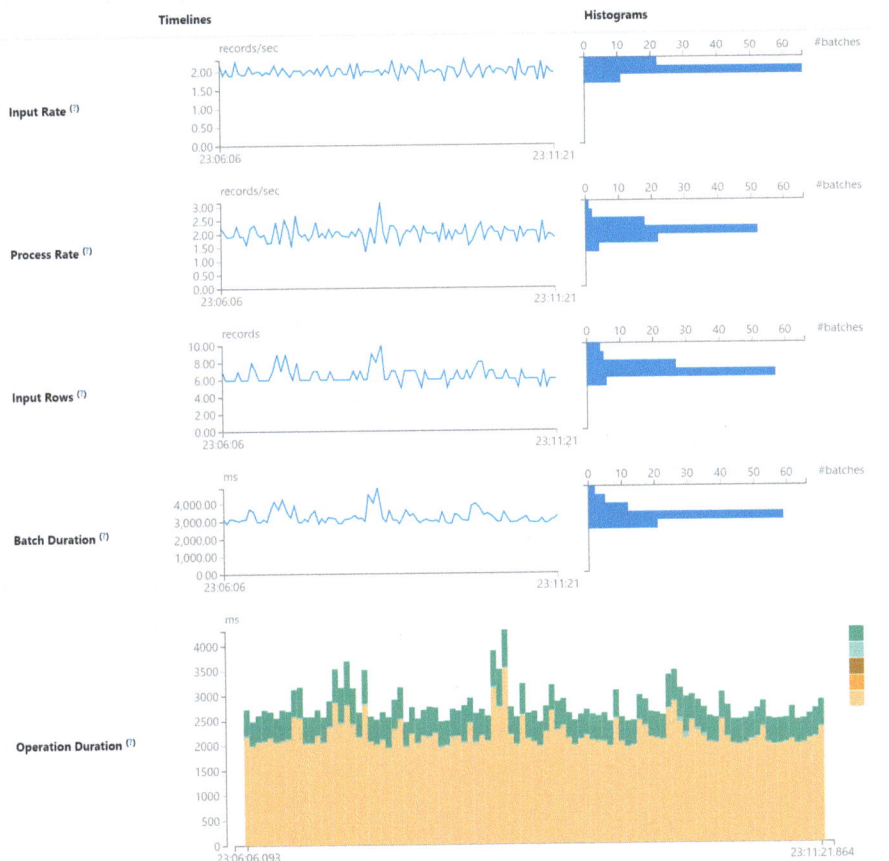

Figure 18-9. *Structured streaming monitor UI*

Conclusion

More than 14 million Structured Streaming jobs run weekly on Databricks, and that number is growing at a rate of more than two times per year (see Figure 18-10).

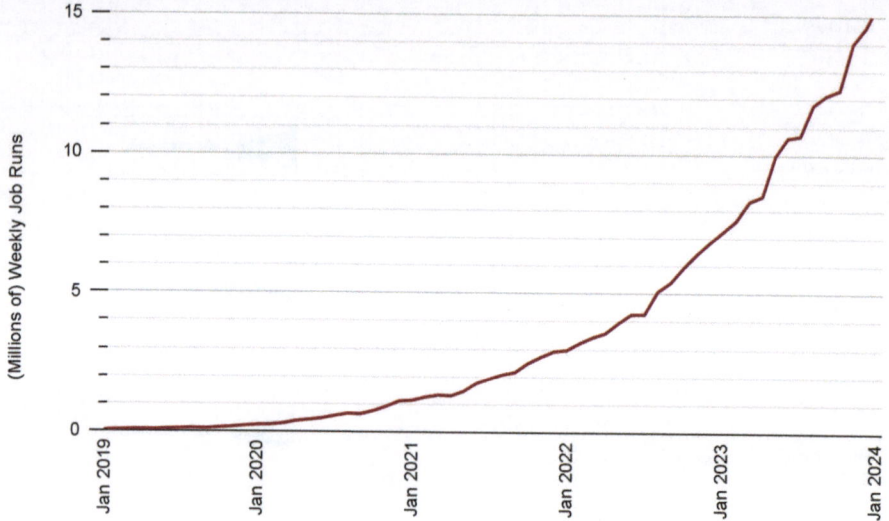

Figure 18-10. *Streaming job runs on Databricks since 2019 (source:*
`https://www.databricks.com/blog/performance-improvements-`
`stateful-pipelines-apache-spark-structured-streaming)`

From Spark Streaming to Structured Streaming, Databricks has evolved on all fronts. From the architecture perspective, it added support for continuous processing and micro-batch processing. It also added support for Python and enhanced state management and watermarking.

The introduction of the Auto Loader when working with Delta Live Tables provides groundbreaking resilience support. It also provides cost savings, comprehensive monitoring, and lineage support.

Further, Project Lightspeed makes a series of enhancements, not only integrating some core Spark innovations into streaming, like Adaptive Query Execution, proving that Spark is fully capable of running at scale and in real time, but it also extends support for different platforms. Asynchronous checkpointing in micro-batch processing is another commitment for Databricks to take streaming more and more seriously.

CHAPTER 19

From Ideation to Creation: A Walk-Through of Building a GenAI Application

In this chapter, we will walk through creating a healthcare and life science application from start to finish. The input is some realistic patient data, but this data was generated by a high-quality data generator, so there are no privacy concerns in this scenario.

We will discuss the downsides of having low-quality data, which will affect downstream data. We will combine the classic machine learning approach and the latest and greatest GenAI techniques for making one great solution. Rest assured, if you are able to master this project, you are already an accomplished data and AI architect.

The Problem Statement

According to the World Health Organization (`https://www.who.int/health-topics/diabetes`),

> *"Diabetes is a chronic, metabolic disease characterized by elevated levels of blood glucose (or blood sugar), which leads over time to serious damage to the heart, blood vessels, eyes, kidneys and nerves. About 422 million people worldwide have diabetes, the majority living in low-and middle-income countries, and 1.5 million deaths are directly attributed to diabetes each year."*

The term *chronic* is the most important in the statement because it means it is a disease that will follow you for a lifetime, and there is no cure for this disease right now. Not only that, the stage of diabetes can range from pre-diabetes to diabetes, and it can also lead to complications later in life if not carefully treated such as blindness, amputation, or even kidney disease.

In this chapter, we will develop a machine learning classification model to classify the severity of diabetes complications using a patient's medical history. We will demonstrate how to use GenAI to give book recommendations to the patient. We do not recommend seeking medical help from GenAI at this stage. That's why this application is meant for enrichment and not medical advice. Using the AI Agent Framework, we will build a chatbot to answer some of the questions related to the complications of diabetes and the ebooks. Finally, we will mimic a real dashboard used in a medical institute to demonstrate that Databricks can build everything from end to end. Figure 19-1 shows this flow in an architecture diagram.

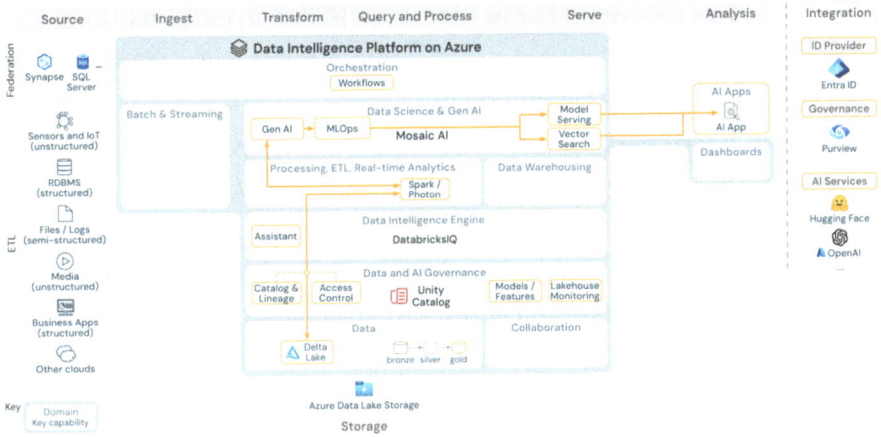

Figure 19-1. *Architecture of a GenAI pipeline*

Data Generation: Source

In this section, we will discuss how to acquire the necessary data for our
use case. Most of the demo code out there leverages Kaggle or open-source
data. Not only is it difficult to acquire high-quality healthcare data, but it is
also impossible to get a large amount of data. Here are some ideas:

Idea 1: Generate data based on a medical journal.

This idea is to leverage pre-existing experiments and reverse engineer
their dataset. Based on a medical journal, generate some random data that
falls within the range and apply some medical knowledge rules on top. For
example, obesity, by definition, has a BMI greater than 30, and our data
will make sure the BMI is greater than 30. This is probably good enough
for a demonstration. The paper "T1DMicro: A Clinical Risk Calculator for
Type 1 Diabetes Related Microvascular Complications" has provided some
insights into what it uses to determine Type 1 diabetes complications.

https://www.ncbi.nlm.nih.gov/pmc/articles/PMC8583376/

Without expert domain knowledge, it is difficult to judge if the data generated makes sense, even at a small scale. Nevertheless, the generated data can be found in the GitHub repo.

Idea 2: Generate data based on a data generator.

Synthea (`https://github.com/synthetichealth/synthea`) is an open-source project that can generate synthetic, realistic (but not real) patient data and associated health records in a variety of formats, including CSV. Using this application, we can generate as many patients as we want along with their medical journey. We didn't think of the later part when generating random data in the first place. Although we could still use the latest record of their hospital visit, it is important to understand that in real life, there will be historical transactions, be it financial records or retail transactions. Having this data will help with different types of modeling.

The detailed data dictionary can be found here: `https://github.com/synthetichealth/synthea/wiki/CSV-File-Data-Dictionary`.

Idea 3: Use the latest and greatest AI model designed for data generation.

> "I believe open source AI will become the industry standard and is the path forward. Partnering with Databricks on Llama 3.1 means advanced capabilities like synthetic data generation and real-time batch inference are more accessible for developers everywhere. I'm looking forward to seeing what people build with this."
>
> —Mark Zuckerberg, founder and CEO, Meta

The is an article that describes how to generate the data using Nvidia GPU and Llama 3.1. If the Synetha data generator doesn't work well, this could be a good idea, but the development process will take time.

`https://developer.nvidia.com/blog/creating-synthetic-data-using-llama-3-1-405b/`

Data Ingestion: Ingest

There are a couple of ways to ingest the data:

- Using the Ingestion UI

- Load data using the Auto Loader

Repeat the same code for the rest of the data. The Databricks Auto Loader will pick up the new files when they arrive the next time the job is run. Hence, we need to ensure that we keep one folder per file type.

Data Transformation: Transform

After getting the data to the Bronze layer, we need to transform the data into features for our machine learning model. We also need to filter our data to ensure we are picking up the latest visit. Data cleansing is also needed to reduce the noise in our model.

1. In the NLM report, different values of HbA1C are calculated as features.

2. For complications, we will take the latest diagnosis and rank the severity.

3. Medication is the third item we are interested in. We want to list all the generic medication names without dosage and represent them in columns. If a patient takes a medication, we will mark this column as 1; otherwise, it will be 0.

4. Following an actual report in the United Kingdom,
    we will build out a sample view and publish it
    to PowerBI: https://digital.nhs.uk/data-
    and-information/publications/statistical/
    national-diabetes-audit-type-1-diabetes/nda-
    type-1-2021-22-detailed-analysis.

5. Finally, we will demonstrate an important feature
    that is a Databricks SQL AI function.

Using Serverless SQL for Transformation

In most cases, we can use serverless SQL for the job. Listing 19-1 shares the
detailed code for the A1C features.

Listing 19-1. Patient_A1C Table

```
create table patient_a1c as
WITH agg_observations AS (
    SELECT
      patient,
      max(value) max_a1c,
      avg(value) avg_a1c,
      stddev(value) std_a1c
    FROM
      observations
    WHERE
      category = 'laboratory'
      AND LOWER(description) LIKE '%a1c%'
    group by patient
)
,filtered_observations AS (
    SELECT
```

```
    patient,
    date,
    description,
    value as current_value,
    LEAD(value) OVER (ORDER BY date desc) AS previous_value,
    ROW_NUMBER() OVER (PARTITION BY patient ORDER BY date
DESC) AS rn
  FROM
    observations
  WHERE
    category = 'laboratory'
    AND LOWER(description) LIKE '%a1c%'

)
SELECT
  latest.*,
  avg_a1c,
  std_a1c,
  max_a1c
FROM
  filtered_observations latest
  JOIN agg_observations a ON latest.patient = a.patient
WHERE rn = 1
```

By doing a little research or by asking an AI program, we can easily find
what the common diabetes complications are. Listing 19-2 shows the code
to group the complications.

Listing 19-2. Patient_complication Table

```
create table patient_complication as
select
patient,
case
when description = 'Kidney Disease' then 'Sev 0'
when description = 'Amputation' then 'Sev 1'
when description = 'Retinopathy' then 'Sev 2'
when description = 'Neuropathy' then 'Sev 3'
when description = 'Hyperglycemia' then 'Sev 4'
when description = 'Proteinuria' then 'Sev 5'
when description = 'Diabetes' then 'Sev 6'
when description = 'Pre-Diabetes' then 'Sev 7'
else description
end as Severity from
(
SELECT
    patient,
    case
    when lower(description) like '%neuropathy%' then
    'Neuropathy'
    when lower(description) like '%retinopathy%' then
    'Retinopathy'
    when lower(description) like '%nephropathy%' then
    'Nephropathy'
    when lower(description) like '%blindness%' then 'Blindness'
    when lower(description) like '%photocoagulation%' then
    'Photocoagulation'
    when lower(description) like '%amputation%' then
    'Amputation'
```

```
    when lower(description) like '%ulcer%' then 'Diabetic
    Foot Ulcer'
    when lower(description) like '%hyperglycemia%' then
    'Hyperglycemia'
    when lower(description) like '%microalbuminuria%' then
    'Microalbuminuria'
    when lower(description) like '%kidney%' then 'Kidney
    Disease'
    when lower(description) like '%proteinuria%' then
    'Proteinuria'
    when lower(description) like '%prediabetes%' then 'Pre-
    Diabetes'
    else 'Diabetes'
    end as description,
    ROW_NUMBER() OVER (PARTITION BY patient ORDER BY start
    DESC) AS rn
FROM
    patients
JOIN
    conditions ON patients.id = conditions.patient
WHERE
    lower(conditions.description) LIKE '%diabetes%'
    OR lower(conditions.description) LIKE '%Diabetic%'
    OR lower(conditions.description) LIKE '%Hyperglycemia%'
    OR lower(conditions.description) LIKE '%Hypoglycemia%'
    OR lower(conditions.description)  like '%neuropathy%'
    OR lower(conditions.description)  like '%retinopathy%'
    OR lower(conditions.description)  like '%nephropathy%'
    OR lower(conditions.description)  like '%blindness%'
    OR lower(conditions.description)  like '%photocoagulation%'
    OR lower(conditions.description)  like '%amputation%'
```

```
    OR lower(conditions.description)  like '%ulcer%'
    OR lower(conditions.description)  like '%hyperglycemia%'
    OR lower(conditions.description)  like '%microalbuminuria%'
    OR lower(conditions.description)  like '%kidney%'
    OR lower(conditions.description)  like '%proteinuria%'
ORDER BY
    patients.id, conditions.start
) where rn = 1
```

Prescribed medication comes with dosage, and it is different for different people or stages of the complication. However, if we want to use these as columns, we need to extract the medication names. Listing 19-3 is an example.

Listing 19-3. Example of Medication Name Standardization

```
emtricitabine 200 MG / tenofovir disoproxil fumarate 300 MG
Oral Tablet → {emtricitabine / tenofovir disoproxil fumarate}
```

But instead of doing manual cleanup, we can leverage Databricks' new AI function, the ai_extract() function, as shown in Listing 19-4.

Listing 19-4. Using AI Function to Extract the Medication Name

```
SELECT distinct description,
    CAST(ai_extract(description, array('medication name without
    dosage')) AS STRING) as med_wo_dosage
from
(
    select distinct description from medications where patient
    in (select patient from diabetes_training)
)
```

Figure 19-2 is the result of the query.

A^B_C description	A^B_C med_wo_dosage
1 atazanavir 300 MG Oral Capsule	{atazanavir}
2 vancomycin 1000 MG Injection	{vancomycin}
3 Diazepam 5 MG Oral Tablet	{Diazepam}
4 NDA020800 0.3 ML Epinephrine 1 MG/ML Auto-Injector	{Epinephrine}
5 benazepril hydrochloride 40 MG Oral Tablet	{benazepril hydrochloride}
6 aspirin 81 MG Oral Tablet	{aspirin}
7 remifentanil 2 MG Injection	{remifentanil}

Figure 19-2. *Result of the AI extract query*

Next, using the `pivot` function, we can transpose the active ingredient
into columns. Our goal is to transpose the rows in Figure 19-3 to columns
in Figure 19-4. Listing 19-5 and Listing 19-6 together will perform this
action. However, because the `PIVOT` function does not allow a dynamic list,
we need to construct a query that is understandable by the engine.

Listing 19-5. Transforming "Active Ingredients" into a List

```
descriptions = [row['desc'] for row in spark.sql("SELECT
DISTINCT `Active Ingredients` as desc FROM medications m join
med_mapping mm on m.description = mm.Prescription").collect()]
```

Listing 19-6. Using the pivot Function to Transform Rows
into Columns

```
# Constructing the dynamic part of the pivot query
pivot_clause = ", ".join([f"'{desc}'" for desc in
descriptions])

# Constructing the full query
query = f"""
SELECT *
FROM (
```

```
SELECT
  m.PATIENT,
  inline(
    collect_list(
      named_struct(
        'description', `Active Ingredients`,
        'stop', CASE WHEN m.STOP IS NULL THEN '1'
        ELSE '0' END
      )
    )
  ) AS (description, stop)
FROM
  medications m
  join med_mapping mm on m.description = mm.Prescription
GROUP BY
  m.PATIENT
) AS subquery
PIVOT (
  MAX(stop)
  FOR description IN ({pivot_clause})
)
"""
```

Figure 19-3 shows what the result looks like.

	PATIENT	Nebivolol	Ferrous Sulfate	Metoprolol	Astemizole	Canagliflo:
1	007a9002-2e48-6355-3381-f4ee4a27e54c	null	1	1	null	null
2	008355cc-1c41-605b-09d3-c215de2b787d	null	null	null	null	null
3	00b87227-cb36-a742-e48f-8c76a9f2f60f	null	null	null	null	null
4	00cd7111-33a9-7b92-7e21-2957abfd48ab	null	null	null	null	null
5	00eeb375-c498-4a0c-9039-e3f5729d3180	null	null	1	null	null
6	00f1164b-0c9c-b3a6-af44-d91cebedc15e	null	null	null	null	null
7	0154b206-3736-a18f-5a86-3500b9768588	null	null	null	null	null
8	016d3877-fa29-54f3-2421-e8faec5d725e	null	null	1	null	null
9	0186a10f-6e39-9344-9c39-0e94f9266ebc	null	null	null	1	null
10	01934c5c-915f-4862-f309-047ac33adceb	null	null	null	null	null

Figure 19-3. *Patient medication table*

Finally, we can combine all the new tables together to create one
training table, as shown in Figure 19-4.

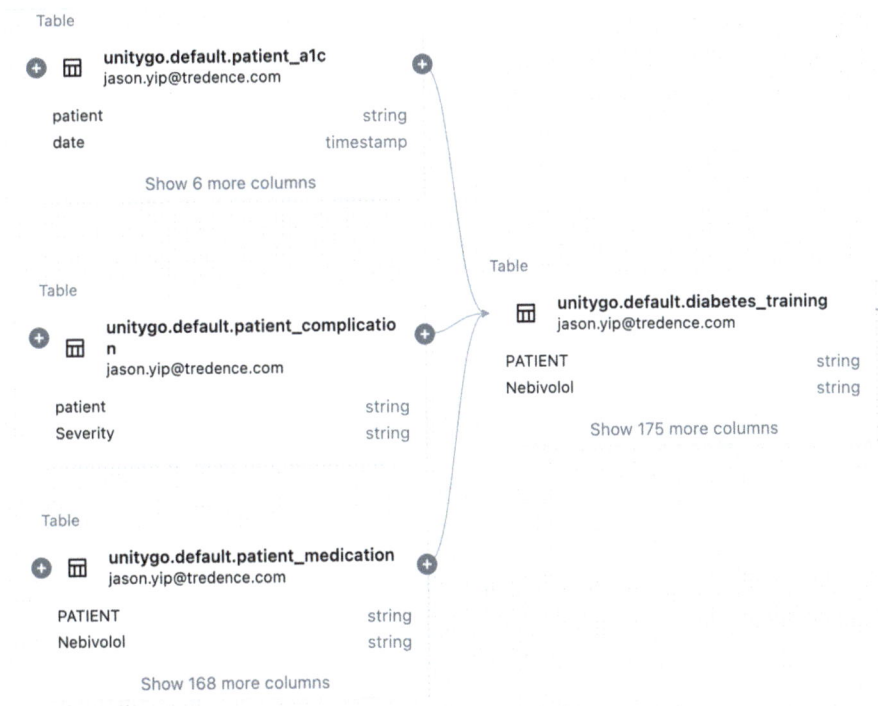

Figure 19-4. *Lineage of the training table*

Machine Learning Model for Diabetes Complication Classification: Query and Process

Typical machine learning projects involve a process called *exploration data analysis* (EDA). But we are going to take a leap of faith and see if our data makes sense by running AutoML on the two datasets that we have. There isn't anything that needs to be done here; just choose the input dataset and select our target variable.

Dataset 1: Randomly generated data

From Figure 19-5, we can tell that the best model scores 0.24. This is not an acceptable base model for fine-tuning, but that's the best we can get with random data.

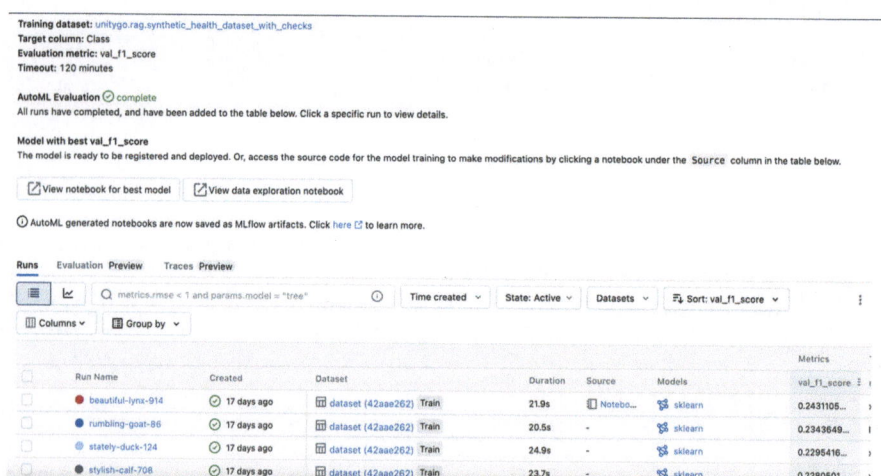

Figure 19-5. *AutoML results for randomly generated data with checks*

Dataset 2: Synthea-generated data

The program has proven track record for generating realistic data. We will give AutoML another try with the diabetes_training data. The result is surprisingly ~0.73 (see Figure 19-6).

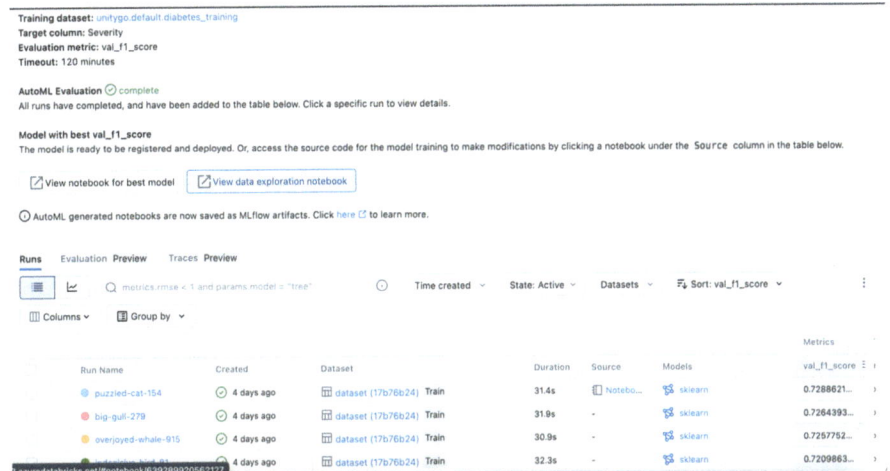

Figure 19-6. *AutoML results for Synthea-generated data with checks*

Generative AI: Serve

This is perhaps one of the most anticipated steps in the whole pipeline. We are using the Mosaic AI stack here. We will first try prompt engineering, and we will move on to a RAG use case by building a simple chatbot. Imagine someone is diagnosed or predicted to have diabetic complications; it can be a daunting task to go through all the self-help guides available. We can first ask an LLM in Databricks' Playground for ebook recommendations for self-help purposes (see Figure 19-7).

445

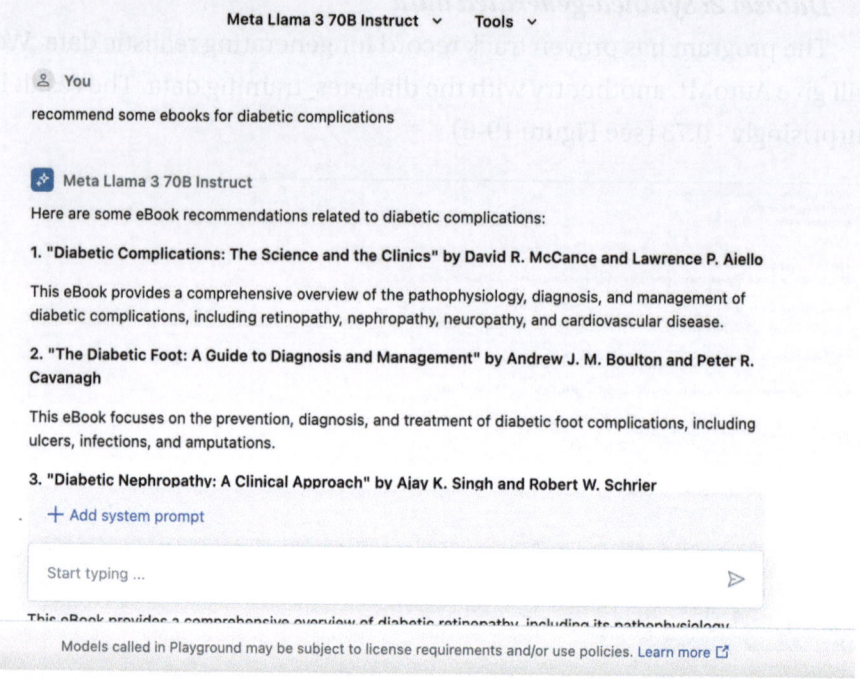

Figure 19-7. *Output from Databricks' GenAI playground*

Next, assuming we have legally acquired all the ebooks, we want to upload them to a volume in Unity Catalog. This step can be done easily via the user interface. We can navigate to a Unity Catalog, choose a preferred database, and click the "Upload to this volume" button, as shown in Figure 19-8.

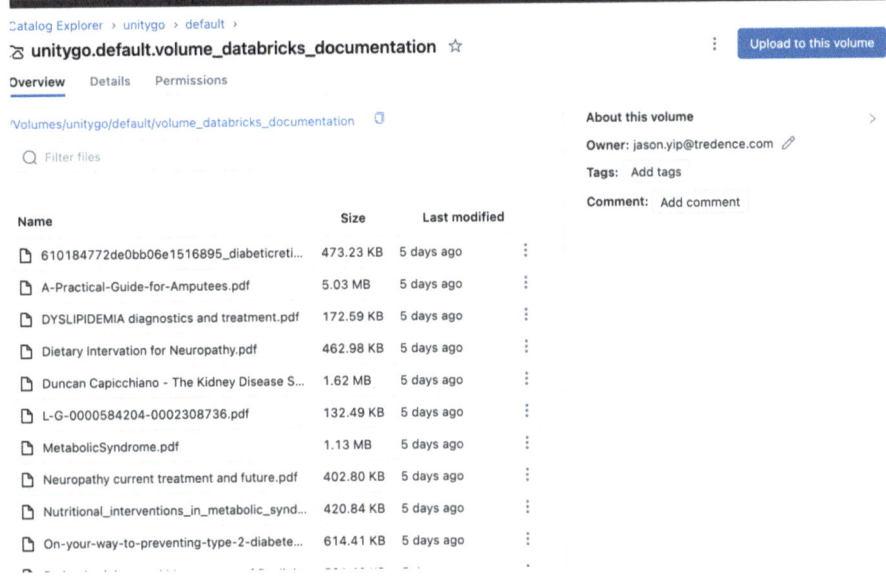

Figure 19-8. *Volume interface from Databricks*

As a starter, we will leverage Databricks' prebuilt template for our proof-of-concept app, and later we will look at techniques to "tune the quality knobs."

☆ **IMPORTANT** ☆

Please bookmark the following site because it contains best practices for the evolving topic of using the AI Agent framework:

```
https://ai-cookbook.io/
```

The following are the prerequisites needed to leverage the AI Agent framework, which can be set up via the user interface:

- **Unity Catalog and Schema:** For storing the parsed/ chunked documents

- **Vector Search Endpoint:** Either a new endpoint or an existing one

- **UC Volume:** An volume that was created using the command and then the documents can be uploaded

- The Foundation Model API is accessible for embedding a calculation

- The MLflow experiment is accessible for model and metric logging

Where Do We Start?

Here are the steps:

1. Verify all the permissions, and deploy something if you are not sure they are correct.

2. Clone the following repo from Databricks into the repo:

   `https://github.com/databricks/genai-cookbook/`

3. Fill in the config in `rag_app_sample_code/00_ global_config`. If you have followed these prerequisites, you should not have any problems filling in the details, but a couple of details that are extremely important not to miss, and can be found in the user interface, include the following:

a. VECTOR_SEARCH_ENDPOINT can be found on the
   Compute tab, as shown in Figure 19-9.

Figure 19-9. *Vector search endpoint UI*

b. Volume can be found under the database of a catalog,
   as shown in Figure 19-10.

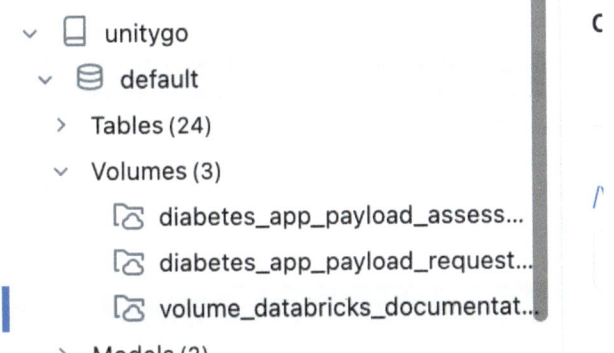

Figure 19-10. *Volumes can be found under a database*

4. To verify everything is set up correctly, run 01_
   validate_config in the corresponding folder of the
   file type of your choice. For example, for PDF files,
   they can be found at the following location:

   /genai-cookbook/rag_app_sample_code/A_POC_app/
   pdf_uc_volume/01_validate_config

   If everything passes, you will see a print message at
   the end of each cell, as shown in Figure 19-11.

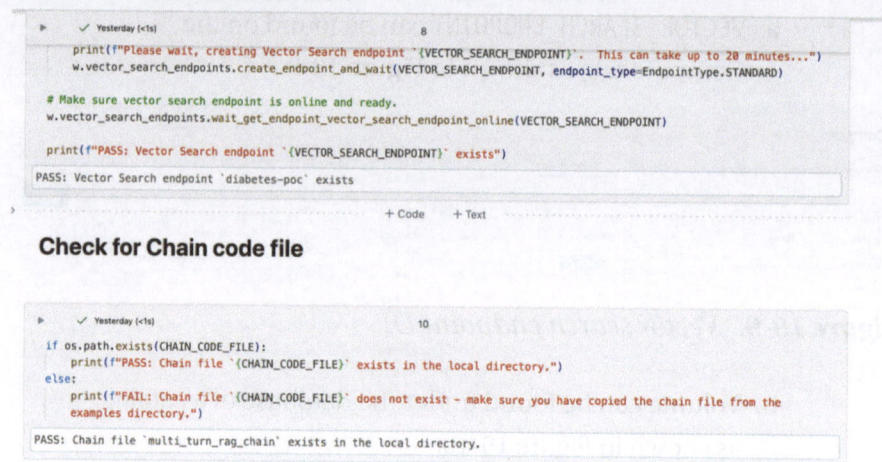

Check for Chain code file

Figure 19-11. *Passing messages at the end of each command*

5. Run 02_poc_data_pipeline, and the result is the
    experiment being logged on the Experiments tab, as
    shown in Figure 19-12.

Experiments

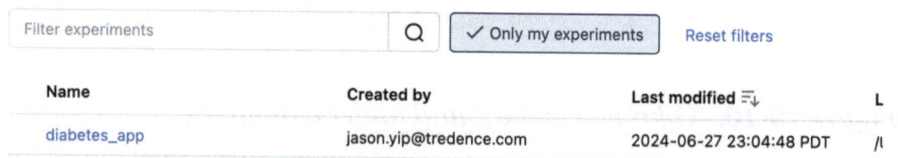

Figure 19-12. *Experiment logged after running the data pipeline*

6. Run 03_deploy_poc_to_review_ap. It will deploy
    an application for you, and you can ask questions
    about your use case. You will notice the relevant
    document is being referenced in the chat, as shown
    in Figure 19-13.

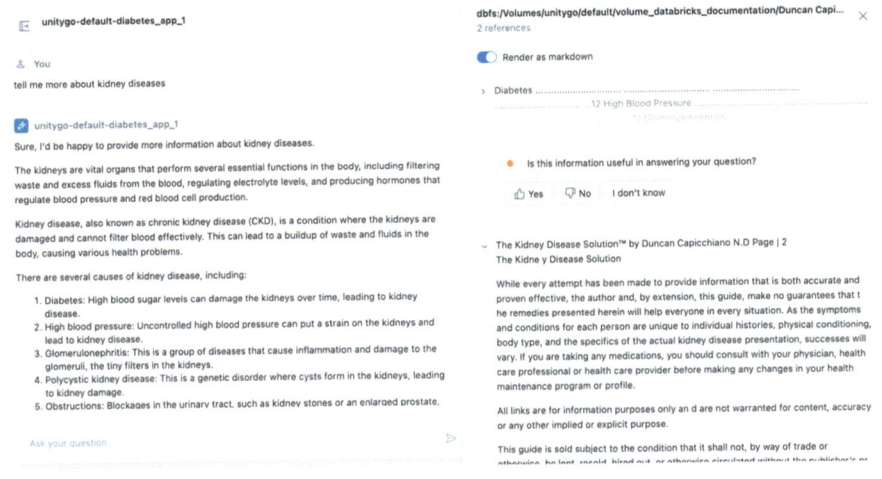

Figure 19-13. *Databricks hosted chatbot interface*

7. Congratulations, you have successfully deployed a chatbot, but the story does not end here.

   a. `04_create_evaluation_set` allows you to create an evaluation dataset. If you know some of the answers that might come from the PDF that you uploaded, you can set up this evaluation dataset so the AI Agent framework can evaluate the accuracy of the output.

   b. `05_evaluate_poc_quality` runs this notebook to evaluate the application.

8. If you have made it this far, I am sure you will appreciate how much work Databricks has done to make it easy, but the story is far from over. Please head over to "RAG quality knobs" section of the GenAI cookbook: `https://ai-cookbook.io/nbs/3-deep-dive.html`. That's where we will learn more about the underlying process of building a RAG application so your application will be future-proof.

Monitoring Dashboard: Analysis

To evaluate hospital treatments, we can build a dashboard similar to
England's National Health Service (see Figure 19-14). We will create a
materialized view for the report so we can refresh the view when the report
is scheduled to refresh, saving time and cost as it can compute incremental
changes.

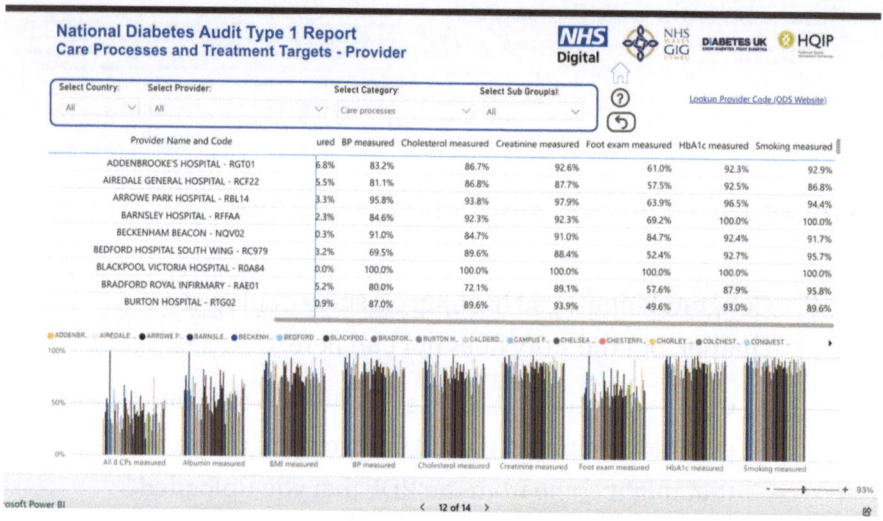

Figure 19-14. *Sample diabetes dashboard from England's National
Health Service (source: https://digital.nhs.uk/data-and-
information/publications/statistical/national-diabetes-
audit-type-1-diabetes/nda-type-1-2021-22-detailed-analysis)*

Listing 19-7 is a query similar to the one on the dashboard in
Figure 19-15.

Listing 19-7. Query for the Health Dashboard

```
create materialized view patient_report as
with encounter_latest as
(
```

```
    select
    *,
    ROW_NUMBER() OVER (PARTITION BY patient ORDER BY start
    DESC) AS rn
    from encounters e
),
observations_new as
(
    select
    distinct
    patient,
    encounter,
    category,
    (case
        when lower(description) like '%cholesterol%' then
        'cholesterol'
        when lower(description) like '%blood%pressure%'
        then 'BP'
        else description end) as description_alias
    from
    observations
)
select o.Name as OrganizationName, o.City as OrganizationCity,
    count(distinct e.patient) as Count_of_Patients,
    count(distinct encounter) as Count_of_Encounters,
    sum( case when category = 'laboratory' and lower
    (s.description_alias) like '%albumin%' then 1 else 0 end )
    as Count_of_Albumin,
    sum( case when category = 'vital-signs' and lower
    (s.description_alias) like '%bmi%' then 1 else 0 end ) as
    Count_of_BMI,
```

```
    sum( case when category = 'vital-signs' and lower(s.
    description_alias) like '%blood%pressure%' then 1 else 0
    end ) as Count_of_BP,
    sum( case when category = 'laboratory' and lower(s.
    description_alias) like '%cholesterol%' then 1 else 0 end )
    as Count_of_Cholesterol,
    sum( case when lower(s.description_alias) like '%foot%'
    then 1 else 0 end ) as Count_of_Foot_exam,
    sum( case when category = 'laboratory' and lower(s.
    description_alias) like '%a1c%' then 1 else 0 end ) as
    Count_of_HbA1C,
    sum( case when category = 'social-history' and lower(s.
    description_alias) like '%smoking%' then 1 else 0 ) as
    Count_of_Smoking
from encounter_latest e
join observations_new s on e.id = s.encounter
join organizations o on e.organization = o.id
where e.patient in (select patient from diabetes_training) and
e.rn = 1
group by o.Name, o.City
```

We can publish this view to a Power BI dashboard, as shown in
Figure 19-15.

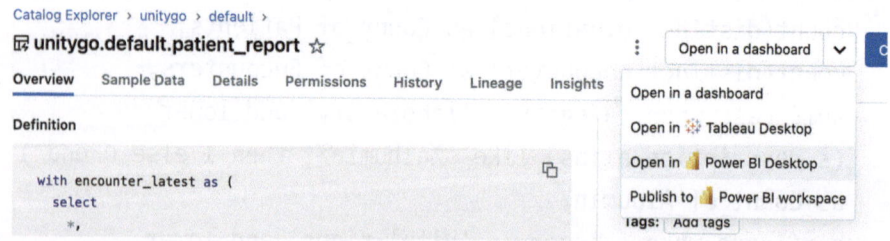

Figure 19-15. *Publishing the materialized view to Power BI Desktop*

Conclusion

The Databricks data intelligence platform not only provides the intelligence to power the next generation application, but it is also built on top of the lakehouse architecture, so it has a strong foundation of supporting any size of workload and complex data transformation. In this chapter, we have demonstrated the ability to create an application from ideation to creation, all within Databricks. While other platforms may come with similar tools in their ecosystem, Databricks' tight integration allows us to stay on the same platform and collaborate closely with the team. Unlocking the GenAI revolution has never been easier!

Index

A

Access control lists (ACLs), 91, 98, 102, 401, 402

Access control mechanisms, 101

Access Tokens, 39, 402

Account admin, 94, 95, 385

ACLs, *see* Access control lists (ACLs)

Add Data UI, 47, 49

Add UI Interface, 47

ADF, *see* Azure Data Factory (ADF)

AI, *see* Artificial intelligence (AI)

AI/BI dashboards, 159, 160

AI/BI Genie, 334, 348–350

ai_extract() function, 440

AI-powered governance, 339

    AI-generated comments enhancements, 340, 341

    AI security filtering, 344–346

    AI security framework, 346

    governance process, 340

    lineage, 341, 342

    PII masking, 342, 343

    Unity Catalog federated governance, 339

ai_query() function, 176

AI2 Wildbench, 276

Amazon S3, 48

API, *see* Application programming interface (API)

Application programming interface (API), 16

Artificial intelligence (AI), 2, 7

    foundation models, 220

    models, 220, 221

Attention mechanism, 318–320

Auto-documentation generation, 221

Auto Loader, 46, 55

    advantages, 422

    Checkpoint, 56

    cloudFiles, 55

    direct listing, 57, 58

    file notification, 58

    manually.inferColumnTypes, 57

    manually.Rescue_Data, 57

    mergeSchema, 56

    Trigger.AvailableNow, 56

Automated machine learning (AutoML), 184–188, 194–196, 198, 199, 444, 455

AutoML, *see* Automated machine learning (AutoML)

Autoregressive, 319

Autoscaling, 9, 16, 19, 25, 28, 30, 133, 144, 145, 147–148, 152, 381, 422
Autotermination, 28
Azure Data Factory (ADF), 47, 51
Azure Data Lake Storage, 48

B

Batch ingestion, 51
Bge-large-en model, 268
BI, *see* Business intelligence (BI)
Bronze layer, 46, 54, 66, 74, 137, 169, 435
Built-in functions, 174
Business intelligence (BI), 2, 3, 13, 153

C

Catalog, 17
Catalog UI, 50
CDC, *see* Change data capture (CDC)
CDF, *see* Change data feed (CDF)
Centralized user management, 100
Chain of thoughts, 258, 265
Chandy-Lamport algorithm, 415
Change data capture (CDC), 74, 133
Change data feed (CDF), 74, 75
Chatbots, 219, 230, 257, 283, 285, 294, 302, 304, 333, 337, 432, 445, 451

ChatGPT, 209, 219, 220, 230
CHECK constraint, 165
Cloud compute costs
    compute cost, 372
    networking cost, 373
    storage cost, 372
cloudFiles, 55, 422
Cloud ingestion, 45
    Add Data UI, 47–50
    file upload UI, 47, 50, 51
    Fivetran, 53
    landing zone, 46
    native cloud tools, 47
    notebook, MySQL to Delta table, 49
    Partner Connect, 52
    via cloud-native tools, 51, 52
    via third-party tools, 52, 53
Cloud object storage layer, 64
Cluster access controls, 31, 381
Cluster access modes, 25
    shared clusters, 25, 26
    standard single-user clusters, 25
    types, 24
Cluster autoscaling, 147, 381
Cluster auto-termination, 381
Cluster best practices, 30, 31
Cluster policies, 22–24, 380
Cluster-scoped libraries, 38
Cluster tags, 28, 30, 383
CMKs, *see* Customer-managed keys (CMKs)
CNNs, *see* Convolutional neural networks (CNNs)

Column masks, 106, 107

Command-line interface (CLI), 16, 17, 39, 102

Conditional execution, 124, 125

Content moderation, 344, 345

Continuous integration/ continuous deployment (CI/CD)

    databricks, 359

    end to end flow, 354

    stages

        build, 357

        deploy, 357

        source, 357

        test, 357

Continuous processing, 414–417, 428

Convolutional neural networks (CNNs), 322, 325

COPY INTO command, 46, 58–60

COPY INTO validate mode, 59

Cosine similarity, 235, 236

CREATE TABLE command, 75

Curated models, 226, 227

Customer-managed keys (CMKs), 398, 408

D

Data access, 90, 104, 107

Data architecture, 2, 4

Databricks

    access modes, 26

    account, 15

AI, 12

algorithms, 196, 197

auto-generated reports, 211

automated root cause analysis, 212

AutoML, 198, 199

AutoML interface, 197, 198

bespoke LLM model, 221

Databricks Asset Bundles, templates, 364, 365

data volume, 232

DBRX, 328, 329

documentation, 41, 198, 385

external feature stores, 239

feature store, 191, 192

    advantages, 192

    AutoML, 194

    dataframe, 194

    delta table, 193

    FeatureEngineeringClient class, 193

    feature table, 193

    primary key, 193

File Upload UI, 47

GenAI, 221, 222

Git folders, 17

infrastructure, 315

ML, *see* Machine learning (ML)

models tab, 201, 202

online tables, 238

production-grade model, 198

registered models, 201, 202

REST API, 40, 41

Terraform, 41, 42

Databricks (*cont.*)
UI, 16, 47
unified monitoring, 211
Unity Catalog, 204
validation, 199
vector database, 231–233
vector search indexes, 234
widgets, 36, 37
websites, 250
workspaces, 16
Databricks all-purpose clusters
access modes (*see* Cluster
access modes)
autoscaling and
autotermination, 28
cluster creation interface, 20
cluster tags, 28
DBR, 27
policies, 22–24
pools, 29, 30
spot instances, 29
Databricks Assistant
autocomplete code/queries, 336
code conversion, 337
code explanation, 337, 338
code fixing, 338, 339
code generation, 335, 336
Databricks command-line interface
(Databricks CLI), 39
Databricks compute/
clusters, 16, 18
all-purpose compute (*see*
Databricks all-purpose
clusters)

job clusters, 19
SQL warehouses, 20
Databricks costs
account tier, 374
Azure, 376
cluster access controls, 381
cluster autoscaling, 381
cluster configuration, 375
cluster policies, 380
cluster tags, 383
compute size and type, 373
instance pools, 382
vs. photon, 378
product SKU, 374
spot instances, 382
SQL warehouse (*see* SQL
warehouse)
Databricks data intelligence
platform, 9–12, 331, 332
AI/BI Genie, 348–351
Databricks IQ, 333
Intelligent Search, 347, 348
Databricks deployment
capacity planning, 391, 392
customer-managed keys,
397, 398
encryptions/auditing, 397
NPIP, 394, 395
private link, 396, 397
VNET injection, 393, 394
Databricks file system (DBFS), 16,
38, 398, 407
Databricks Foundation Model API,
232, 233

Databricks IQ, 12, 333
  AI/BI Genie, 334
  AI-powered governance, 333,
      339 (*see also* AI-powered
      governance)
  automated job tuning, 334
  Databricks Assistant, 333, 334
      (*see also* Databricks
      Assistant)
  requirements, 333
  search and discovery, 334
Databricks job
  job-level parameters
      continuous, 122
      file arrival, 120, 121
      job parameters, 123
      job tags, 122
      scheduled, 119
      table update, 122
  task-level parameters
      cluster, 117, 118
      create, 116
      dependent libraries, 118
      duration threshold, 119
      notifications, 118
      parameters, 118
      source, 117
      task retries, 119
      type, 116
Databricks lakehouse
  cloud object storage, 7
  core concepts and
      capabilities, 8, 9
  Delta Lake, 7

enterprise-scale
    implementations, 7
  lakehouse architecture, 7
  platform, 6
Databricks MLOps stack
  CI/CD, 366
  CI/CD workflow, 369
  code review, 368
  develop model, 368
  initialize, 367
  setting CI/CD, 368
  YAML file, 368
Databricks open sharing
    protocol, 108
Databricks Partner
    Connect, 52
Databricks platform, 7
  account, 15
  catalog, 17
  cluster, 16 (*see also* Databricks
      compute/cluster)
  cluster best practices, 30, 31
  cluster creation interface, 21
  DBFS, 16
  DBR, 18
  DBU, 18
  Delta Lake, 18
  external connectivity (*see*
      External Databricks
      connectivity)
  folder, 17
  libraries, 17
  notebooks, 16 (*see also*
      Notebooks)

Databricks platform (*cont.*)
    workflows, 17
    workspaces, 16
Databricks reference architecture,
        ingestion, 46
Databricks repo, 358
    branch, 362
    checkout, 362
    clone, 361
    Cloud Git providers, 360
    commit, 362
    commit & push, 363
    Git repositories, 361
    merge, 363
    on-premises Git providers, 360
    pull, 363
    push, 363
    rebase, 363
    reset, 363
Databricks runtime (DBR), 18, 24,
        27, 38, 148
Databricks SQL (DBSQL)
    architecture diagram, 153
    BI tools, 154
    constraints
        enforced, 165, 166
        informational, 165–167
    definition, 154
    materialized view (*see*
        Materialized views)
    streaming tables,
        168, 169
Databricks-to-Databricks
        sharing, 108

Databricks unit (DBU), 18, 373,
        374, 377, 379, 388
Databricks workflows, 17, 115
    advanced features
        cluster reuse, 124, 125
        conditional execution, 125
        late jobs, 127, 128
        modularize jobs, 128, 129
        repair/re-run, 125
        running jobs, 129
    end-to-end architecture, 114
    Job Matrix View, 131
    Job Run dashboard
        error types, 130
        finished run chart, 130
        jobs list, 130
        monitoring, 131
    task types, 124
Data cleansing, 435
Data drift, 204, 206, 209
Data engineers, 113–152, 155, 168,
        181, 182, 218, 331, 378, 412
Data exploration
    pandas profiling, 188–190
    summarization, 191
Data generation, 433, 434
Data ingestion, 45–60, 135–137, 435
Data lakehouse, 4, 5, 61, 171
Data lakes, 2–5, 61–88
Data lineage, 102, 103
Data owners, 94, 349
Data scientists, 181, 184, 186, 188,
        189, 191, 195, 207, 212
Data search, 104

Datasets, 249, 271, 444, 445

Data sharing, 104, 107, 108, 110

Data swamps, 2

Data transformation, 134, 173, 435, 455

Data warehouses, 3–7, 13, 45, 60, 61, 76, 153–179

DBFS, *see* Databricks file system (DBFS)

DBR, *see* Databricks runtime (DBR)

DBU, *see* Databricks unit (DBU)

DBRX, 252, 318

  benchmarks, 315

  cost of pre-training, 313, 314

  end-to-end capabilities, 315

  enterprises, 312

  fine-tuning, 315

  inference performance, 317, 318

  instruct, open models, 315, 316

  Open AI, 316

  vs. open-source models, 311, 312, 330

  open-sourcing

    model, 314

    tooling, 314

  production, 313

  vs. prominent models, 317

  Pytorch FSDP, 327

  transformer architecture, 320–322

DBSQL, *see* Databricks SQL (DBSQL)

DDP, *see* DistributedData Parallel (DDP)

Decision trees, 197

Deep clones, 72, 73

Deep learning, 322, 334

Deep neural networks (DNNs), 321, 322

Defragmentation, 85

Delta bronze layer, 46, 137

Delta ingestion, 46, 54

  via Auto Loader, 54–58

  via COPY INTO, 54, 58–60

Delta Lake, 7, 18, 46–48, 53–56

  challenges, other storage formats, 61

  change data feed (CDF), 74–76

  clone delta tables, 71–73

  components, 62, 63

    cloud object storage layer, 64

    Delta log, 63

    Parquet files, 63

  definition, 62

  Delta format, 86

  Delta optimization, 80

    optimize, 80, 81

    partitioning, 80

    vacuum, 82

    Z-ordering, 81

  generated columns, 73

  key features

    ACID transactions, 64

    compliance, 65

    schema enforcement, 64

    time travel, 65

Delta Lake (*cont.*)
    unified batch and
      streaming, 64
    version control, 64
  limitations, 84
  liquid clustering, 82–84
  Medallion (*see* Medallion
    architecture)
  MERGE SQL operation, 68
  ML/AI to rescue, 88, 89
  predictive I/O, 85
  schema evolution, 69–71
  storage protocol, 61
  time travel feature, 70, 71
  Universal Format, 77–80
  Update and Delete
    commands, 68
Delta Lake format, 7, 13
Delta Live Tables (DLT), 116,
    133, 206
  CDC, 137
  create pipeline, 142, 143
  DAG, 145
  data ingestion, 135, 137
  enhanced autoscaling
    algorithm, 147
  expectations, 139, 151
  logs, 146
  materialized view, 135
  metrics, 141
  monitor, 146
  parameters
    compute, 144
    pipeline mode, 144

    product edition, 143
    source code, 143
  retail sales pipeline (*see* Retail
    sales pipeline)
  streaming table, 135
  views, 135
  PowerBI Desktop, 178, 179
Delta log, 63, 70
Delta Sharing
  confirmation, 110
  customers, 110
  differentiators/benefits, 108
  working, 108, 109
Delta UniForm, 77
Deploy model
  model overview, 200
  model registration, 200, 201
  Model Serving/inferencing,
    202, 203
    inference tables, 203
    query endpoint button,
      203, 204
    Unity Catalog, 203
    workflow, 204
Descriptive analytics, 2
Dev environment, 98
Digital transformation, 1
DistributedDataParallel (DDP), 326
Distributed training libraries, 326
DLT, *see* Delta Live Tables (DLT)
DNNs, *see* Deep neural
    networks (DNNs)
Document Q&A, 229, 230
Dynamic view, 107

E

E-commerce system, 424, 425

Embeddings, 230–235, 240, 241, 259, 261, 267–270

Encoder and decoder network, 319

Enforced constraints, 165, 166

Enhanced Autoscaling algorithm, 147

Epoch markers, 415

ERP/CRM systems, 45

Evaluation-driven approach, 305–307

Exploratory data analysis (EDA), 186, 258, 444

External Databricks connectivity
    Databricks CLI, 39, 40
    Databricks Terraform, 41, 42
    REST APIs, 40, 41

External tables, 93

F

Feature engineering, 187, 188, 194, 195, 256

File arrival triggers, 120, 121

File upload UI, 47, 50, 51

Fine-tuning, 10, 220, 247, 248

Fivetran, 47, 50, 53

Foundation Model API, 328, 330

Fully Sharded Data Parallel (FSDP), 326, 327

G

Gartner Magic Quadrant, 182

GenAI, *see* Generative artificial intelligence (GenAI)

GenAI techniques
    architecture, 433
    dashboard, 452, 454
    output, 446
    steps, 448–451
    user interface, 448
    volume interface, 447

Generative artificial intelligence (GenAI), 9–12, 219, 253
    content, 220
    embeddings, 231
    journey, 223, 224, 247
    knowledge graph, 231
    pricing, 250
    vs. ChatGPT, 220, 221

Genie space, 349–351

Glass Box approach, 184

Gold layer, 67, 75

H

Hadoop systems, 2, 45

Hierarchical Navigable Small World (HNSW), 235, 236

HNSW, *see* Hierarchical Navigable Small World (HNSW)

Hybrid PaaS, 390

I

Identity and access control
    ACL, 401, 402
    best practices, 407
    IP access lists, 399–401
    MFA, 399
    SAT, 404–406
    SSO, 399
    Token Management API, 402–404
If/else condition task type, 126
Informational constraints, 165–167
Ingestion partners, 52
Instance pools, 382
Interactive clusters, 21, 53, 117,
    152, 157, 379, 388

J

Job clusters, 19, 81, 117, 122, 157,
    378, 380
Job parameters, 123, 124
Job tags, 122
Jupyter notebooks, 31

K

Kafka, 48, 51, 135–137, 169,
    170, 410

L

Lakehouse, 5
    architecture, 3
    data architecture paradigm, 4

    object stores/cloud-based
        storage, 4
    paradigm, 6
Lakehouse monitoring
    drift analysis metrics, 206, 207
    inference profile, 207
    metrics tables, 210
    model metrics, 209
    monitor, 205, 206
    notebooks, 214
    nyctaxi_trips, 213
    one-stop interface, 209
    output table, 214
    queries and dashboard, 214, 215
    reports, 209, 211
    Responsible AI, 208
    reusable format, 212
    set up, 205
    snapshot profile, 206
    table schema, 209
    tables documentation, 214
    tables relationship, 213
    tables usage, 214
    time series profile, 206
LangChain model, 280–283
LangChainTool, 242
Large language model operations
    (LLMOps)
    advantages, 260
    bge-large-en model, 268
    calculate embeddings, 269, 270
    components, 258
    considerations, 256
    create vector search, 267

feedback mechanism
    model packaging, 262
    Model Serving and
        interference, 263
    index creation, 266
    judging, 278, 279
    model evaluation, 262
    model fine-tuning, 261, 270
    model pre-training, 262, 271
    prompt engineering
        chain of thoughts, 265
        templates, 263–265
Large language models (LLMs), 9,
        10, 12, 220, 222, 224, 232,
        242, 244, 247, 253, 311
Late job, 127, 128
Legacy dashboards, 159
Libraries, 17, 38, 39
LightGBM, 197
Liquid clustering, 61, 82–84
Llama, 253, 314, 434
Llama 2, 252
Llama-2-70B-Chat model, 226
Llama Guard, 345
LLMs, *see* Large language
        models (LLMs)
Logistic regression, 197
Long-Term Support (LTS)
        versions, 27, 117

M

Machine learning (ML), 2, 3, 13, 182
    best practices, 215–218

black boxes, 184
components, 182
engineers, 181, 218
experiments, 183, 184
flight delay/cancellation, 185
    data preparation, 186
    EDA, 186, 187
    feature engineering, 187, 188
lifecycle, 185
problem types, 195, 196
randomly generated data, 444
Synthea-generated data, 445
user personas, 181
Machine learning
        operations (MLOps)
    deployment server, 273
    interfaces for LLM, 280
    LangChain, 280–283
    roles/responsibilities, 255
Massive Text Embedding
        Benchmark (MTEB), 269
Materialized views
    AI functions, 173
    BI tools, 176
    create, 170
    custom models, 176
    Lakehouse Federation,
        171, 173
    LLM models, 174, 175
    PowerBI, 177, 178
    refresh, 171
Medallion architecture, 65
    Bronze layer, 66
    Gold layer, 67

Medallion architecture (*cont.*)
  main layers, 65
  as multihop architecture, 65
  Silver layer, 66
Merge conflict, 363
Metastore admin, 94, 98, 386
Micro-batch processing, 410, 413,
    416, 422, 428, 429
Mistral, 252, 253
Mixtral-8x7B Instruct model, 226
Mixture of Expert (MoE), 316
  fine-grained, 324, 330
  MegaBlocks/Dropless MoE
    blocks, 324
  neural networks
    architecture, 322
  traditional architecture,
    322, 323
MLflow Deployments Server
  advantage, 274
  AI2 Wildbench, 276
  create serving endpoint, 274
  EluetherAI LM Evaluation
    Harness, 276
  Mosaic Model Gauntlet, 276
  saving credentials, 275
MLOps, *see* Machine learning
    operations (MLOps)
MoE, *see* Mixture of Expert (MoE)
Model architecture, 250, 272
Model fine-tuning, 261, 270
Model pre-training, 262, 271
Model Serving, 174, 176, 201–204,
    248, 263, 313, 327, 333

Model training, 185, 188, 191, 195,
    221, 249, 250, 256, 262,
    270, 272
Modern data platforms, 3
Mosaic AI agent framework, 285
  deployment
    LangChain model
      registration, 302
    one-line deployment
      command, 303
    retrieving review app
      URLs, 303
    sharing permission, 304, 305
    status, 303
  features, 286
  installations, 287
  LangChain, 309
  LangChain parameterization,
    287, 288
  MLflow evaluation
    array structure, 292, 293
    custom metrics, 290
    databricks-agent, 290
    input dataset schema, 292
    metrics, 293, 294
    mlflow.evaluate, 291
    model metrics tab, 289
    principles, 308
    YAML file, 291
  model development
    basic chain, 296
    chat endpoint, configuration
      file, 299, 301
    components, 297

LangChain pipeline, 295–297
LangChain trace
    interface, 302
log_model() function, 295
mlflow.langchain.log_
    model() function, 294
RAG artifacts *vs.* MLflow
    artifacts, 295
rag_chain_config.yaml, 298
testing, 301
YAML configurations, 297
    workflow, 285, 286
Mosaic AI Agent Framework,
    231, 285–310
Mosaic AI vector Search, 235, 240
MosaicML, 12, 248, 325, 326, 330
MPT, 252
MPT-7B-8K-Instruct model, 227
MPT-30B-Instruct model, 227
MTEB, *see* Massive Text
    Embedding
    Benchmark (MTEB)
MySQL, 48–49, 53, 172, 239

N

Native cloud tools, 47
Natural language processing
    (NLP), 318
Network security group (NSG), 393
Neural networks, 67, 270, 315, 318,
    320–322, 368
NLP, *see* Natural language
    processing (NLP)

Notebooks, 16
    attach SQL warehouse, 36
    cell debugging, 34
    code-first development
        tool, 31
    collaborative, 32
    Databricks widgets, 36, 37
    debugging, 33, 34
    Jupiter notebooks, 31
    multiple language
        support, 32
    reproducible, 33
    sample Databricks
        notebook, 31, 32
    scheduled, 33
    serverless, 35
    visualizations, 33
NOT NULL constraints, 165
NSG, *see* Network security
    group (NSG)

O

Observability, 371–388, 423
OLMo, *see* Open Language
    Model (OLMo)
Open AI, 247, 248, 253, 329
Open Language Model (OLMo),
    249, 250, 271
    dataset, 271
    hardware, 272
    model architecture, 272
    model training, 272
Output modes, 417, 420

P, Q

PAT, *see* Personal access tokens (PAT)

Pandas Profiling, 186–191, 212

Parameter-efficient fine-tuning (PET), 271

Parquet, 46, 54, 55, 58, 61–64, 76, 77, 86, 189

Partitioning, 5, 72, 79–82

Partner Connect, 47, 52, 53, 179

Personal access tokens (PAT), 39, 40, 402, 403

Personal Compute, 24

Personalized models, 230

Personally identifiable information (PII) masking, 342, 343, 346

PET, *see* Parameter-efficient fine-tuning (PET)

Photon, 28, 144, 157, 158, 374, 377–379

PII masking, *see* Personally identifiable information (PII) masking

PIVOT function, 441

PowerBI workspaces, 177, 178

Predictive I/O, 85, 334

Pre-training, 223, 247, 248, 257, 261, 262, 271, 313, 314

Project Lightspeed, 423–425, 429

Prompt engineering, 224, 258

    curated models, 227

    key points, 224, 225

    types, 225

Python debugger, 34

R

RAG, *see* Retrieval augmented generation (RAG)

RAG Studio, 10

Random forests, 197

RDD, *see* Resilient Distributed Dataset (RDD)

readStream, 56

Real-time streaming sources, 45

Recurrent neural networks (RNNs), 318, 319

Resilient Distributed Dataset (RDD), 410, 412

Retail sales pipeline

    data lineage, 150, 151

    data quality, 151

    data validation, 149–151

    streaming, 149

Retrieval augmented generation (RAG), 220, 259, 285

    creation, 237

    embeddings/sync, 240, 241

    feature and function serving

        external feature stores, 238, 239

        online feature stores, 240

        online tables, 238

    LangChainTool, 242

    MLflow LLM evaluation

        data, 246

        metrics, 244, 245

        predefined model types, 242–244

RNNs, *see* Recurrent neural networks (RNNs)
Row filters
    apply to table, 105
    create, 105
"Run if" dependencies, 126

S

SAT, *see* Security Analysis Tool (SAT)
SCD, *see* Slowly Changing Dimensions (SCD)
Scheduled trigger, 119, 120
Schema evolution, 58, 59, 68, 69
Secure Cluster Connectivity, 394–396
Security Analysis Tool (SAT), 404–406
Sentiment analysis, 174–176, 228
serverless SQL, 436, 437, 440
Serverless workflows, 118
Shallow clones, 72, 73
Shared clusters, 25, 26
Silver layer, 66, 137
Slowly Changing Dimensions (SCD), 138, 152
SME, *see* Subject-matter expert (SME)
SMoE, *see* Sparse mixture of experts model (SMoE)
Snowflake, 48, 172, 173
SpaceX spaceship, 11
SpaceX's Starship, 10, 12

Spark Streaming
    components, 410
    data engineering, 412, 413
    Scala, 411
    workflow, 411
Sparse mixture of experts model (SMoE), 226
Spot instances, 16, 29, 382
SQL editor, 154, 158, 159, 335, 336
SQL warehouses, 20
    AI/BI dashboards, 159, 160
    alerts, 161
    AWS, 380
    Azure, 380
    Canvas tab, 160
    classic, 155, 379
    create new, 379
    Data tab, 160
    GCP, 380
    parameters
        cluster size, 156
        scaling, 157
    photon, 157
    pro, 155
    query history, 161, 162
    query profile, 162, 163
    serverless, 155, 379
    serverless compute, 163, 164
    setting up, 156
    SQL editor, 158, 159
Standard runtime version, 27
Standard single-user clusters, 24, 25

Star schemas, 164
State management, 411, 412, 414,
        418–419, 423, 424, 428
Streaming tables, 135, 168,
        169, 179
Structured Streaming
    best practices, 426
    challenges, 414
    group aggregations, 418
    monitor UI, 427
    state management, 418, 419
    watermark, 420, 421
Subject-matter expert (SME), 12
Summarization, 191, 226,
        227, 229
Super Heavy booster, 10
System tables
    audit logs, 386
    billable usage, 386–388
    catalogs, 386
    Databricks documentation, 385
    table lineage, 387
    Unity Catalog, 385
    use cases, 383, 384
    uses, 384
    workflow, 387

T

Table update trigger, 122
Tokenizers, 251–253

Tokens, 251–253
Transformer model, 319, 320
Trigger modes, 161, 416

U

UC, *see* Unity Catalog (UC)
Unified monitoring, 211
UniForm (Universal Format),
        77–79, 88
UNIQUE constraint, 165
Unity Catalog (UC), 7, 8
    admin roles
        account admin, 94
        create metastore,
            95–97
        data owners, 94
        metastore admin, 94
        workspace admin, 94
    capabilities, 90
    catalog, 92
    column masks, 106
    data sharing, 107
    definition, 90
    dev environment, 98
    dynamic view, 107
    –enabled clusters, 25
    external tables, 93
    features
        access control
            mechanisms, 101

centralized user
  management, 100
data access, 104
data lineage, 102, 103
data search, 104
granting permissions, 102
SCIM synchronization, 100
hierarchy, 92
managed tables, 93
metastore, 92
row filter (*see* Row filters)
schema, 93
SDLC workspaces, 99
structure, 98
support, 26
tables, 93
volumes, 93
website, 98
with/without, 91
workspace setup, 99
Unstructured text
    parsing, 228

V

Variable Explorer, 34, 35
Vector database, 231–233, 237,
        240–241, 259, 260, 266,
        269, 270
Vector index, 231, 233, 259, 260,
        266, 283
Vector library, 260, 266
Vector Search, 10, 233–235, 260,
        267, 269, 296, 297
Virtual machines (VMs), 16, 18,
        28, 29, 372
VMs, *see* Virtual machines (VMs)

W, X, Y

Workspace admin, 94, 401

Z

Z-ordering, 81